Chaos in
Biological Systems

NATO ASI Series

Advanced Science Institutes Series

A series presenting the results of activities sponsored by the NATO Science Committee, which aims at the dissemination of advanced scientific and technological knowledge, with a view to strengthening links between scientific communities.

The series is published by an international board of publishers in conjunction with the NATO Scientific Affairs Division

A	Life Sciences	Plenum Publishing Corporation
B	Physics	New York and London
C	Mathematical and Physical Sciences	D. Reidel Publishing Company Dordrecht, Boston, and Lancaster
D	Behavioral and Social Sciences	Martinus Nijhoff Publishers
E	Engineering and Materials Sciences	The Hague, Boston, Dordrecht, and Lancaster
F	Computer and Systems Sciences	Springer-Verlag
G	Ecological Sciences	Berlin, Heidelberg, New York, London,
H	Cell Biology	Paris, and Tokyo

Recent Volumes in this Series

Series A: Life Sciences

Chaos in Biological Systems

Edited by

H. Degn
Odense University
Odense, Denmark

A. V. Holden
The University of Leeds
Leeds, United Kingdom

and

L. F. Olsen
Odense University
Odense, Denmark

Plenum Press
New York and London
Published in cooperation with NATO Scientific Affairs Division

Proceedings of a NATO Advanced Research Workshop on
Chaos in Biological Systems,
held December 8–12, 1986,
at Dyffryn House, Cardiff, Wales

Library of Congress Cataloging in Publication Data

NATO Advanced Research Workshop on Chaos in Biological Systems (1986:
  Cardiff, South Glamorgan)
  Chaos in biological systems.

  (NATO ASI series. Series A, Life sciences; vol. 138)
  "Proceedings of a NATO Advanced Research Workshop on Chaos in
Biological Systems, held December 8–12, 1986, at Dyffryn House, Cardiff,
Wales"—T.p. verso.
  "Published in cooperation with NATO Scientific Affairs Division."
  Includes bibliographies and index.
  1. Chaotic behavior in systems—Congresses. 2. Physiology—Congresses. I.
Degn, Hans. II. Holden, Arun V., 1947–   . III. Olsen, L. F. (Lars F.) IV. North
Atlantic Treaty Organization. Scientific Affairs Division. V. Title. VI. Series:
NATO ASI series. Series A, Life sciences; v. 138. [DNLM: 1. Biochemistry—
congresses. QU 34 N279c 1986]
  QP33.6.C48N38   1986   574   87-7203
  ISBN 0-306-42685-4

PREFACE

In recent years experimental and numerical studies have shown that chaos is a widespread phenomenon throughout the biological hierarchy ranging from simple enzyme reactions to ecosystems. Although a coherent picture of the fundamental mechanisms responsible for chaotic dynamics has started to appear it is not yet clear what the implications of such dynamics are for biological systems in general. In some systems it appears that chaotic dynamics are associated with a pathological condition. In other systems the pathological condition has regular periodic dynamics whilst the normal non-pathological condition has chaotic dynamics. Since chaotic behaviour is so ubiquitous in nature and since the phenomenon raises some fundamental questions about its implications for biology it seemed timely to organize an interdisciplinary meeting at which leading scientists could meet to exchange ideas, to evaluate the current state of the field and to stipulate the guidelines along which future research should be directed.

The present volume contains the contributions to the NATO Advanced Research Workshop on "Chaos in Biological Systems" held at Dyffryn House, St. Nicholas, Cardiff, U.K., December 8-12, 1986. At this meeting 38 researchers with highly different backgrounds met to present their latest results through lectures and posters and to discuss the applications of non-linear techniques to problems of common interest. In spite of their involvement in the study of chaotic dynamics for several years many of the participants met here for the first time. As organizers we feel that the outcome lived up to our expectations, and we should like to thank all participants for the enthusiasm they carried with them and for contributing to this volume. We also acknowledge the staff at Dyffryn House and at the Microbiology Department, University College, Cardiff for their help in the organization. Special thanks to Mrs. Nia Pagett for her assistance before, during and after the meeting and Mrs. Lynn Buckley and Mrs. Jackie Hill, Leeds University for typing the manuscripts. Finally, we thank the NATO Scientific Affairs Division, Brussels and the Commission of the European Communities, Directorate-General for Science Research and Development, Brussels for financial support of the meeting. Without the joint support from both parties this meeting could not have been held.

March 1987

Hans Degn
Arun V. Holden
David Lloyd
Lars F. Olsen

CONTENTS

τὰ μὲν ἄλλα παντὸς μοῖραν μετέχει, νοῦς δέ ἐστιν
ἄπειρον καὶ αὐτοκρατὲς καὶ μέμεικται οὐδενὶ χρήματι,
ἀλλὰ μόνος αὐτὸς ἐφ' ἑαυτοῦ ἐστιν. εἰ μὴ γὰρ ἐφ'
ἑαυτοῦ ἦν, ἀλλά τεωι ἐμέμεικτό ἄλλωι, μετεῖχεν ἄν
ἁπάντων χρημάτων, εἰ ἐμέμεικτό τεωι· ἐν παντὶ γὰρ
παντὸς μοῖρα ἔνεστιν, ὥσπερ ἐν τοῖς πρόσθεν μοι
λέλεκται· καὶ ἄν ἐκώλυεν αὐτὸν τὰ συμμεμειγμένα, ὥστε
μηδενὸς χρήματος κρατεῖν ὁμοίως ὡς καὶ μόνον ἐόντα ἐφ'
ἑαυτοῦ. ἔστι γὰρ λεπτότατόν τε πάντων χρημάτων καὶ
καθαρώτατον, καὶ γνώμην γε περὶ παντὸς πᾶσαν ἴσχει καὶ
ἰσχύει μέγιστον· καὶ ὅσα γε ψυχὴν ἔχει καὶ τὰ μείζω
καὶ τὰ ἐλάσσω, πάντων νοῦς κρατεῖ. καὶ τῆς περιχωρήσ-
ιος τῆς συμπάσης νοῦς ἐκράτησεν, ὥστε περιχωρῆσαι τὴν
ἀρχήν. καὶ πρῶτον ἀπό του σμικροῦ ἤρξατο περιχωρεῖν,
ἐπὶ δὲ πλέον περιχωρεῖ, καὶ περιχωρήσει ἐπὶ πλέον. καὶ
τὰ συμμισγόμενά τε καὶ ἀποκρινόμενα καὶ διακρινόμενα
πάντα ἔγνω νοῦς. καὶ ὁποῖα ἔμελλεν ἔσεσθαι καὶ ὁποῖα
ἦν, ἄσσα νῦν μὴ ἔστι, καὶ ὅσα νῦν ἔστι καὶ ὁποῖα ἔσται,
πάντα διεκόσμησε νοῦς, καὶ τὴν περιχώρησιν ταύτην, ἣν
νῦν περιχωρεῖ τά τε ἄστρα καὶ ὁ ἥλιος καὶ ἡ σελήνη καὶ
ὁ ἀὴρ καὶ ὁ αἰθὴρ οἱ ἀποκρινόμενοι. ἡ δὲ περιχώρησις
αὕτη ἐποίησεν ἀποκρίνεσθαι. καὶ ἀποκρίνεται ἀπό τε τοῦ
ἀραιοῦ τὸ πυκνὸν καὶ ἀπὸ τοῦ ψυχροῦ τὸ θερμὸν καὶ ἀπὸ
τοῦ ζοφεροῦ τὸ λαμπρὸν καὶ ἀπὸ τοῦ διεροῦ τὸ ξηρόν.
μοῖραι δὲ πολλαὶ πολλῶν εἰσι. παντάπασι δὲ οὐδὲν
ἀποκρίνεται οὐδὲ διακρίνεται ἕτερον ἀπὸ τοῦ ἑτέρου πλὴν
νοῦ. νοῦς δὲ πᾶς ὅμοιός ἐστι καὶ ὁ μείζων καὶ ὁ
ἐλάττων. ἕτερον δὲ οὐδέν ἐστιν ὅμοιον οὐδενί, ἀλλ'
ὅτων πλεῖστα ἔνι, ταῦτα ἐνδηλότατα ἓν ἕκαστόν ἐστι καὶ
ἦν.

Anaxagoras, Fragment No. 12

     Whereas all other things contain a portion of everything, the Mind
(Nous) is unlimited and autonomous, mixing with nothing but being alone
by itself. For it it were not self-contained but mixed with anything
else, it would partake with everything, since in everything there is a
part of everything as I said before. Whatever were intermingled with it
would prevent it from having power over anything, of the kind it has now
being alone by itself. It is the finest of all things and the purest,
haing every knowledge about everything and exerting the greatest power.
Whatever, great or small, exists (has psyche), is controlled by the Mind.
The whole recurrence (perichoresis) is under the control of the Mind,
such that it set into recurrent motion the beginning. First from the
small, the recurrence was started. Then more got involved in the about-
moving (perichoretic) process, and even more will be involved in it.
What is mixed together, and what is separate and distinguished, was all
known to the Mind. All that was to come into being - all that has been
but no longer is, all that is now, and all that there will be - got neatly
led out by the Mind. Even the recurrent motion that governs the stars
and the sun and the moon and the air and the aether, separate as they are
to date, was included. The recurrence in turn generated the separation
process, From the thin there got separated the dense, from the cold the
warm, from the dark the glowing, from the moist the dry. Many parts are
there of the many things. For all of them it holds true that nothing
gets separated off and distinguished, the one from the other, except
through the Mind. The Mind as a whole is self-similar no matter whether
it refers to the large or the small. Of the other things, however, none
is similar to any other: always that which is the most strongly
represented in each thin determines its properties.

CHAOS IN PHYSIOLOGY: HEALTH OR DISEASE?

Ary L. Goldberger[1] and Bruce J. West[2]

[1]Harvard Medical School
Cardiovascular Division
Beth Israel Hospital
330 Brookline Avenue
Boston
Massachusetts 02215

[2]Division of Applied
Nonlinear Problems
La Jolla Institute
10280 N. Torrey Pines Road
Suite 260
La Jolla
California 92037

The universality of 'chaotic' dynamics in mathematical and physical systems [1-4] has prompted renewed interest in the application of non-linear analysis to biological processes [4,5]. Attention has also focused on the physiological and medical implications of these concepts [4,6-11]. The prevailing viewpoint is that the dynamics of health are ordered and regular and that a variety of pathologies represent a bifurcation to chaos [6,9,12]. For example, Smith and Cohen [9] advanced the hypothesis that ventricular fibrillation, the arrhythmia most commonly associated with sudden cardiac death, is a turbulent process (cardiac chaos) that may result from a subharmonic bifurcation (period-doubling) mechanism.

We [8,13-16] have proposed an alternative viewpoint, contrary to this notion of chaotic disease. In particular, we have suggested that chaos is useful in modelling the 'constrained randomness' [17] inherent in the healthy function of physiological systems. This countervailing hypothesis is supported by the following lines of evidence

1.   The frequency spectra associated with healthy dynamics in a variety of apparently unrelated settings (beat-to-beat heart rate variability, peripheral white blood cell fluctuations, neuronal responses) show a broadband ('noisy') profile with $1/f$-like (inverse power-law) distribution (Fig. 1) [18-20]. Broadband spectra of this kind are typical of chaotic (fractal) processes. Such processes manifest self-similar fluctuations across multiple temporal orders of magnitude and, therefore, do not have a single, characteristic frequency [21].

2.   Many anatomical structures (tracheo-bronchial tree, His-Purkinje system, chordae tendineae, biliary network, vascular tree, urinary collecting system) demonstrate a fractal architecture [14,15,22-26] (Fig. 2). The ubiquity of these fractal anatomies suggests that their morphogenesis is encoded by strange attractors or renormalization group algorithms [24] that iteratively generate complex, self-similar forms lacking a character-istic scale of length. In the case of the lung tree, for example, the contribution of multiple scales provides a mechanism for both the order and variability apparent in normal bronchial architecture. The order is

HEART RATE DYNAMICS

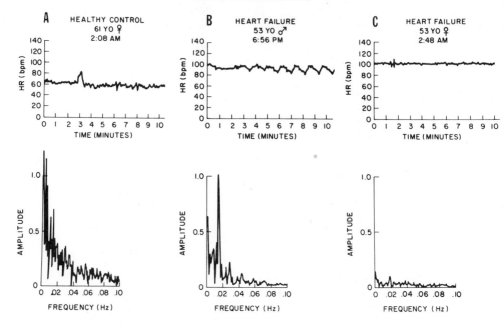

A HEALTHY CONTROL
61 YO ♀
2:08 AM

B HEART FAILURE
53 YO ♂
6:56 PM

C HEART FAILURE
53 YO ♀
2:48 AM

Fig. 1. Healthy subjects (A) show considerable beat-to-beat heart rate variability (top panel) with a 'noisy', 1/f-like spectrum (lower panel). Loss of spectral reserve in heart failure is evidenced by either (B) the dominance of relatively low frequency (∿.02 Hz) oscillations that correlate with Cheyne-Stokes breathing or (C) a marked overall reduction in sinus rhythm variability. HR (bpm) = heart rate in beats/min. Adapted from [15].

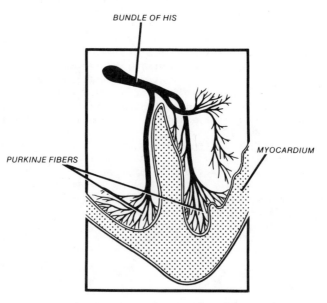

Fig. 2. Many anatomical structures, such as the His-Purkinje conduction network in the heart, show a fractal-like architecture. Adapted from [23].

reflected in the harmonically-modulated, inverse power-law scaling that governs the decrease in mean bronchial dimensions from one level (generation) of branchings to the next smaller one in man and other mammals [24]. The variability is evidenced in the multiplicity of tube dimensions within any single generation of branchings.

3.   A wide class of pathologies is characterized by a loss of broadband stability and the emergence of highly periodic behaviour [8,27]. Perhaps most surprising is the finding that ventricular fibrillation has a relatively narrowband spectral representation, an observation inconsistent with the widely-held view of fibrillation as cardiac chaos [16]. The spectral data, however, are in accord with recent mapping studies from the epicardial and endocardial surfaces of the ventricles showing unexpected evidence of temporal and spatial organization during canine ventricular fibrillation [28,29]. Highly periodic oscillations of the electrocardiogram are also a defining feature of *torsades de pointes*, another cardiac arrhythmia associated with sudden death [30].

4.   Period-doubling phenomenology appears to be widespread in pathophysiology. When excessive fluid collects in the pericardial sac and compresses the heart, for example, the frequency of the cardiac swinging motion may decrease abruptly to one-half of its initial value [31]. Behaviour suggesting a subharmonic bifurcation mechanism has also been described with cardiac electrical dysfunction [7,13,14,32,33]. However, there are no clinical data to support the suggestion [9] that a cascade of period-doublings ever leads to a broadband (chaotic) disease state, analogous to the subharmonic bifurcation route to chaos in physical systems [2,3]. Instead, the dynamics observed in sudden cardiac death syndromes suggest an inverse bifurcation [16] -- from healthy variability (chaos) to pathological order (periodicity).

5.   This bifurcation from a healthy broadband state to pathological periodicity in cardiovascular dynamics is complemented by a recent analysis of electroencephalographic time series data. Babloyantz and Destexhe [11] reported a marked reduction in the dimensionality of the brain attractor during epilepsy as compared with normal function.

The notion of the chaotic (fractal) nature of healthy dynamics supports the more general concept of health as an 'information-rich' (broadband) state [16]. Highly periodic behaviour, in contrast, seen in a variety of pathologies [8,27], reflects a loss of physiological information and is represented by a relatively narrowband spectrum. As a corollary, time series analysis of electrocardiographic data may provide a new means of monitoring patients at high risk for sudden death who would be predicted to show alterations in the healthy, 1/f-like, broadband interbeat interval spectrum [8,14]. Loss of sinus rhythm heart rate variability, sometimes associated with low frequency oscillations, has already been observed both in adults with severe heart failure [8] (Fig. 1) and in the foetal distress syndrome [34]. Loss of broadband spectral reserve may also be a marker of pathological dynamics in other diseases, including certain malignancies [35]. Whether periodic attractors of this sort [8,13,27] are universal in disease remains unanswered. The conjectured importance of chaos to healthy structure and function suggests additional applications of abstract nonlinear models to bedside medicine.

REFERENCES

[1]   R.M. May, <u>Nature</u> (London) 261:459-467 (1976).
[2]   M.J. Feigenbaum, <u>Commun. Math. Phys.</u> 77:65-86 (1980).

[3]   J-P. Eckmann, Rev. Mod. Phys. 53:643-654 (1981).

[4]   A.V. Holden, ed.,"Chaos", University Press, Manchester and Princeton (1986).

[5]   L.F. Olsen and H. Degn, Quart. Rev. Biophys. 18:165-225 (1985).

[6]   M.C. Mackey and L. Glass, Science, 197:287-289 (1977).

[7]   M.R. Guevara and L. Glass, J. Math. Biol. 14:1-23 (1982).

[8]   A.L. Goldberger, L.J. Findley, M.R. Blackburn and A.J. Mandell, Am. Heart J. 107:612-615 (1984).

[9]   J.M. Smith and R.J. Cohen, Proc. Natl. Acad. Sci. U.S.A. 81:233-237 (1984).

[10] M. Sernetz, B. Gelleri and J. Hofmann, J. Theor. Biol. 117:209-230 (1985).

[11] A. Babloyantz and A. Destexhe, Proc. Natl. Acad. Sci. U.S.A. 83:3513-3517 (1986).

[12] D. Ruelle, Math. Intelligencer 2:126-137 (1980).

[13] A.L. Goldberger, V. Bhargava, B. West and A.J. Mandell, Physica 17D:207-214 (1985).

[14] A.L. Goldberger, B.J. West and V. Bhargava, in "Proc. 11th International Modeling and Computers in Simulation" World Congress, Oslo, Norway, Vol. 2, eds. B. Wahlstrom, R. Henriksen and N.P. Sundby, 239-242, North Holland Publishing Co., Amsterdam (1985).

[15] A.L. Goldberger, in "Temporal Disorder in Human Oscillatory Systems", eds. L. Rensing, U. An der Heiden and M. Mackey, Springer, Berlin, in press.

[16] A.L. Goldberger, V. Bhargava, B.J. West and A.J. Mandell, Physica 19D:282-289 (1986).

[17] A.J. Mandell, S. Knapp, C.L. Ehlers and P.V. Russo, in "Neurobiology of the Mood Disorders", eds. R.M. Post and J.C. Ballenger, 744-776, Williams & Wilkins, Baltimore (1983).

[18] M. Kobayashi and T. Musha, IEEE Trans. Biomed. Eng. 29:456-457 (1982).

[19] A.L. Goldberger, K. Kobalter and V. Bhargava, IEEE Trans. Biomed. Eng. 33:874-876 (1986).

[20] T. Musha, Y. Kosugi, G. Matsumoto and M. Suzuki, IEEE Trans. Biomed. Eng. 28:616-623 (1981).

[21] E.W. Montroll and M.F. Shlesinger, J. Stat. Phys. 32:209-230 (1983)

[22] B.B. Mandelbrot, "The Fractal Geometry of Nature", W.H. Freeman, New York (1982).

[23] A.L. Goldberger, V. Bhargava, B.J. West and A.J. Mandell, Biophys. J. 48:525-528 (1985).

[24] B.J. West, V. Bhargava and A.L. Goldberger, J. Appl. Physiol. 60:1089-1097 (1986).

[25] J. Lefevre, J. Theor. Biol. 102:225-248 (1985).

[26] B.J. West and A.L. Goldberger, Am. Scientist, in press.

[27] H.A. Reimann, "Periodic Diseases", F.A. Davis, Philadelphia (1963).

[28] R.E. Ideker, G.J. Klein, L. Harrison et al., Circulation 63:1371-1379 (1981).

[29] S.J. Worley, J.L. Swain, P.G. Colavita et al., Am. J. Cardiol. 55:813-820 (1985).

[30] V. Bhargava, A.L. Goldberger, D. Ward and S. Ahnve, IEEE Trans. Biomed. Eng. 33:894-896 (1986).

[31] A.L. Goldberger, R. Shabetai, V. Bhargava, B.J. West and A.J. Mandell, Am. Heart. J. 107:1297-1299 (1984).

[32] M.R. Guevara, L. Glass and A. Shrier, Science 214:1350-1353 (1981).

[33] A.L. Ritzenberg, D.R. Adam and R.J. Cohen, Nature (London), 307:159-161 (1984).

[34] H.D. Modanlou and R.K. Freeman, Am. J. Obstet. Gynecol. 142:1033-1038 (1982).

[35] H. Vodopick, E.M. Rupp, C.L. Edwards, F.A. Goswitz and J.J. Beauchamp, N. Engl. J. Med. 286:284-290 (1972).

PATTERNS OF ACTIVITY IN A REDUCED IONIC MODEL OF A CELL FROM THE

RABBIT SINOATRIAL NODE

Michael R. Guevara[1], Antoni C.G. van Ginneken[2] and
Habo J. Jongsma[2]

[1]Department of Physiology
McGill University
3655 Drummond Street
Montreal, Quebec
Canada H3G 1Y6

[2]Fysiologisch Laboratorium
Universiteit van Amsterdam
Academisch Medisch Centrum
Meibergdreef 15
1105 AZ Amsterdam, Nederland

ABSTRACT

Numerical simulation of an ionic model of an isolated nodal cell
produces patterns of activity similar to those seen in the Belousov-
Zhabotinsky reaction and in other physical and chemical systems in which
chaotic dynamics is said to exist. However, no evidence of chaotic dynamics has yet been found in the modelling work. Recent experimental
results on the sinatrial node reinforce this conclusion.

I INTRODUCTION

The heart normally beats with a regular rhythm. In fish, amphibians,
and reptiles, the natural pacemaker of the heart is a separate chamber of
the heart, the sinus venosus, which drives the rest of the heart and so
sets the overall heart rate. In mammals, the pacemaking centre is the
sinoatrial node (SAN), a small pale area of tissue located in the right
atrium near the point where the venae cavae enter the heart. The main
function of the SAN is to spontaneously generate a periodic train of
action potentials, which propagate out of the SAN into the muscle of the
right atrium, and thence to the rest of the heart.

There are several disease processes that affect the SAN in human
beings. In some instances, a cardiac arrhythmia called atrial standstill
is seen, in which the electrocardiogram shows that the atria are not
being activated as usual by the output of the SAN. If a subsidiary
pacemaker, located elsewhere in the heart, does not then take over the
pacing function of the SAN, the heart will stop pumping blood, and the
individual will die. In fact, sinoatrial dysfunction is presently a
leading indication for the surgical implantation of electronic cardiac
pacemakers, which are man-made substitutes for the SAN.

It is generally thought that standstill of the atria comes about as
a result of one of two main causes: either (i) the SAN stops generating
action potentials (sinoatrial arrest), or (ii) there is a block of
propagation from the SAN to the right atrium (sinoatrial exit block).

It is thus of some interest to consider how cells within the SAN might be made to cease their spontaneous activity. This can be done by carrying out experiments on isolated right-atrial preparations in which, for example, the normal constitution of the medium bathing the preparation is changed. An alternative approach, which gives insight into the ionic mechanisms underlying the maintenance and abolition of spontaneous activity, is to carry out computer simulations using ionic models of the SAN.

The electrical activity of cardiac cells is due to the flow of ions of various species (e.g. Na^+, K^+, Ca^{++}, Cl^-) through channels in the cell membrane. Voltage-clamp experiments, in which the transmembrane potential difference (V) is clamped to a fixed potential using an electronic circuit, have been carried out on small effectively isopotential pieces of tissue cut out of the SAN. These experiments result in quantitative descriptions of the various membrane currents. A model of the SAN cell membrane can then be assembled which incorporates these descriptions.

We consider here the model of Irisawa and Noma [21], which incorporates five currents: the fast inward sodium current (I_{Na}), the slow inward current (I_s), the potassium repolarizing current (I_K), the hyperpolarizing-activated pacemaker current (I_h), and the time-dependent background or leakage current (I_L). The model is formulated as a seven-dimensional set of ordinary differential equations, with the seven variables being the transmembrane potential V, the activation and inactivation variables of I_{Na} (m, h resp.) and I_s (d, f resp.), and the activation variables of I_K(p) and I_h(q). We show that this model can be reduced to a three-dimensional model, without impairing its impulse-generating properties, and demonstrate three qualitatively different ways in which spontaneous activity can be abolished.

II METHODS

We carry out simulations by numerically integrating the Irisawa-Noma equations using an efficient variable-time-step algorithm, which can be proven to be convergent [50]. When, in iterating from time t to time t+Δt, the change in V (ΔV) is greater than 0.4 mV, the time step Δt is successively halved and the calculations redone until ΔV < 0.4 mV. In addition, when ΔV < 0.2 mV at any iteration, the value of Δt is doubled for use at the next iteration. In this way, by allowing Δt to take one of nine values in the range 0.032 ms $\leqslant \Delta$t $\leqslant$ 8.192 ms, ΔV is kept below 0.4 mV. The formula appearing in footnote (2) of Victorri et al. [50] is used to calculate the contribution of the total transmembrane current to ΔV. Single precision arithmetic is used throughout ($\sim$7 significant decimal digits in FORTRAN on a Hewlett-Packard model 1000F minicomputer) and l'Hôpital's rule for indeterminate forms is applied when necessary. Initial conditions are V = -60.000 mV, m = 0.064260, h = 0.92720, d = 0.030477, f = 0.86991, p = 0.20890, q = 0.012767, with Δt = 4.096 ms.

III RESULTS

Figure 1A shows spontaneous activity in the Irisawa-Noma model. The periodic activity in V (and in the other six variables m, h, d, f, p and q) suggests that there is a (stable) limit cycle present in the phase space of the system. The initial conditions chosen closely approximate a point on this limit cycle. Inspection of the waveforms of the currents underlying the spontaneous activity of Fig. 1 (see Fig. 8 of [21]) reveals that the contribution of the currents I_{Na} or I_h to total transmembrane current is not very significant, since these currents are much smaller

6

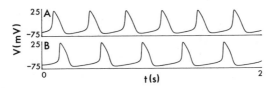

Fig. 1. Spontaneous activity in the full 7-dimensional Irisawa-Noma model (A) and in the reduced 3-dimensional model with $I_{Na} = I_h = 0$ and $d = d_\infty$ (B).

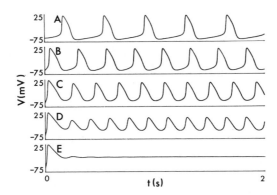

Fig. 2. Effect on spontaneous activity of injecting a constant depolarizing bias current. $I_{bias} = 0.0$(A), -0.5(B), -1.0(C), -1.5(D), and -2.0(E) $\mu A/cm^2$.

in magnitude than I_s, I_K, or I_L. In fact, it is well known that in the central part of the SAN, blockade of I_{Na} with tetrodotoxin makes little difference to spontaneous activity [25]. In addition, in some specimens cut out of the SAN, I_h is either completely absent or is present to a negligible extent [2,33]. Thus, removing I_{Na} and I_h from the model causes only small changes in the voltage waveform. One is thus left with a model containing only the three currents I_s, I_K, and I_L; the model is then four-dimensional, with variables V, d, f, and p.

The equation governing d is given by

$$\dot{d} = -\frac{1}{\tau_d}(d-d_\infty),\tag{1}$$

where the overdot denotes differentiation with respect to time, and τ_d and d_∞ are functions of V alone. The model can be further reduced in dimension by noting that during spontaneous activity the variable d is not far removed from its asymptotic value d_∞ at all times. Thus, d can be replaced with d_∞, and we are left with a three-dimensional model of a cell of the SAN, with the three variables being V, f, and p. Figure 1B shows spontaneous activity in the reduced three-variable model, and demonstrates that the reduced model produces a voltage waveform quite similar to that of the original seven-variable model (Fig. 1A).

We have investigated the effect of changing, one at a time, many of the parameters in our three-dimensional model. In several instances, when a parameter is pushed far enough, spontaneous activity is abolished. For example, Fig. 2 shows the effect of injecting a constant depolarizing current I_{bias}. As I_{bias} is increased, the action potential falls in amplitude until quiescence eventually occurs (Fig. 2E). Over the range of

7

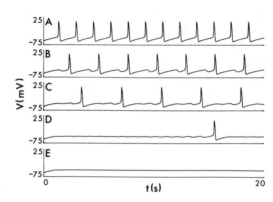

Fig. 3. Effect on spontaneous activity of multiplying I_L by a constant
factor c. c = 1.86(A), 1.87(B), 1.885(C), 1.9(D), and 1.905
(E). Note the appearance of afterpotentials in B-D, and
quiescence in E.

I_{bias} used in Fig. 2, there is only one equilibrium point in the system.

A qualitatively different sequence of patterns is seen in the model
when I_L is increased by multiplying its usual value at any voltage V by
a constant factor, c. Instead of a gradual continuous reduction in
action potential amplitude as in Fig. 2, small amplitude events ('skipped
beats' or 'delayed afterdepolarizations'[9]) are seen interspersed among
the action potentials (Fig. 3B, C, D). These afterpotentials increment
in amplitude until an action potential is produced. As c increases, the
frequency of occurrence of these afterpotentials relative to action
potentials also increases. Eventually a damped oscillation (barely
visible at the start of Fig. 3E) is seen that results in quiescence.
Again, there is only one equilibrium point present in the system over the
parameter range used in Fig. 3.

IV CONCLUSIONS

The sequence of waveforms shown in Fig. 2 is compatible with that
seen in experimental work in which depolarizing current is injected into
a cell in a small piece of tissue from the SAN [29]. Activity similar to
that shown in Fig. 2 can also be seen in experiments in which I_s is
blocked using a calcium-channel blocker [4] or in which Ba^{++} is applied
[35]. In addition, gradual reduction of g_s, the maximal conductance of
I_s, produces a similar sequence of waveforms in the reduced three-
dimensional model.

The situation is quite different when I_L is gradually increased (Fig.
3): a qualitatively different sequence of waveforms is seen. Patterns
of activity containing afterdepolarizations (similar to Fig. 3B, C, D)
have been described several times in the SAN in response to a variety of
interventions [3,24,30,31,32,51]. A sequence of patterns essentially
equivalent to the entire sequence shown in Fig. 3 has been recently
described in experimental work on the SAN as the K^+ concentration in the
medium bathing the preparation is gradually elevated [19]. However, in
that case, a maintained small-amplitude subthreshold oscillation can be
observed at a parameter value intermediate to those at which waveforms
similar to those shown in Fig. 3D (c = 1.9) and 3E (c = 1.905) are seen.
Oscillatory activity of low amplitude can indeed be seen at parameter

values between c = 1.9 and 1.905 in our reduced model. However, there are two factors that complicate the interpretation of this finding. First, since our simulations are carried out for only finite times, there is no guarantee that an oscillation of seemingly constant amplitude is not in reality growing very slowly, eventually resulting in a skipped beat pattern of very long period (>> 10 seconds). Alternatively, the oscillation might eventually damp out after a very long time [28]. Secondly, one might expect that a skipped-beat run of very long period would be converted into a maintained small-amplitude oscillation due to the finite precision used in our computations. In experimental work, membrane voltage noise would have a similar effect. In fact, a maintained small-amplitude subthreshold oscillation has been described but rarely in the SAN; indeed, in our own experimental work, the external K^+ concentration must be exquisitely adjusted in order to obtain a maintained small amplitude oscillation [19]. This agrees with the results of Fig. 3, which suggest that if a small-amplitude oscillation does exist, it would be sustained over only a very narrow parameter range.

Traces similar to some of those shown in Fig. 3 can be seen in experimental work on Purkinje fibre and on aggregates of embryonic heart cells (see refs. in [16]). In fact, a sequence of patterns analogous to that shown in Fig. 3 has recently been described in an ionic model of Purkinje fibre [16]. There is one major difference however: the current I_{K2} (which replaces I_h in the SAN model considered above) is directly implicated in generating the delayed afterdepolarizations and the maintained subthreshold oscillation seen in that case. This is not the case in Fig. 3, where I_h is identically zero. A different sequence of patterns, which is in many ways the mirror image of that shown in Fig. 3, has been described in Purkinje fibre and ventricular muscle and their ionic models (see [6] and other refs. in [16]). In these cases, there are 'early afterdepolarizations' in the plateau range of potentials due to failures of repolarization, and not delayed afterdepolarizations in the pacemaker range of potentials due to failures of depolarization as in Fig. 3. In addition, the small-amplitude oscillation lies in the plateau, and not the pacemaker, range of potentials, with the membrane potential eventually coming to rest in the plateau range of potentials as a parameter is changed.

We have described above two qualitatively different ways in which spontaneous activity in the SAN can be abolished (Figs. 2,3). There is only one other way of which we are aware in which this can occur: injection of a brief pulse of current at the right point in the cycle can terminate spontaneous activity. This 'annihilation' does not occur in the standard Irisawa-Noma model or in our three-dimensional reduction of that model, since the only equilibrium point present in either of those two models is unstable. This agrees with the experimental finding in small pieces cut from the rabbit SAN that clamping the membrane to the voltage corresponding to that point for some time and then releasing the voltage clamp results in a resumption of spontaneous activity [32]. However, annihilation has been reported once in the SAN, in strips of kitten atrium containing nodal tissue [22]. The different findings in these two experimental studies could be due to a difference in species, age, or experimental protocol. However, changing a parameter in the full seven-dimensional Irisawa-Noma model can lead to a situation in which annihilation can occur [18]. In fact, a similar statement holds for the reduced three-dimensional model that we have studied above. For example, injection of a constant hyperpolarizing bias current can lead to the creation of a new pair of equilibrium points via a saddle-node bifurcation. Since the equilibrium point lying at the most negative potential is then stable over a range of I_{bias}, annihilation can then occur.

Waveforms analogous to those shown in Fig. 3 have also been described in experimental [23,27,34,36-38,41,42,44,46,49] and modelling [1,
6,10-12,14,15,20,26,27,34,37-41,45,47-49] studies on several biological,
chemical, and physical systems. In the majority of these reports,
chaotic dynamics is said to exist at parameter values intermediate to
those at which periodic behaviours occur ('alternating periodic-chaotic
sequence'). In some instances, chaotic behaviour arises out of a cascade of period-doubling bifurcations [11,12,14,46]. We have not observed such bifurcations in our work on the reduced model, nor have we
observed apparently non-periodic dynamics. In some systems, a cascade of
period-doubling bifurcations of a small-amplitude orbit has been shown to
exist at parameter values close to that at which a homoclinic orbit is
found.[11,12,14]. The waveform shown in Fig. 3D is close to being a
homoclinic orbit, and, as noted above, a maintained small-amplitude
oscillation might exist nearby. This would suggest that the system might
indeed be chaotic at parameter values differing slightly from those used
to obtain Fig. 3D.

However, two other factors should be kept in mind. First, the
existence of a homoclinic orbit does not necessarily guarantee the
existence of chaotic dynamics; for chaotic dynamics to exist in a neighbourhood of a homoclinic orbit in a three-dimensional system, the
Šil'nikov condition on the eigenvalues of the equilibrium point must be
satisfied [45]. In addition, in a recent experimental report on a
chemical system in which waveforms similar to those shown in Fig. 3 were
described, no evidence of period-doubling bifurcations or chaotic
dynamics could be found [27]. Similarly, in a numerical study of a
simple three-variable oscillator in which waveforms similar to those
shown in Fig. 3 were seen, the Liapunov exponents were always negative,
indicating the absence of chaotic dynamics [20]. Secondly, the importance of a homoclinic orbit as an organizing feature of chaotic dynamics
only becomes evidence when one examines the dynamics as a function of
two parameters [11,12,14]. Thus, at this point in time, we do not know
whether or not chaotic dynamics is present in Fig. 3. To make such a
decision based on simulation runs alone would require carrying out
further computations (in double precision) in which c would be changed
by amounts smaller than we have employed so far. The fact that very
small changes would have to be made to perhaps see chaotic dynamics
leads us to suggest that period-doubling bifurcations and chaotic
dynamics might not be seen in the corresponding experiment.

In several systems in which chaotic dynamics is said to exist, it is
possible to obtain a one-dimensional description of the dynamics by constructing a first-return Poincaré map [5,7,37-40,46]. This can be done
by, for example, plotting the amplitude of each maximum of a time series
as a function of the immediately preceding maximum. We have attempted to
do this with the waveforms of Fig. 3. However, the absence of maxima in
the voltage range. -50 mV $< V < 0$ mV has led us to a situation in which
the map is undefined over the major part of the range of the variable V.
We believe that the absence of spikes of intermediate amplitude is due
to the very fast (indeed, infinitely fast) activation of I_s in our model.
There is thus an effectively all-or-none threshold for initiation of an
action potential at $V \cong -45$ mV. Once again, computation on a much finer
scale should produce events of intermediate size, and so allow the map to
be determined over its full range (for an analogous situation in neural
tissue, see [8]).

However, the return map must possess a maximum, since consecutive
afterpotentials increment in amplitude and a large-amplitude action
potential is followed by a low-amplitude afterpotential (e.g. Fig. 3D).
In some systems where patterns of activity similar to those shown in

Fig. 3 are found, the return map actually has two extrema [11,12,38].
In that case, it should be possible to see one of two different periodic
orbits at a fixed value of the parameter, depending on initial conditions
[12,13,43]. We as yet have no evidence for such bistability in Fig. 3,
since we have always started simulations from one set of initial conditions. Note that, as in the case of homoclinic orbits mentioned above,
bistability is best appreciated when a two-parameter analysis is carried
out. In fact, two-parameter analysis has demonstrated intimate con-
nections between bistability, period-doubling bifurcations, chaotic
dynamics, homoclinic orbits, and cusp catastrophes in several different
systems in which study of the dynamics can be reduced to the study of a
one-dimensional map [11-14,17,43].

Waveforms such as those shown in Fig. 3B,C,D can exist only in
systems of dimension higher than two; they cannot be seen in two-
dimensional systems, since trajectories would then have to cross each
other, violating uniqueness of solution. There are now several modelling
studies on three-dimensional systems in which waveforms similar to those
shown in Fig. 3 are to be seen. A common characteristic of these systems
is the existence of a two-dimensional S-shaped slow manifold. The two
attracting leaves of this manifold are involved in the generation of the
large-amplitude relaxation oscillation. An S-shaped two-dimensional slow
manifold also exists in our three-dimensional model and is given by the
solution of the equation $V = 0$. As initially pointed out by Rössler [40],
the existence of displaced reinjection of trajectories is possible when
the slow manifold is S-shaped; this then leads to return maps having at
least one extremum.

In conclusion, we find it somewhat surprising that the activity of
an SAN cell can be well represented by a model of dimension as low as
three. In addition, this minimal model displays the three qualitatively
different ways of abolishing spontaneous activity that have been demonstrated so far in experimental work on the SAN. Finally, the possible
existence of period-doubling bifurcations, homoclinic orbits, bistability
and chaotic dynamics needs further investigation.

V   ACKNOWLEDGEMENTS

We thank Sandra James for typing the manuscript and Robert Lamarche
for photographing the figures. MRG would like to thank the Canadian
Heart Foundation and the Natural Sciences and Engineering Research
Council of Canada for postdoctoral fellowship support (1984-86).
Supported by grants to HJJ from ZWO (Dutch Organization for Pure Research,
grant no. 900516-80) and to MRG from the Medical Research Council of
Canada.

REFERENCES

[1]   A. Arneodo, P. Coullet and C. Tresser, J. Stat. Phys. 27:171 (1982).
[2]   H. Brown, J. Kimura and S. Noble, in "Cardiac Rate and Rhythm",
         eds. L.N. Bouman and H.J. Jongsma, Martinus Nijhoff, The Hague,
         pp. 53-68 (1982).
[3]   E. Bozler, Am. J. Physiol. 138:273 (1943).
[4]   C. McC. Brooks and H.-H. Lu, "The Sinoatrial Pacemaker of the Heart',
         Thomas, Springfield, pp. 89 (1972).
[5]   T.R. Chay, Physica 16D:233 (1985).
[6]   T.R. Chay and Y.S. Lee, Biophys. J. 47:641 (1985).
[7]   T.R. Chay and J. Rinzel, Biophys. J. 47:357 (1985).
[8]   J.R. Clay, J. theor. Biol. 64:671 (1977).

[9]   P.F. Cranefield, Circ. Res. 41:415 (1977).
[10] P. Gaspard, Phys. Lett. 97A:1 (1983).
[11] P. Gaspard and G. Nicolis, J. Stat. Phys. 31:499 (1983).
[12] P. Gaspard, R. Kapral and G. Nicolis, J. Stat. Phys. 35:697 (1984).
[13] L. Glass, M.R. Guevara, J. Belair and A. Shrier, Phys. Rev. 29A: 1348 (1984).
[14] P. Glendinning and C. Sparrow, J. Stat. Phys. 35:645 (1984).
[15] A. Goldbeter and O. Decroly, Am. J. Physiol. 245:R478 (1983).
[16] M.R. Guevara, in Proceedings of Conference "Temporal Disorder in Human Oscillatory Systems", Bremen, 1986, Springer, Heidelberg (in press).
[17] M.R. Guevara and L. Glass, J. Math. Biol. 14:1 (1982).
[18] M.R. Guevara and H.J. Jongsma: unpublished.
[19] M.R. Guevara, T. Op't Hof and H.J. Jongsma: unpublished.
[20] J. Honerkamp, G. Mutschler and R. Seitz, Bull. Math. Biol. 47:1 (1985).
[21] H. Irisawa and A. Noma, in "Cardiac Rate and Rhythm", eds. L.N. Bouman and H.J. Jongsma, Martinus Nijhoff, The Hague, pp. 35-51 (1982).
[22] J. Jalife and C. Antzelevitch, Science 206:695 (1979).
[23] C.D. Jeffries, Physica Scripta T9:11 (1985).
[24] I. Kodama and M.R. Boyett, Pflüg. Arch. 404:214 (1985).
[25] D. Kreitner, J. mol. cell. Cardiol. 7:655 (1975).
[26] R. Lozi, C.R. Acad. Sci. Paris 294:21 (1982).
[27] J. Maselko and H.L. Swinney, Physica Scripta T9:35 (1985).
[28] G. Matsumoto and T. Kunisawa, J. Phys. Soc. Japan 44:1047 (1978).
[29] A. Noma, Jap. J. Physiol. 26:619 (1976).
[30] A. Noma and H. Irisawa, Jap. J. Physiol. 24:617 (1974).
[31] A. Noma and H. Irisawa, Pflüg. Arch. 351:177 (1974).
[32] A. Noma and H. Irisawa, Jap. J. Physiol. 25:287 (1975).
[33] A. Noma, M. Morad and H. Irisawa, Pflüg. Arch. 397:190 (1983).
[34] L.F. Olsen and H. Degn, Biochim. Biophys. Acta 523:321 (1978).
[35] W. Osterrieder, Q.-F. Yang and W. Trautwein, Pflüg. Arch. 394:78 (1982).
[36] L.-Q. Pei, F. Guo, S.-X. Wu and L.O. Chua, IEEE Trans. Circuits and Syst. CAS-33:439 (1986).
[37] A.S. Pikovsky, Phys. Lett. 85A:13 (1981).
[38] A.S. Pikovsky and M.I. Rabinovich, Physica 2D:8 (1981).
[39] A.S. Pikovsky and M.I. Rabinovich, Sov. Phys. Dokl. 213:183 (1978).
[40] O.E. Rössler, Z. Naturforsch. 31a:259 (1976).
[41] O.E. Rössler and K. Wegmann, Nature (Lond.) 271:89 (1978).
[42] J.-C. Roux, R.H. Simoyi and H.L. Swinney, Physica 8D:257 (1983).
[43] M. Schell, S. Fraser and R. Kapral, Phys. Rev. 28A:373 (1983).
[44] R.A. Schmitz, K.R. Graziani and J.L. Hudson, J. Chem. Phys. 67:3040 (1977).
[45] L.P. Sil'nikov, Math. USSR Sbornik, 10:91 (1970).
[46] R.H. Simoyi, A. Wolf and H.L. Swinney, Phys. Rev. Lett. 49:245 (1982).
[47] K. Tomita and I. Tsuda, Phys. Lett. 71A:489 (1979).
[48] J.J. Tyson, J. Math. Biol. 5:351 (1978).
[49] J.S. Turner, J.-C. Roux, W.D. McCormick and H.L. Swinney, Phys. Lett. 85A:9 (1981).
[50] B. Victorri, A. Vinet, F.A. Roberge and J.-P. Drouhard, Comp. Biomed. Res. 18:10 (1985).
[51] T.C. West, in "The Specialized Tissues of the Heart", eds. A.P. DeCarvalho, W.C. DeMello and B.F. Hoffman, Elsevier, New York, pp. 81-94 (1961).

OSCILLATORY PLANT TRANSPIRATION: PROBLEM OF A MATHEMATICALLY CORRECT

MODEL

I. Gumowski

University of Toulouse 3
118, route de Narbonne
31062 Toulouse

ABSTRACT

Oscillatory water transpiration is known to occur in many plant
species with a repetition rate of the order of 30 minutes. The transpiration rhythm is directly related to a cyclic variation of stomatal
apertures, and indirectly to the water content of guard, subsidiary and
mesophyll cells. There exists a large number of accurately measured
oscillation wave forms, some of which appear to be chaotic. A lumped-
element hydraulic elastic model of the water transport was formulated in
1972, without specifying explicitly the forces responsible for the upward
movement of water. An analysis of the implications of this model has
disclosed an agreement with some observed properties, in particular the
frequency of periodic oscillations, and a qualitative and quantitative
disagreement with others, in particular the oscillation wave forms, as
well as the response of the latter to various physical and chemical
perturbations. The argument is re-examined and an improved model is
proposed.

I SIMPLIFYING ASSUMPTIONS LEADING TO A LUMPED-PARAMETER MODEL

Combined research results of many experimental and theoretical
workers have led to the conclusion that water transport in plants, from
the roots to the leaves and finally to the atmosphere, can be described
by four 'collective' macroscopic variables: water potential, rate of
water flow, resistance to water flow and storage capacity of water volume.
Osmotic pressure can be expressed by an equivalent potential. The
observed phenomenological interrelationships between these four variables
give rise to a lumped-element hydraulic-elastic model of water flow,
whose detailed structure depends on the intended purpose. Two predominant
purposes have emerged: (i) study of water flow on a time scale of days,
and (ii) study of water flow on a time scale of hours. The former is
intended essentially for the study of plant growth, the effect of water
stress, of artificial irrigation, and other long-term perturbations [1],
[2], [3], and the latter for a study of oscillatory transpiration [4].
Since lumped electrical networks are easier to interpret and analyse than
hydraulic networks, it is convenient to use electrical equivalents.
Water potentials are represented by electrical potentials v(t), rates of

water flow by currents i(t), resistances to water flow by resistances R, or conductances g, water volumes by electric charges w(t), i=dw/dt, and water storage capacities by capacitors C. Subscripts refer to macroscopic plant parts or to their combinations.

Groupings of plant parts into relevant lumped elements are not quite the same for the models (i) and (ii), because some parts, like for example the water storage capacity of the root and soil system, do not have the same effect on slowly and rapidly evolving water transport rates i(t). The i(t) and v(t) can be separated into constant (DC) and variable (AC) parts, the former representing suitably chosen time-averages of the latter. This separation permits the determination of an equivalent network for each component containing fewer relevant elements, which results in a considerable simplification of analysis. The equivalent 'dynamic' or 'AC' network resulting from the Cowan model [4] is shown in Fig. 1. Its components are identified in the figure caption. The reference for the incremental elements is a constant (DC) state. The resistances and capacitances are assumed to be linear, with constant values of R and C depending on the DC state. The central feature of the model of Fig. 1 is the ideal current generator $i(t)=f(w_g(t),w_s(t)$, representing the dependence of the stomatal transpiration rate on the water content of guard and subsidiary cells. The form of the function $i=f(w_g,w_s)$ is defined by the mechanical structure of the stomatal complex, consisting of the particular assembly of guard and subsidiary cells around a stomatal aperture. The ideal current generator i=f provides a source of positive feedback without which the existence of self-excited and self-sustained oscillations would be impossible. The energy source responsible for the widening and narrowing of the size of the stomatal aperture appears neither in the dynamic model of Fig. 1 nor in the full model [4].

II COMPARISON OF THEORETICAL AND EXPERIMENTAL DATA

The practical usefulness of the model of Fig. 1 can be easily evaluated in principle, because highly accurate measurements of oscillatory transpiration and of simultaneous water absorption by the roots are available for some plant species. The most complete observations, including the effect of various perturbing chemical and physical agents, have been published for young oat plants (Avena sativa) grown in a controlled environment [9]. The initial justification of the model of Fig. 1 was based on the observation that for plausible parameter values a simulation on an analogue computer produced oscillations with a frequency which matched rougly the measured oscillation frequency of a cotton leaf on an intact plant [4]. A further exploration of the model of Fig. 1, with the effect of mesophyll cells neglected, has disclosed a qualitative disagreement between simulated and observed oscillation wave forms [10].

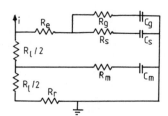

Fig. 1. Dynamic model of oscillatory water transport. Subscripts: e epidermis, g guard cells, ℓ leaf, m mesophyll cells, r roots + xylem, s subsidiary cells; i=f ideal current generator.

A more detailed exploration, with the effect of mesophyll cells taken into account, managed to reproduce some wave forms observed under particular conditions, but in general it confirmed the existence of a fundamental discrepancy [5], [6]. The wave forms deduced from the model of Fig. 1 did not change in the same way as the observed ones under the influence of perturbations. The latter were applied one at a time, and the lumped parameters were modified to reflect the known qualitative influence. Quantitative analysis has thus shown that the model of Fig. 1 is fundamentally unsound. A posteriori it is possible to identify the reason of the unsoundness without a repetition of the detailed comparison of computed and observed wave forms. By looking at Fig. 1, or the original model [4], it is sufficient to note that water can flow from guard, subsidiary and mesophyll cells directly to the soil, bypassing both the stem of the plant and the atmosphere. The same type of bypass exists also in the long-time model [1], [2], [3]. A more thorough examination of the forces and energy sources responsible for the upward water transport in a plant suggests a way to remove the unrealistic bypass.

III   THE MECHANISM OF WATER TRANSPORT

What forces or combination of forces are responsible for the water transport from the soil to the leaves of the plants, where a part of the water is evaporated? The ducts through which water flows against the force of gravity are rather narrow, the cross-sectional diameters are 0.1 mm or less. Present knowledge suggests that root pressure, capillarity and transpiration-cohesion are likely candidates in that order of increasing importance. Root pressure appears to play a minor role, because cut flowers and twigs maintain their turgor and transpiration for a considerable time when the cut stems are immersed in water.

The force of capillarity is well understood in principle. For narrow tubes it is governed by the relation

$$h = (2\sigma\cos\theta)/(g\rho r) \tag{1}$$

where g = acceleration of gravity, ρ = density of water, r = internal radius of the tube, σ = surface tension, θ = contact angle at the tube boundary, and h = the height over the reservoir to which water rises inside the tube. The relation (1) is weakly temperature-dependent. In order to illustrate the modus operandi of the transpiration-cohesion force assume that the upper end of the capillary tube is capped by, say, a horizontal platform somewhat larger than the tube area, and that the tube can be telescopically shortened or lengthened without any change of the internal radius. If the tube is slowly shortened till its length $h_1 < h$, a drop of water will appear on the platform. If the tube is then slowly lengthened, it is possible to obtain $h_1 > h$ without losing completely the water drop on the platform. The weight of the excess water column of height $h_1 - h$ is maintained by the increased surface tension force: drop-platform + drop-air. This force is transmitted to the water column by intermolecular cohesion. The water lost by evaporation from the outer surface of the drop is replaced internally in the same way as if capillarity were acting alone, i.e. by free motion of molecules inside the liquid. It would be helpful to have an explicit expression like (1) for the excess height corresponding to a given geometrical configuration. Unfortunately, the theoretical problem of fluid mechanics involves a free boundary, and remains at present unexplored. The force of transpiration-cohesion allows water molecules to escape into the atmosphere through the liquid-air boundary layer dominated by the surface tension force, which is an energy expending process. The necessary energy is taken from the heat available near the liquid-air boundary layer, i.e. it is the atmosphere, the plant environment and the incident heat radiation which does the work nominally ascribed to the force of transpiration-cohesion. Summarising the physical argument it is possible to affirm that capillarity and transpiration-cohesion are sufficient to assure the water transport in small plants, whose height above the soil water reservoir is, say, 20 cm or less. The plants used for the experiments reported in the publications [9] had a height of about 5 cm. The simulation of the stomatal water transport force by means of an ideal current generator appears therefore to be qualitatively and quantitatively justified, provided one is not interested in the reaction of the atmosphere on the internal state of the plant. If the internal reaction is not negligible, the ideal current generator must be completed by an internal impedance, and the resistance or impedance of the atmosphere-soil path must be added to the equivalent electrical network.

IV MODIFICATION OF THE MODEL

A re-examination of the argument leading to the model of Fig. 1 in the light of quantitative data suggests that all its lumped elements have a role to play, but that their interconnections should be rearranged. The simplest rearrangement which appears to reproduce better the structure of an actual plant is shown in Fig. 2. If one writes out the equations of water flow, one obtains an implicit operator system. Without loss of generality it is possible at first to neglect R_e. One obtains then

$$i_s + i_g = (1 + \frac{R_g\,C_s\,s + \alpha}{R_s\,C_s + 1})\,i_g = f(w_g, w_s), \quad s = d/dt$$

$$w_s = \frac{R_g\,C_s\,s + \alpha}{R_s\,C_s\,s + 1}\,w_g, \qquad \alpha = C_s/C_g, \qquad i_g = s\,w_g \tag{2}$$

which can be converted into a single nonlinear integro-differential equation. It is noteworthy that the system (2) does not depend on the root, stem and mesophyll cell parameters. This intuitively surprising property is a consequence of the simplifying assumption that i=f is an ideal current generator having an infinite internal impedance. The elements missing in the system (2) happen to be in series with this impedance, and are therefore automatically eliminated. In excised plants a compression of the stem is equivalent to the survival of the ideal current generator. A practical study of the system (2) requires the know-ledge of the shape of the function f. This shape is not the same for all plant species, because it depends on the specific structure of the stomatal complex. In the absence of the corresponding physical information it is necessary to assume some plausible shapes inferred from qualitative

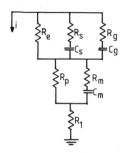

Fig. 2. Simplest rearrangement of the model of Fig. 1. Absence of a parasitic water path. $R_p = \frac{1}{2}R_\ell$, $R_1 = R_r + R_x + R_p$

observations, and to compare the resulting waveforms to those observed experimentally. In linear approximation it is plausible to use the relation [4]

$$f = aw_g + bw_g, \quad ab < 0, \tag{3}$$

where a,b are constants. Combining the relations (2), (3) and rearranging one obtains the second order equation

$$(R_g + R_s)C_s\ddot{w}_g + [1 + \alpha - (aR_sC_s + bR_gC_s)]w_g - (a + b\alpha)w_g = 0, \quad = d/dt \tag{4}$$

The steady state solution $w_g(t) = 0$ admits complex conjugate eigenvalues with a positive real part, provided

$$-(a + b\alpha) > 0 \quad \text{and} \quad 1 + \alpha - aR_sC_s - bR_gC_s < 0 \tag{5}$$

The first condition is satisfied when a>0, b<0 and $-bC_s > aC_g$. Since in most plants subsidiary cells are larger than guard cells, $C_s > C_g$, this condition is realistic. The second condition requires that aR_s be somewhat larger than $-bR_g + (C_s + C_g)/(C_sC_g)$. It is satisfied when subsidiary cells are less permeable than guard cells, i.e. when $R_s > R_g$. The conditions (5) are changed only in a minor way when the resistance R_e is not negligibly large.

If increasing oscillatory transients are to terminate on periodic solutions of the limit cycle type, nonlinear terms must be added to the rhs of equation (4). From an inspection of the symmetry properties of experimentally observed wave forms it is possible to infer that these terms must be predominantly even. The simplest possibility is that they be quadratic. In such a case the periodic solution is not of a standard type, but it can be constructed just as easily [7], [8]. No matter how many nonlinear algebraic terms are added, the oscillatory steady state solution will remain periodic. In other words, a nonlinear algebraic generalisation (free of integral, higher order differential or delayed terms) of equation (4) cannot account for the observed apparently chaotic oscillations. Such oscillations occur under low illumination levels and necessarily involve mesophyll cells.

An influence of mesophyll cells can be secured in several ways. The most obvious one consists in the replacement of the ideal current generator i=f by one with a finite internal impedance. The resulting equations become, however, much more complex, and they appear very difficult to analyse. An alternative, qualitatively less exact way, is to keep the ideal current generator, and to modify only the structural positions of R_m and C_m. A plausible configuration is shown in the equivalent network

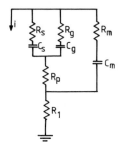

Fig. 3. A plausible variant of the model of Fig. 2.

of Fig. 3. With R_e initially neglected one obtains then the operator system

$$(1 + \frac{R_gC_s \ s + \alpha}{R_sC_s \ s + 1} + \frac{C_mZ_1s}{R_mC_ms+1} \cdot \frac{R_sC_s \ s + 1 + \alpha}{R_sC_s \ s + 1})i_g = f(w_g,w_s), \ s = d/dt,$$

$$w_s = \frac{R_gC_s \ s + \alpha}{R_sC_s \ s + 1} \ w_g, \ \alpha = C_s/C_g, \ i_g = sw_g, \ Z_1 = \frac{A_o s^2 + A_1 s + A_2}{R_1C_1s^2 + s}, \quad (6)$$

$$R_1 = R_g+R_s, \ A_1 = R_p+R_g \ \frac{C_1}{C_s} + R_s \ \frac{C_1}{C_g}, \ C_1 = \frac{C_gC_s}{C_g+C_s}, \ A_o = (R_pR_1+R_sR_g)C_1,$$

$$A_2 = \frac{C_1}{C_gC_s}$$

which reduces to the system (2) as $R_m \to \infty$ or $C_m \to 0$. The effective differential order of this system being higher than two, it may admit chaotic oscillatory solutions, at least for an appropriate choice of the function f. Whether it actually does so in a quantitatively satisfactory manner, remains to be seen. Although a quantitative analysis of the system (6) is rather laborious, the results will either confirm the validity of the model of Fig. 3, lead to further modifications, or suggest additional experiments. In all cases a quantitative analysis will contribute to a better understanding of the mechanism of oscillatory transpiration.

REFERENCES

[1]   A. Berger, Actes Coll, Uppsala 1970, UNESCO 5: 201-212 (1973).
[2]   A. Berger, Bull. Soc. Bot. de France 125: 159-176 (1978).
[3]   A. Berger, La Houille Blanche 3/4: 227-233 (1978).
[4]   I.R. Cowan, Planta (Berl) 106: 185-219 (1972).
[5]   I. Gumowski, Cycle Res. 12: 273-291 (1981).
[6]   I. Gumowski, Cycle Res. 14: 33-41 (1983).
[7]   I. Gumowski, Comptes rendus Acad. Sc. Paris 295(I): 483-486 (1982).
[8]   I. Gumowski, Bull Acad. Roy. Belge. Cl. Sc. 5e S. 69: 204-209
        (1983).
[9]   A. Johnsson et al., Physiol. Plant. Part I 28: 40-50 (1973);
        Physiol. Plant. Part II 28: 341-345 (1973);
        Physiol. Plant. Part III 31: 112-118 (1974);
        Physiol. Plant. Part IV 31: 311-322 (1974);
        Physiol. Plant. Part V 32: 256-267 (1974);
        Physiol. Plant. Part VI 42: 379-386 (1978).
        Planta (Berl) 124: 99-103 (1975).
        Z. Pflanzenphysiol. 80: 251-260 (1976).
[10] A. Johnsson, Bulletin IMA (UK) 12(1): 22-26 (1976).

BIFURCATIONS IN A MODEL OF THE PLATELET REGULATORY SYSTEM

Jacques Bélair

Département de mathématiques et de statistique and
Centre de recherches mathématiques
Université de Montréal
C.P. 6128, Succursale A
Montréal
Québec H3C 3J7
Canada

ABSTRACT

Modelling of the control mechanism for the regulation of platelet production leads to a functional differential equation with two time delays, one accounting for the senescence time of platelets, and the other one due to the maturation time of megakaryocytes. Local stability analysis and numerical simulations are performed to evaluate the possible behaviours of the solutions as clinically relevant, physiologically realistic parameters are varied. In particular, possible mechanisms for the onset of cyclic thrombocytopenia and idiopathic thrombocytopenic purpura are discussed.

I   INTRODUCTION

Dynamical diseases have been defined [5] as the manifestation, in an intact physiological regulatory system, of the control parameters taking values outside their normal range, the latter being operationally defined as the range of values for which the system can be considered 'healthy'. This concept has led to the interpretation of various haematological and respiratory disorders in terms of bifurcations occurring in the corresponding mathematical systems.

In this short paper, we present preliminary analysis of a model for the regulation of platelet production: this model can be reduced to a differential equation containing two time delays. The main (unusual) feature of our model is the presence of a second time delay, in addition to the one usually incorporated in most population models. This significantly complicates the analysis, even the local stability of the stationary solutions being difficult to assess in all generality. The results of numerical integration of our equation are shown: the main parameter we vary is a factor of "random" destruction which we suppose small in normal conditions, and leads to singular, although periodic, oscillations.

II MODEL AND ANALYSIS

Normal mammalian thrombopoiesis is accepted to be organised as follows [7]. A self-maintaining population of pluripotential stem cells gives rise to more mature stem cells committed to the eventual production of platelets. The signal responsible for triggering cells in the pluri-potential stem cell population into the committed platelet series is unknown. From these committed cells derive the first morphologically recognisable platelet precursors, the megakaryocytes. The latter form intracellular cytoplasmic platelet units which are subsequently released into the circulation as platelets. Unlike the precursors of the erythro-cytes and neutrophils, maturation of the megakaryocytes is not accompanied by cellular division. The total time elapsed between the appearance of a recognisable megakaryocyte and when it may start to produce platelets is about 9 days in the normal human [8]. Once released into the circulation, platelets normally disappear primarily through senescence, living for approximately 10 days [6]. There is, however, evidence for the random destruction of platelets at a low level in normal humans, and this may be exacerbated in certain disease states.

The regulation of platelet production is accomplished via the humoral stimulator thrombopoietin, analogous to the erythropoietin present in erythropoiesis. The control of thrombopoiesis is somewhat less well understood than that of erythropoiesis. What is clear, though, is that a fall in platelet number leads to an increase in thrombopoietin levels and a subsequent increase in platelet production [7], thought to be primarily accomplished by an increase in the flux of cells entering the megakaryocyte compartment from the committed stem cell population.

Let $P(t)$ denote the total number of circulating platelets of all ages, T_M stand for the maturation time of a megakaryocyte (from the time it becomes recognisable to the time it may start to produce platelets), and T_s represent the age of death, due to senescence, of a platelet. If, in addition, we let γ be an age-independent rate of random destruction of platelets, then we may derive [2] the equation

$$dP/dt = -\gamma P(t) + \beta(P(t-T_M)) + \beta(P(t-T_M-T_s))\exp(-\gamma T_s) \qquad (1)$$

where the function β reflects the feedback ruling the influx of cells into the recognisable megakaryocyte compartment. For definiteness purposes, we let $\beta(u) = \beta_0 u \theta^n/(\theta^n+u^n)$. The parameters β_0, θ and n will be evaluated from clinical data.

Equation (1) possesses the stationary solution $P_1 = 0$ for all values of the parameters. Another steady state appears at $P_2 = \theta[\beta_0(1-\exp(-\gamma T_s))/\gamma - 1]^{1/n}$, when $\beta_0 > \gamma/(1-\exp(-\gamma T_s))$, and the trivial solution P_1 is locally asymptotically unstable for the parameter values satisfying this last relation. We have been able to locate geometrically a curve of Hopf bifurcations in the plane of the parameters γ and β_0 [1]: these occur when the characteristic equation of equation (1) linearised around P_2 possesses a pair of pure imaginary roots, and all other solutions are in the left half of the complex plane.

We now consider, by numerical integration, the influence of an increase in the level of random destruction of the platelets, corresponding to a variation in the values of the parameter γ. To find appropriate estimates for the other parameters appearing in the feedback function, we have used published clinical data [3] to obtain $\beta_0 = 37$, $\theta = 0.068$ and n = 2.2. The sequence of bifurcations we obtain do not seem to depend on the precise values of these parameters, although large deviations would undoubtedly lead to different dynamics.

We show in Fig. 1 the results from one set of time series once transients have (apparently) vanished, along with the value of the parameter γ. For very low values of γ, the steady state $P_2 = 1$ is stable, and, for large values of γ, the stationary solution $P_1 = 0$ attracts all solutions: this is consistent with the role played by an additional destruction factor such as γ. In the intermediate range, however, an interesting sequence of oscillating solutions can be observed.

As γ is decreased, a significant proportion of each cycle is spent at very low levels of concentration of the platelets, of the order of 1% of the normal value. Simultaneously, the amplitude of the oscillation is increased to more than six times the normal concentration. The period of the oscillations is also increasing. These types of oscillations are reminiscent of the platelet levels observed in both cyclic thrombocytopenia and idiopathic thrombocytopenic purpura [4]: periodicity in platelet levels and very low platelet concentration are fundamental features of these diseases.

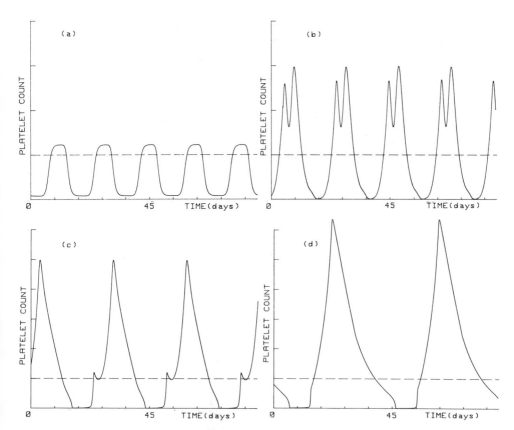

Fig. 1. Platelet count as a function of time. In all cases, n = 2.2, $\theta = 0.068$ and $\beta_0 = 37$, as fitted from [3]. Duration of each frame is 95 days; the dashed line indicates the normal level (1.8E+5/μl). The parameter γ is varied: (a) $\gamma = 1$; (b) $\gamma = 0.3$; (c) $\gamma = 0.1$; (d) $\gamma = 0.05$.

III CONCLUSION

The onset of clinical thrombocytopenia and idiopathic thrombocytopenic purpura may be explained by variations in the controlling parameters, putting the intact regulatory mechanisms in a range of values giving rise to these pathologies. They could be considered as dynamical diseases.

IV ACKNOWLEDGEMENTS

This study has been supported by NSERC (Canada), FCAR and MES (Quebec). We have benefitted from indispensable conversations with Michael Mackey.

REFERENCES

[1] J. Bélair, "Stability of a differential-delay equation with two time lags", CMS Proceedings, to appear (1987).
[2] J. Bélair and M.C. Mackey, "Oscillatory pathologies of platelet control", in preparation (1987).
[3] I. Branehög, J. Kutti and A. Weinfeld, Brit. J. Haematol. 27: 127-143 (1974).
[4] T. Cohen and D.P. Cooney, Scand. J. Haematol. 12: 9-17 (1974).
[5] L. Glass and M.C. Mackey, Ann. N.Y. Acad. Sci. 316: 214-235 (1979).
[6] L.A. Harker, J. Clin. Invest. 48: 963-974 (1969).
[7] L.A. Harker, Clin. in Haematol. 6: 671-693 (1977).
[8] U. Müller, J. Kulina, J. Luber, H.-P. Hebestreit, M. Mayer, U. Kempgens and W. Quisser, Blut 34: 31-38 (1977).

CHAOS IN A SYSTEM OF INTERACTING NEPHRONS

Klaus Skovbo Jensen[1], Niels-Henrik Holstein-Rathlou[2],
Paul P. Leyssac[2], Erik Mosekilde[1] and Dan Rene Rasmussen[1]

[1]Physics Laboratory III   [2]Institute for Experimental
Technical University of Medicine
Denmark   University of Copenhagen
DK-2800 Lyngby   DK-2100 Copenhagen
Denmark   Denmark

ABSTRACT

Urine is formed from blood by a process of filtration followed by
selective reabsorption and secretion. The basic functional unit is the
nephron of which there are about 30,000 in a rat kidney (1 million in a
human kidney).

Recent experimental results have revealed an oscillatory behaviour
of the proximal intratubular pressure in rat nephrons with a typical
period of 20-30 sec. The oscillations are assumed to arise from temporal
self-organisation in the glomerular filtration of the individual nephron.
While for normal rats these oscillations have a well-defined frequency and
amplitude, highly irregular oscillations are observed for spontaneously
hypertensive rats.

We present some simulation results from models of non-interacting as
well as coupled nephrons and compare these with results from analysis of
experimental data.

I INTRODUCTION

Recent experimental investigations performed at the Institute for
Experimental Medicine, University of Copenhagen, have revealed an
oscillatory behaviour of the proximal intratubular pressure in rat nephrons [4]. For normal, healthy rats, the oscillations have a well-defined
frequency and amplitude. For rats with genetically elevated blood pressure, however, the oscillations are highly irregular. Figure 1 shows
typical experimental results. The basic cycles (0.4 - 0.6/min) are about
5 mmHg peak to peak around a mean value of approximately 12 mmHg.

The above investigations are contributions to a forthgoing research
programme, the aim of which is to elucidate the pathophysiological changes
which lead to chronically elevated blood pressure, one of the most common
diseases in the Western World. A growing body of evidence points to a
close connection between essential hypertension and kidney function.

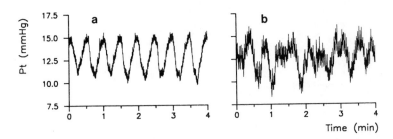

Fig. 1. Typical oscillations in the proximal intratubular pressure, P_t, as measured in (a) a normal rat and (b) a spontaneously hypertensive rat.

It is likely that maintenance of a high blood pressure in hypertensive persons is due, at least partly, to a malfunction of offset in the regulatory mechanisms which control salt and water excretion. One such mechanism is the tubuloglomerular feedback loop involved in regulation of pressures and flows in individual nephrons. Our model presented here is concerned with the dynamics of that particular loop.

II THE SINGLE-NEPHRON SYSTEM

In the upper end, the nephron contains 20-40 capillary loops arranged in parallel. This configuration, the glomerulus, is supplied with blood through the afferent arteriole, see Fig. 2.

The glomerulus acts as a filter through which low-molecular blood constituents can pass into the proximal tubule. Typically 25-25% of the plasma is filtered out as the blood passes the capillary system. Blood cells and proteins are retained within the glomerular capillaries, and therefore the blood viscosity and the colloid osmotic pressure increase. Through selective reabsorption and secretion, the constituents of the filtrate are changed as it passes through the various segments of the tubule: the proximal tubule, the loop of Henle, the distal tubule, and the collecting duct.

It has long been recognised that the kidneys are capable of compensating for changes in arterial blood pressure, and that this ability partly rests with feedback mechanisms in the individual nephrons. By adjusting mainly the flow resistance (R_a) in the afferent arteriole, the pressure (P_t) in the proximal tubule can thus be maintained at a relatively constant level.

For simplification, we consider the venous pressure (P_e) and the efferent flow resistance (R_e) as constants, and the glomerular hydrostatic pressure (P_g) is then effectively determined by the arterial pressure (P_a) and the flow resistance in the afferent arteriole.

The difference between glomerular filtration pressure (hydrostatic pressure minus colloid osmotic pressure) and intratubular pressure controls the filtration rate (F_{filt}) into the proximal tubule. The filtration again influences the two pressures by changing the stores of capillary plasma and filtrate, respectively. The intratubular pressure determines the flow of filtrate into the loop of Henle (F_{hen}), and via the juxtaglomerular apparatus, the Henle flow again determines the afferent arteriolar resistance. If the Henle flow increases, the afferent

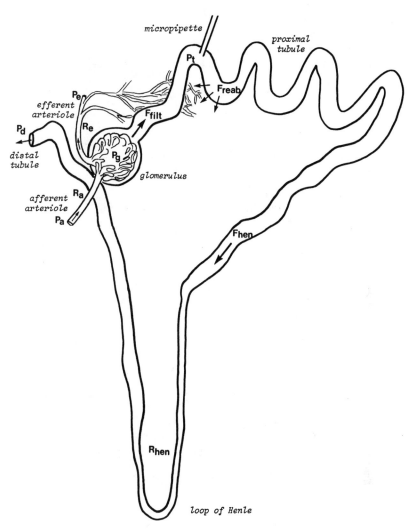

Fig. 2. The nephron. Some of the variables and constants used in the model are indicated corresponding to their physical meaning.

arteriolar resistance also increases, and the glomerular filtration pressure is reduced. Altogether, the feedback is thus negative.

The physiological mechanisms responsible for adjusting the afferent arteriolar resistance are not known in detail. It is likely that the arteriolar wall is stimulated by a signal arising from specialised cells situated where the terminal part of the loop of Henle passes the afferent arteriole (the juxtaglomerular apparatus). Some of these cells, the macula densa cells, are assumed to be sensitive to Na^+ and Cl^- ions. Due to the complicated balances of absorption and secretion processes in the loop of Henle, the Na^+ and Cl^- concentrations in the exiting fluid depend upon the flow rate, and the activity of the macula densa cells therefore effectively measures the Henle flow.

Due to finite transit time through the loop of Henle, the Na^+ concentration at the macula densa cells does not change immediately upon a change in the Henle flow. An additional delay is introduced in transmission of the signal from the macula densa cells to the arteriolar wall, presumably because this transmission involves a series of subsequent processes.

With respect to the dynamical response of the afferent arteriole only limited information is available. It has been observed, however, that stepwise changes in vasoconstrictor nerve activity can give rise to large-amplitude damped oscillations with typical frequencies around 0.03 Hz [7]. A simple equation capable of describing this behaviour is a second order ordinary differential equation corresponding to a non-linear, damped oscillator. For physiological reasons, we assume that both the force constant (K) and the damping increase when the diameter of the arteriole approaches its two extrema (Fig. 5B).

Variation of parameters which are likely to differ between normal and SH rats causes our single-nephron model (described in [5]) to develop through a cascade of period-doubling bifurcations to reach a chaotic regime. This route to chaos is indeed supported by experimental findings (Fig. 3).

The "regular" in vivo pressure oscillations, however, show amplitude modulation and slow fluctuations in base line, probably caused by drifting in biological constants and/or interaction between nephrons.

III COUPLED NEPHRONS

Due to the very dense packing of nephrons in the kidney, one may expect a certain degree of interaction between neighbouring nephrons. Thus, in a recent study [3] pairs of nephrons were found in which the intratubular pressure oscillations had identical frequency and phase relations. Furthermore, in such pairs, perturbation of the flow in one nephron could elicit changes in the pressure oscillation of the other nephron. Nephrons with a 180° phase lag between their oscillations have also been observed (not published). The interaction is likely to occur as a sum of several possible coupling mechanisms, one of them being simultaneous action on a common vascular segment through the feedback systems of neighbouring nephrons.

Based on a slight modification of the single-nephron model in [5], we made a set-up of 256 nephrons on the parallel computer PALLAS [1], each nephron being influenced by its four neighbours in a square lattice. Eqs. (1)-(10) in Fig. 4 describe one nephron, including the coupling

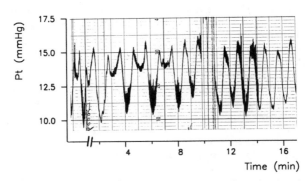

Fig. 3. Proximal intratubular pressure recording from a normal rat showing the transition from a period-2 to a period-1 limit cycle. The oscillations are very slow due to cooling (25°C) of the kidney surface.

$$(1) \quad \frac{dP_t}{dt} = \frac{1}{C_{tub}} \left(\frac{1}{R_a}(1 - H_a)(P_a - P_g)(1 - \frac{C_a}{C_e}) - F_{reab} - \frac{P_t - P_d}{R_{hen}} \right)$$

$$(2) \quad \frac{dP_g}{dt} = \frac{1}{C_{glo}} \left(\frac{P_a - P_g}{R_a} [1 - (1 - H_a)(1 - \frac{C_a}{C_e})] - \frac{P_g - P_e}{R_e} \right)$$

$$(3) \quad \frac{dR_a}{dt} = R_{ad}$$

$$(4) \quad \frac{dR_{ad}}{dt} = \omega^2 K(R_a) [R_{a0} \Psi(\frac{3X_3}{T} + \Delta) - R_a] - 2d\omega R_{ad} \sqrt{K(R_a)}$$

$$(5) \quad \frac{dX_1}{dt} = \frac{P_t - P_d}{R_{hen}} - \frac{3X_1}{T}$$

$$(6) \quad \frac{dX_2}{dt} = \frac{3(X_1 - X_2)}{T}$$

$$(7) \quad \frac{dX_3}{dt} = \frac{3(X_2 - X_3)}{T}$$

$$(8) \quad \frac{dX_{coup}}{dt} = K_{coup} (\frac{3X_3}{T} - X_{coup})$$

$$(9) \quad \text{where} \quad C_e = \sqrt{\left(\frac{a}{2b}\right)^2 + \frac{P_g - P_t}{b}} - \frac{a}{2b}$$

$$(10) \quad \text{and} \quad \Delta = \varepsilon \sum_{i=1}^{4} \{X_{coup}(i) - F_{hen0}\}$$

Fig. 4.   Set of eight ordinary differential equations describing a single nephron and its coupling to nearest four neighbours. The two one-argument-functions, ψ, and K, used in Eq. (4) are shown in Fig. 5. In all simulations we use C_{glo} = 0.15 nl/mmHg and K_{coup} = 0.1/sec. The other constants are the same as in [5].

mechanism. The magnitude and the degree of delayed action of the coupling are given by ε and K_{coup}.

All nephrons have the same parameter values except for the resonance frequency, ω, of the arteriolar resistance. ω is chosen at random from an interval (± 30%) around the nominal value and assigned to each nephron before the simulation.

One purpose of this lattice model was to study possible phase transitions in temporal and spatial structures as the coupling parameter values are changed. We also wanted to see the effect of such transitions on macroscopic variables such as the total renal filtration fraction. Here, however, we concentrate on the impact of coupling on the tubular pressure oscillation in the individual nephron.

Several physiological constants in the model are effective bifurcation parameters, such as the steepness of the feedback function, the time delay at macula densa, the "force constant" and the resonance frequency of the afferent arteriole, etc. In [5] we showed how the single-nephron model produces a chaotic attractor very similar to a Rössler band when the time delay, T, is increased. Here, we arbitrarily choose P_a to be the bifurcation parameter (Fig. 6), since it is not certain to which of the just mentioned parameters the difference between normal rats and SH rats shall be ascribed.

Figure 7 shows the phase portraits of a simulated pressure oscillation

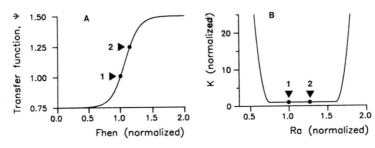

Fig. 5. Feedback function, ψ, and "force constant", K, used in the nephron model. Two different operating points (unstable equilibrium points) are shown corresponding to an arterial pressure, P_a, of 100 mmHg (1) and 120 mmHg (2), respectively.

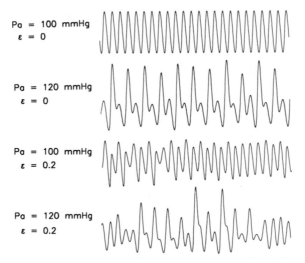

Fig. 6. Simulation results showing the pressure, P_t, in a single nephron out of 256 nephrons with nearest-neighbour coupling. The two top rows display tubular pressure oscillations for two different arterial pressures when there is no coupling to neighbours ($\varepsilon = 0$). Introducing a weak coupling, $\varepsilon = 0.2$, leads to more irregular oscillations (the two bottom rows).

(with coupling parameter $\varepsilon = 0.2$) and an irregular oscillation from a SH rat. The phase space coordinates are created from a single time series as described by Takens [9]: $[V(t),V(t+\tau),...,V(t+(m-1)\tau)]$, where m is the embedding dimension. To make the visualisations comparable, we adjust the time delay, τ, to the mean period of each time series as 1 to 5.5. Before embedding, all experimental data are bandpass filtered (9 - 90 mHz) to remove the strong ripple originating from respiration and heart beating.

When finding the correlation dimension [2], D_2, a time delay of 1.2-2.4 sec is used for the experimental data, see Figs. 8 and 9. All time series used for estimation of D_2 have a length of 5000-7000 points which corresponds to a number of 35-50 cycles.

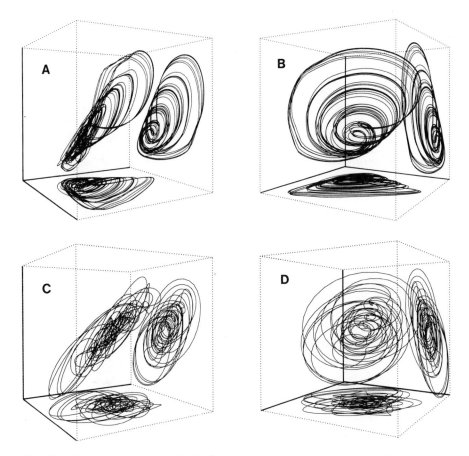

Fig. 7. Reconstruction of 3-dimensional phase portraits from a
single time series using Takens' scheme. (A) Simulation as
in Fig. 6, bottom row. A time delay of 5.0 sec was used for
the reconstruction. (B) Same object as in A, but viewed
from another angle. (C,D) Irregular oscillations in a SH
rat. Time delay is 3.5 sec. The time series was bandpass
filtered before making the phase portrait.

IV DISCUSSION

     From visual inspection of time series and phase portraits, and
estimation of the correlation dimension, we have found some similarity
between the dynamics of our model of coupled nephrons and the chaotic-
looking tubular pressure oscillations in SH rats. Here, we would like to
point out some problems and aspects concerning the interpretation of these
findings.

     First of all, the existence of a plateau in Fig. 8(D) does not itself
prove the existence of low-dimensional deterministic chaos in the
experimental system. Since the experimental data are rather noisy due to
components from respiration and heart beating as well as biological
fluctuations and measurement problems, filtration of the data is essential.
In general, the correlation integral should be calculated from raw data,
because filtration and smoothing operations increase the correlation
between the data points. In this case, however, high-frequency
oscillations transmitted via the blood pressure are merely added without

29

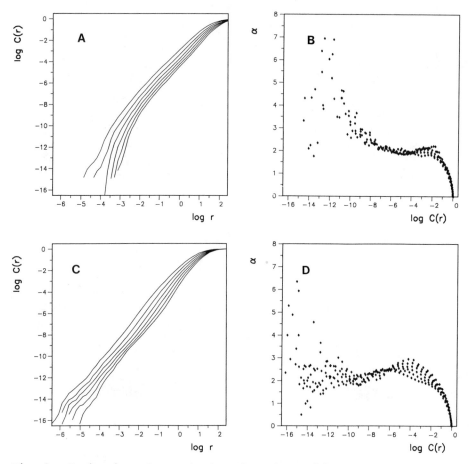

Fig. 8. Estimation of correlation dimension. (A) Correlation
integral for the simulated attractor in Fig. 7 with
embedding dimension m=4 to m=9. A time delay of 2.0 sec
was used for the phase space reconstruction. (B)
Corresponding plot of the local slope, α = d log C(r)/d log r.
From the plateau between -7.0 and -3.5 we find the correlation
dimension, D_2, to be approximately 2.1. (C,D) Similar plots
for the experimental data from Fig. 7, but now using a time
delay of 1.2 sec. Estimation of α between -8.5 and -6.0 gives
$D_2 = 2.3$.

disturbing the dynamics of the basic feedback system which makes
separation by filtration possible. An attempt to estimate the correlation
dimension from unfiltered data failed, as the correlation integral scaled
as r^m, m being the embedding dimension.

When comparing experimental data with simulation in Figs. 7 and 8,
one should also have in mind that the model describes only one (although
in this context the most prominent one) of several feedback regulations
in the nephron. Feedback loops involving the tubular reabsorption and the
flow resistance in the efferent arteriole could very well give rise to a
more complicated behaviour than seen in Fig. 7(A,B). Furthermore,
biological parameters that are global to the nephrons (plasma hormone
concentrations, mean arterial blood pressure, etc.) inevitably fluctuate
during an experiment (typical duration 20-25 min). Figure 9 illustrates

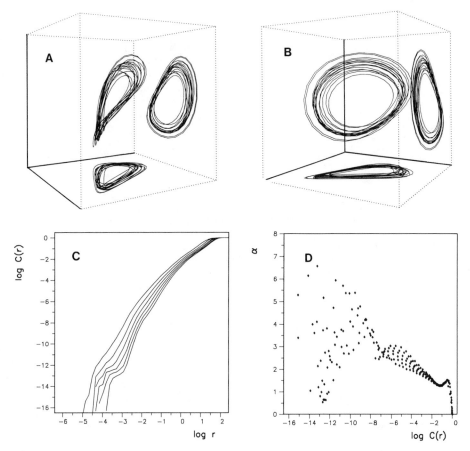

Fig. 9.   Regular oscillations from a normal rat. (A,B) Phase portrait
          viewed from different angles. Time delay between coordinates
          is 5.4 sec. Compared to SH rats this portrait is very flat,
          without scrolls or foldings as in Figs. 7C and 7D. (C,D)
          Correlation integral and local slope for the regular
          oscillations. Embedding dimension m=4 to m=9. The time
          delay is 2.4 sec. The slopes do not seem to converge or
          coincide except in a very narrow interval around log C(r) =
          1.5. Here we find α = 1.2.

the combined effect of coupling between nephrons, additional feedback
loops in the same nephron, and drifting in the parameters. According to
the single-nephron model, the phase portrait in Fig. 9 corresponds to a
limit cycle, and should therefore have the correlation dimension D_2 = 1.

By using a modification [6] of the algorithm described in [8], we
tried to estimate the Lyapunov spectra from experimental as well as
simulated time series. The estimates, however, depended strongly on the
parameter values used in the algorithm, and no reliable results were
obtained.

Taking into account the above mentioned conditions, we conclude that
the irregular intratubular pressure oscillations in SH rats are likely to
reflect a chaotic attractor inherent in the autonomous nephron system, but
under influence of slow variations in internal parameters.

ACKNOWLEDGEMENTS

We would like to thank Henrik Bohr for giving us the possibility to use the parallel computer PALLAS. We also gratefully acknowledge the work done by Thor Petersen and Bo Rathjen: the implementation of the nephron model on PALLAS had been impossible without their expertise.

REFERENCES

[1]   H. Bohr, T. Petersen, B. Rathjen, E. Katznelson and A. Nobile, "Parallel Computations of Lattice Models in Physics", Computer Physics Communications, November (1986).
[2]   P. Grassberger and I. Procaccia, Physica D 9: 189-208 (1983).
[3]   N.-H. Holstein-Rathlou, Pflügers Arch. 408: 438-443 (1987).
[4]   N.-H. Holstein-Rathlou and P.P. Leyssac, Acta Physiol. Scand. 126: 333-339 (1986).
[5]   K.S. Jensen, E. Mosekilde and N.-H. Holstein-Rathlou, Mondes en Developpement 14: No. 53 (1987).
[6]   D.R. Rasmussen, Master thesis (unpublished), University of Copenhagen (1986).
[7]   M. Rosenbaum and D. Race, Am. J. Physiol. 215(6): 1397-1402 (1968).
[8]   M. Sano and Y. Sawada, Phys. Rev. Lett. 55(10): 1082-1085 (1985).
[9]   F. Takens, Lecture Notes in Mathematics 898 (1981).

POLYPEPTIDE HORMONES AND RECEPTORS: PARTICIPANTS IN AND PRODUCTS OF A

TWO PARAMETER, DISSIPATIVE, MEASURE PRESERVING, SMOOTH DYNAMICAL SYSTEM

IN HYDROPHOBIC MASS ENERGY

Arnold J. Mandell

Laboratory of Biological Dynamics and
Theoretical Medicine (M-003)
University of California
La Jolla
CA 92093

ABSTRACT

Loss of a complex Fourier spectrum and the emergence of a single
periodic mode indicate loss of information transport capacity, desensitisation, in the dynamics of signal-sensitive neurobiological systems.
These findings are consistent with coding theorems requiring the growth
rate of orbits (topological entropy) of a channel to be equal to or
exceed that of its source of information. The problem becomes that of
developing a coding scheme consistent with protein motions characteristic
of chaotic dynamical systems.

Integrating the Cartwright-Littlewood differential equation representing forced dissipation in hydrophobic free energy in the parameter
regions of its homoclinic bifurcations, digitising the resulting time
series, and autocovariance and Fourier transformation generated a full
range of polypeptides resembling those found in the brain.

Using the same analytic scheme, similar patterns in hydrophobic
broad bands were found between several neuroendocrine polypeptide ligands
and their protein receptors.

I. POLYPEPTIDE AND PROTEIN THERMODYNAMICS IN HYDROPHOBIC MASS-ENERGY

Intuition easily grasps the frictional heat of both rough surfaces
of a wooden block and table as the forced slide of one upon the other
decelerates to rest. It is more difficult to imagine an inverse expression of this thermodynamic-mechanical conservation law: the relative

motion of the two participants driven by their spontaneous cooling. Outside of the relativistic realm, in our daily world of Galilean invariance and speeds much less than light, this interdependence of mass, m, and energy, U, along with their constant of proportionality, k,

$$U = mk \tag{1}$$

seems irrelevant. This assumption may not be true for living systems which conserve rather than maximise entropy.

In 1935, Edsall [7] observed that with respect to the more polar metabolic intermediates such as sugars, carboxy acids, and urea, molal heat capacities, Cp (generally, the amount of heat in Joules/kg/T° K required to raise the temperature of one kilogram, kg, of a substance one degree Kelvin, K), were about the same when compared in the crystalline and aqueous state. In contrast, the Cp of amino acids composing the dynamical "worker ants" of biology, such as the catalytic enzymes and binding and transport proteins, deviated positively from this ideal solution law. Increases in heat capacity, Cp, inverse to changes in enthalpy H, (the heat content such that $\Delta Cp > 0$, $\Delta H < 0$) occurred roughly at the rate of 20 cal/mol of added hydrophobic methylene group, $-CH_2-$. The thermodynamic law involving organic compounds in aqueous solution

$$\Delta U = \Delta H - T°\Delta S, \quad \Delta U = 0 \tag{2}$$

where S is the entropy, has been called entropy-enthalpy compensation [10,11]. Empirically, it successfully predicts that from 250 - 315° absolute temperature, T° (spanning more than the range of all living things), the Gibbs free-energy relationships are linear when studied across changes in hydrophobic mass and/or geometry for the rate and equilibrium constants of a particular class of water-solvated organic chemical reactions, enzymatic catalysis or protein ligand binding. Said another way, changes in energies of activation and heat capacity occur in parallel with changes in the dynamics of hydrophobic mass energy, dU/dt.

The introduction of a single amino acid into aqueous solution induces countervailing influences within the 109.5° angled cage structure of molecular water: maximisation of the number of hydrogen bonds neutralising the unbonded electron pairs of water (5 - 7 kcal/mol versus van der Waals interactions of <0.3 kcal/mol) versus an entropic force seeking its maximum to return lost solvent degrees of freedom. These represent two nearly equivalent dynamical influences of opposite sign. This cage strain, as seen in surface-minimising, strong attraction between hydrophobic moieties, is transported as a long range force with distance, $D \sim 15Å$, which is within the range of a neuronal gap junction. It decays exponentially as exp(-D/1.4) [16]. Here the analogy with the thermomechanical wooden block and table conservation law becomes relevant: Frank and Evans [9] were the first to describe the dynamical transconfirmation of autonomously "flickering" transient clusters of water molecular aggregates of variable size, $[H_2O]_n$, into hierarchies of "icelike lattices" around hydrophobic substances introduced into aqueous solution. A broad distribution of slowed and stretched NMR water relaxation times in protein solutions has been documented [2]. Metabolic heat driven, incessant cycles of hydrophobic polypeptide-water system melting-moving-freezing-arrest are an instance of entropy (here a probability more like Shannon's entropy $S = -k\Sigma_i p_i \ln p_i$ than the inefficiency protrayed in the Clausius relation, $dS = dQ/T°$) being conserved like the other extensive quantities of thermodynamics). Some of the mysterious energetics of the reaction eliminating molecular water in peptide bond formation against the enormous concentration gradient of solvent water can be rationalised in this context. The C-N bond formed is equally single and double bonded in character, reducing entropy demands on surrounding water due to its

inability to ionise or be proteinated within the range of physiological
pH. The solvent gains entropy via the new "free" molecule of metabolic
water as the polypeptide loses it through the relative rigidity of the
C-N bond.

Tanford [22] discovered the elements of a sequential amino acid (AA)
code, in hydrophobic mass energy, U, by studying their individual
partition kinetic energies,

$$U = kT°[AA]_O/[AA]_W, \text{ kcal/mol} \qquad (3)$$

as the product of the Boltzmann constant k, the absolute temperature, and
the ratio of the equilibrium concentrations of AA in organic solvents, O,
and water, W (Table 1):

Table I: Hydrophobic Mass Energy, kcal/mol

Ala	0.87	Glu	0.67	Leu	2.17	Ser	0.07
Arg	0.85	Gln	0.00	Lys	1.64	Thr	0.07
Asn	0.09	Gly	0.10	Met	1.67	Trp	3.77
Asp	0.66	His	0.87	Phe	2.87	Tyr	2.76
Cys	1.52	Ile	3.15	Pro	2.77	Val	1.87

These equilibrium concentration measurements transformed as free energies
can be viewed in more dynamical terms as relative "escape velocities"
from water into hydrophobic solvent or toward each other in nucleotide-
amino acid binding, protein folding, and ligand-receptor interactions.
The entropy-enthalpy compensation phenomena of organic chemistry are
demonstrated equally well by rate and by equilibrium constants [10]. This
generates a sequence code, $dU/dt \sim U_i$, for polypeptides and proteins which
generalises to the nucleotides for which the middle member of the triples
accounts for 70 -80% of the peptide sequence variance and uridine/adenine
at this position codes for the high values of U_i and cytosine/guanine for
the low ones [6].

When an amino acid polymer, a polypeptide or protein, is introduced
into water, another and more energetic competition ensues. The motions
of hydrophobic mass energy, dU/dt, here analogous to a surface tension,
are minimised by close packing [3], the proportionality constant k
computed to be 20 - 25 cal/mol/Å^2 of spherical surface at 25° [18]. If
this thermodynamic property were unitary, the spatially scaled ($R^3 \rightarrow R^2$)
dynamical conservation law would be

$$\frac{dU}{dt} \sim (m^{2/3}k)^{-1} \qquad (4)$$

where mass m is in gram molecular weight. However, this stabilising effect
is balanced almost precisely by destabilising losses in chain entropy
engendered by hydrophobic aggregation as calculated from ΔCp during
temperature-induced folding-unfolding transitions [17]. The two very
large and equal dynamical influences on stability of opposite sign for a
150 - 200 residue protein have been estimated to be in the range of 300
kcal/mol. Small variations in energy, dU/dt, around an attractive-
repelling critical point, dU/dt = 0, determine the changes in global
behaviour of these hydrophobic mass-energy dynamical systems seen in the
form of bifurcations of the equations of motion and resulting conformational transitions. Cp at constant volume is related to the mean square
fluctuations in mass energy

$$\langle(\Delta U)\rangle^2 = kT^2mCp \tag{5}$$

which for an average protein would amount to $3 \cdot 10^{-9}$ J/molecule times Avogadro's number (assuming the quasi-coherence of the large breathing motions of proteins [24]); about 160 kJ/mol. This very small proportion of the total energy is sufficient to induce folding or unfolding as documented by microcalorimetry [17]. For example, it is well known that energies within this range (2 - 5 kcal/mol) drive the conformational transition between the higher and lower oxygen and carbon dioxide affinity states of haemoglobin.

The nonunitary status of dU/dt with the solutions of the equation of motion for hydrophobic mass energy U appearing as dynamic instability, macromolecular motion, results in energy available for the forward vectors of conformational transition. Hydrophobic mass energy oscillates autonomously around the attractive-repelling fixed point, dU/dt = 0. This can be seen as a decomposition of the thermodynamic manifold into stabilising (convergent) and destabilising (divergent) submanifolds transverse to the forward derivative of thermomechanical action; this diffeomorphic system is suspended as a continuous time hyperbolic flow in R^3. Conditions on the use of these features to justify the use of the theory of quasi-Anosov diffeomorphisms and their flows [21] include uniform hyperbolicity assumed here to be true to within some e with respect to smooth measure at the relevant time and space scales of peptide and protein dynamics - from the methylene group, $-CH_2-$, of a single amino acid to protein polymers composed of thousands - due to the unique entropy-enthalpy compensation property of peptides in water. The rate of divergence of nearby orbits tangent to the flow called pressure, entropy, and/or free energy in the abstract thermodynamic formalism of hyperbolic theory [20a] becomes a physical reality in the dynamical behaviour of hydrophobic mass energy in biological macromolecules in water.

II. HYPERBOLIC DYNAMICS OF HYDROPHOBIC MASS-ENERGY dU/dt; SOME GLOBAL GEOMETRIC AND STATISTICAL PROPERTIES OF HOMOCLINIC TANGENCY AND BIFURCATION

The dynamics of polypeptides in aqueous solution require the introduction of a dissipative/forcing term into Hamiltonian-like equations of motion. A dissipative analogue of the restricted three body problem, its solutions which include unstable periodic and aperiodic recurrence motivated the context within which Poincaré discovered the dynamical complexity of homoclinic orbits. If $\phi^t U:M \to M$ is a diffeomorphism of compact manifold M (a discrete time process in R^2), and p represents a fixed point for it, we say that p has a homoclinic orbit through U if $\lim \phi^t U = p$ for both $t \to \infty$ and $t \to -\infty$. When p is periodic of period n, we study the orbits of $\phi^{nt}U$. For amino acids and polypeptides in solution, p is an attractive repeller and $D\phi^t U(p)$ has no eigenvalue, $\bar{\lambda}$, of norm one - a definition of hyperbolicity. The set of points, $U_i \, \epsilon \, M$, is such that $\phi^t U_i \to p$ for $t \to \infty$ constitutes motions on the immersed stabilising submanifold of p, $W^s(p)$ representing aqueous spherical surface minimising hydrophobic packing (4); the set of points, $U_i \, \epsilon \, M$, such that $\phi^t U \to p$ for $t \to -\infty$ compose the destabilising submanifold of p, $W^u(p)$, inclined toward regaining lost polypeptide chain entropy. A suspension of this diffeomorphism composing a continuous time flow [21] includes a third submanifold supporting the forward orbits of macromolecular function. Hyperbolic dynamics of hydrophobic mass energy both drive and sculpt the time-dependent pattern of critical variations in macromolecular observables on submanifolds perpendicular to the direction of the flow. We focus on the transverse interaction, $W^s \cdot W^u$, across the directional vector of action which serves as a "carrier current" energised and modulated by homoclinic dynamics on these competing submanifolds. Homoclinic orbits of hydrophobic mass energy, dU/dt, $\gamma(U)$, associated to p are the intersection (transversal or not) of $W^s(p)$ and $W^u(p)$. Birkhoff [1] showed that every transversal homoclinic orbit is accumulated by hyperbolic periodic points, and in this context Smale [21] proved that every transversal homoclinic orbit is part of a ϕ^t-invariant compact hyperbolic set, Λ. The behaviour of $\gamma(U) \in \Lambda$ combines (1) unpredictability called sensitive dependence on initial conditions: there exists a $\delta > 0$ such that for any neighborhood Y of U and $U \in R$, there exists $\tilde{U} \in Y$ and $t \geqslant 0$ such that $|\phi^t U - \phi^t U| > \delta$; (2) topological transitivity, indecomposability of the point set: for any open sets $U \in (X,Y)$, there exists a $t > 0$ such that $\phi^t U(X) \cap Y \neq \emptyset$ and (3) regularity in the behaviour of γU manifested by a dense set of periodic orbits; there exists an $n > 0$ such that $\phi^{t+s}(U) = \phi^t(U)$ for all $s \geqslant n$. These complicated attractors have been called strange because the asymptotic geometric behaviour of $\gamma(U)$ under the action of $\phi^t(U)$ is topologically distinct from either a periodic orbit or the invariant, attracting simple closed curves of ordinary differential equations.

At first glance, such messy mathematical objects representing hydrophobic mass energy in motion appear inappropriate to an understanding of the processes of specific signal generation and reception required by the work of amino acid polypeptides and proteins (and their parental, RNA, and grandparental DNA polymer chains composed of the nucleotides). Remarkably, for a particular set of parameters, the behaviour generated by, say, a one parameter group of transformations representing the dynamics of hydrophobic mass energy in co-dimension two, in the region of homoclinic tangency and bifurcation and within a critical range, $r_c < r < r_c'$,

$$\phi^t_{r,\lambda}(U):M \to M \tag{6}$$

has precisely the topological and statistical stability as well as sensitivity to small changes in parameters, r and λ, required by the job of biological dynamic coding. The dense set of periodic orbits (modes, resonances) components of the ϕ^t-invariant set is such that $\phi^t(\Lambda) = \Lambda$ and the tangent bundle $T_\Lambda M = W^s \cdot W^u$ where W^s and W^u are invariant by $D\phi^t: ||D\phi^t|W^s|| < \tilde{\lambda} < 1$ and $||D\phi^{-t}|W^u|| < \tilde{\lambda} < 1$. Through the points of Λ there exist ϕ^t-invariant leaves of the stable and unstable foliations, and the hyperbolic structure of Λ_ϕ is persistent: if $\tilde{\phi}$ is close to ϕ in the C^r topology, $r \geqslant 1$, then there is a $\Lambda_{\tilde{\phi}}$ such that $\phi\Lambda$ and $\tilde{\phi}\Lambda$ have the same dynamics. There is a homeomorphism $h:\Lambda_\phi \to \Lambda_{\tilde{\phi}}$ such that $h\phi(U) = \tilde{\phi}h(U)$, a definition of topological stability.

Statistical descriptions of $\gamma(U)$ generated by $\phi^t_{r,\lambda}(U)$ can be found in the form of an invariant measure, $\hat{\phi}^t_{r,\lambda}(U)$. We can place a hyper-surface, Σ, on M, transversal to flow $\phi^t_{r,\lambda}(U)$, and compute the time average of the Dirac deltas at the points $\gamma(U)_t \in \Sigma$ formed by its inter-sections (within the limits of a realistic biological sample length, t_n)

$$\hat{\phi}^t_{r,\lambda}(U) = \lim_{t_0 \to t_n} \frac{1}{T} \int_0^T dt \delta U_t \tag{7}$$

Measure $\hat{\phi}^t_{r,\lambda}(U)$, is invariant under the time evolution of $\phi^t_{r,\lambda}(U)$, that is

$$\hat{\phi}^t_{r,\lambda}(U)[f \cdot \phi^t_{r,\lambda}(U)] = \hat{\phi}^t_{r,\lambda}(U)(f) \tag{8}$$

a definition of statistical stability.

In place of the consistency afforded by biologically unrealistic asymptotic solutions to (8) ($t_0 \to t_\infty$ instead of $t_0 \to t_n$) we compute an autocovariance transformation on $\gamma(U)$

$$\phi^t UU'(\hat{t}) = \frac{\sum\limits_{i=1}^{n-k} (U_i - \bar{U})(U_{i+k} - \bar{U})}{\sum\limits_{i=1}^{n} (U_i - \bar{U})^2} \tag{9}$$

in which $n = t_n - t_0$, $k = n/2$, $\bar{U}$ is the mean value of dU/dt, and $\hat{t}$ represents the arbitrary time step of the partition. This transformation generates a new, discrete orbit $\hat{\gamma}(U)$ composed of the k autocovariance coefficients, c_k. The Fourier transformation of the sequence, $c_k, \phi^t UU'(\omega)$, generates a power spectrum which protrays both the mode resonances of $\phi^t_{r,\lambda}(U)$, the dense set of periodic orbits contained in $\gamma(U) \varepsilon \Lambda (\phi^{t+s}(U)' = \phi^t(u))$ as well as the more amorphous surrounding regions of probability – continuous distribution of power – representing the underlying "rug" (dense set) of topologically indecomposable points $(\phi^t(U)(X) \cap Y \neq \emptyset)$. The extent of the latter property can be estimated as

$$\text{Var } \phi^t UU'(\omega) \qquad \frac{\sum\limits_{i=1}^{n}(\Sigma p_i)}{1\sum\limits_{i=1}^{n} p_i^2} \tag{10}$$

where p is the probability squared ("power") at a given mode and 1 is the number of modes sampled [8].

Another statistical characterisation of the discrete time series, c_k, $\varepsilon \hat{\gamma}(U)$ can be achieved by assigning the hypersurface Σ to the zero set of $\phi^t UU'(\hat{t})$ and computing a 2 x 2 matrix of transition probabilities of the intersection of $\phi^t UU'(\hat{t})$, M, with Σ (zero crossing) and then the sum of its diagonal entries, a topological invariant, the trace, Tr M. We could also have used the determinant of M since in 2 x 2 matrices of transition probabilities, Tr M - Det M = 1.0.

With respect to the required parameter sensitivity [15], when dynamical system (6) is moved from r < 0 without homoclinic orbits or strange attractors, through r ~ 0 where a homoclinic tangency generically unfolds along with its disjoint intervals along the r-axis arbitrarily close to zero with infinitely many periodic attractors and repellers [13], and then becomes transversal at r > 0, we are adding at least one strange attractor to the dynamics of (6). With it comes inifinitely many more hyperbolic period points and stable and unstable foliations. After this, for infinitely many values of parameter r, new orbits of tangency between stable and unstable leaves representing periodic and aperiodic orbits emerge.

Global stability along with this parameter sensitivity is maintained because generically, large portions in terms of Lebesgue measure of parameter space of (6) are persistently hyperbolic as in the region of the first homoclinic tangency and bifurcation [15].

III. AN AREA-PRESERVING DISSIPATIVE DYNAMICAL SYSTEM ENCODING POLYPEPTIDE HORMONES AND RECEPTORS, dU/dt

The following theory derives from the above considerations along with the seminal x-ray crystallographic discoveries of Rose [20] that the locations among peptide chains in globular proteins in which dU/dt = 0 marked the beginning and end of helical loops of varying amino acid residue "frequencies", λ, and Chothia et al. [4] that almost all helices of varying λ were in close apposition, "communicating," in the interiors of folded proteins. We numerically simulate the temporal evolution of these hydrophobic mass energy dynamical interactions integrating a generic, Lienard-type non-linear, integrodifferential equation in the parameter region of homoclinic tangency and transversality representing (6). This equation of motion for hydrophobic mass energy, (11a), describes equally well nucleotide-nucleotide, nucleotide-polypeptide or polypeptide-receptor information transport processes in the form of hydrophobic mass energy dynamics:

$$\frac{dU}{dt} = \mu U(1 - \int U^2 dt) - \int U dt + \mu r \cos \lambda t \tag{11a}$$

$$U = \frac{r\mu \sin \lambda t}{\lambda} - \frac{t^2 U}{2} + \mu U \left(t - \frac{t^3 U^2}{3} \right) \tag{11b}$$

r represents the hydrophobic mass energy parameter and λ a multiple of the amino acid residue or nucleotide hydrophobic sequence "frequency" of dU/dt ∿ U_i, the two informational parameters imposed upon the receptive, autonomously "breathing" [14] macromolecule in the dynamics of the DNA → RNA, RNA → polypeptide or polypeptide → receptor interactions. μ is the global, environmental aqueous field dissipative forcing parameter representing the myriad of "background" body fluid constituents which is held constant, μ = 5; r ε [0,1] is varied from 0.15 to 0.60, and λ ranges from 0.1 - 2.0 Hz with the autonomous breathing motions of the receptive mechanism within the realistic range of the larger quasi-coherent peptide and nucleotide macromolecular motions, 0.07 Hz [24].

Figure 1 portrays the geometry of representative orbits on the phase plane, dU/dt vs U around the attractive-repeller centre, dU/dt = U = 0, across changes in r and λ. The transitions λ,0.1 → 0.2 (r = 0.2, 0.5 and 0.6) and r,0.2 → 0.15 and 0.25 (λ = 0.1) portray the orbital geometry of homoclinic bifurcation of dU/dt to complex attractors. In some abstract sense, these objects represent the dynamical equivalent of the x-ray crystallographic maps of the α-carbons of macromolecular structure. Periodic, unstable periodic, aperiodic and long transient orbits are in evidence as are parameter-dependent, quasi-systematic changes in mode and residue structure. Increasing r across λ = 0.5 demonstrates the nonlinear effect of hydrophobic mass energy on dynamical polymer complexity. Increasing spiral order (less "random" chain), called percent helix in circular dichroism and other studies of the optical activity of proteins in aqueous solution [9a], changes in a nonlinear manner over increasing λ at all r's. Figure 1 portrays the orbital geometry in dU/dt versus U phase space of measure-preserving, dissipative dynamical system (11a), the smooth hyperbolic dynamics of hydrophobic mass energy which are subsequently partitioned by discrete time steps into coded sequences. The result of diffeomorphic transformations of dynamical system (11a), $\int dU/dt → \Sigma dU/dt$ ∿ U_i, are seen in Fig. 2: 2x2 transition probability matrices, M, and their traces, Tr M, on c_k, k = 500, from autocovariance transformation, $\phi^t UU'(\hat{t})$ across the same values in co-dimension 2 in which n = 1000 and $\hat{t}$ = 0.50 Hz^{-1}. We note that for some λ, intervals across r demonstrate the nonlinear appearance and disappearance of high spiral mode densities; higher and lower values of Tr M indicate dominance by long and short wave-

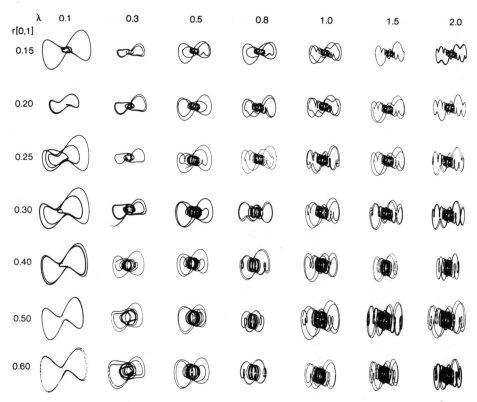

Fig. 1. The dynamics of hydrophobic mass energy in the dU/dt vs U
phase plane demonstrate complex orbital geometries around
their attractive-repelling critical point, dU/dt = 0. These
abstract dynamical analogues of the x-ray crystallographic
maps of macromolecular structure demonstrate homoclinic
bifurcation and quasi-systematic increases in complexity
induced by changes in the parameters representing hydrophobic
mass energy (r) and hydrophobic mass energy frequency (λ) of
the biological information being transported.

lengths respectively with values in between, ∿1.0, reflected by a flatter
power spectrum resembling white noise. One might anticipate such parameter sensitivity in regions of tangent bifurcations and the "wild hyperbolic sets" of Newhouse [13]. On the other hand, consistencies are found
in the spectral locations of the spiral mode densities for a given λ when
comparing r = 0.50 with r = 0.60 (see Fig. 3).

Figure 3 examines the wavelengths and broadband densities of these
hydrophobic mass energy dynamics, spiral helical modes in dU/dt ∿ U_i via
Fourier transformation of the autocovariance transformation, the poles of
$\phi^t UU'(\omega)$ and using Var $\phi^t UU'(\omega)$ (indicated here as a decimal fraction
≤ 1.0) to estimate the extent of their dispersion. The representative
peptide chain examined was at 625 - 675/1000. In the language of the
hydrophobic mass energy dynamical spiral modes observed in polypeptides
and proteins [12]: λ = 1.0, r = 0.15, 0.20, 0.30, 0.50, 0.60 and λ = 0.8,
r = 0.20, 0.40, 0.50, 0.60 are the β-strands of about 2.2 residues seen in
corticotropin releasing factor, CRF, and ACTH; α-helices of about 3.6
residue wavelength as in calcitonin can be seen at λ = 0.5, r = 0.15,
0.25, 0.30, 0.50, 0.60; the 4.0 residue mode of growth hormone releasing
factor, GRF, its competitive inhibitor somatostatin, and glucagon are seen
at λ = 1.5, r = 0.25, 0.40, 0.50, 0.60; the 4.4 residue mode of the
polypeptide opiates, at λ = 2.0, r = 0.20 and λ = 0.8, r = 0.20; the

$r[0,1]$ \ λ	0.1	0.3	0.5	0.8	1.0	1.5	2.0
0.15	$\begin{bmatrix}0.841 & 0.158\\0.157 & 0.842\end{bmatrix}$ Tr: 1.684	$\begin{bmatrix}0.804 & 0.195\\0.193 & 0.807\end{bmatrix}$ 1.611	$\begin{bmatrix}0.478 & 0.522\\0.518 & 0.482\end{bmatrix}$ 0.959	$\begin{bmatrix}0.192 & 0.808\\0.828 & 0.172\end{bmatrix}$ 0.364	$\begin{bmatrix}0.052 & 0.947\\0.944 & 0.056\end{bmatrix}$ 0.109	$\begin{bmatrix}0.938 & 0.060\\0.064 & 0.936\end{bmatrix}$ 1.875	$\begin{bmatrix}0.807 & 0.193\\0.075 & 0.925\end{bmatrix}$ 1.732
0.20	$\begin{bmatrix}0.841 & 0.159\\0.161 & 0.839\end{bmatrix}$ Tr: 1.680	$\begin{bmatrix}0.815 & 0.186\\0.202 & 0.798\end{bmatrix}$ 1.611	$\begin{bmatrix}0.479 & 0.520\\0.528 & 0.471\end{bmatrix}$ 0.951	$\begin{bmatrix}0.169 & 0.830\\0.801 & 0.198\end{bmatrix}$ 0.367	$\begin{bmatrix}0.090 & 0.909\\0.954 & 0.045\end{bmatrix}$ 0.137	$\begin{bmatrix}0.500 & 0.500\\0.508 & 0.492\end{bmatrix}$ 0.992	$\begin{bmatrix}0.920 & 0.080\\0.038 & 0.962\end{bmatrix}$ 1.882
0.25	$\begin{bmatrix}0.845 & 0.154\\0.161 & 0.839\end{bmatrix}$ Tr: 1.684	$\begin{bmatrix}0.788 & 0.212\\0.221 & 0.779\end{bmatrix}$ 1.567	$\begin{bmatrix}0.478 & 0.522\\0.535 & 0.465\end{bmatrix}$ 0.943	$\begin{bmatrix}0.696 & 0.303\\0.448 & 0.552\end{bmatrix}$ 1.248	$\begin{bmatrix}0.940 & 0.060\\0.065 & 0.934\end{bmatrix}$ 1.874	$\begin{bmatrix}0.489 & 0.510\\0.502 & 0.498\end{bmatrix}$ 0.988	$\begin{bmatrix}0.957 & 0.043\\0.046 & 0.954\end{bmatrix}$ 1.911
0.30	$\begin{bmatrix}0.842 & 0.158\\0.162 & 0.838\end{bmatrix}$ Tr: 1.680	$\begin{bmatrix}0.673 & 0.327\\0.321 & 0.679\end{bmatrix}$ 1.342	$\begin{bmatrix}0.478 & 0.522\\0.531 & 0.469\end{bmatrix}$ 0.947	$\begin{bmatrix}0.558 & 0.441\\0.520 & 0.480\end{bmatrix}$ 1.038	$\begin{bmatrix}0.040 & 0.960\\0.980 & 0.020\end{bmatrix}$ 0.060	$\begin{bmatrix}0.940 & 0.060\\0.064 & 0.935\end{bmatrix}$ 1.874	$\begin{bmatrix}0.975 & 0.024\\0.033 & 0.967\end{bmatrix}$ 1.942
0.40	$\begin{bmatrix}0.614 & 0.386\\0.453 & 0.546\end{bmatrix}$ Tr: 1.160	$\begin{bmatrix}0.680 & 0.320\\0.320 & 0.680\end{bmatrix}$ 1.360	$\begin{bmatrix}0.486 & 0.514\\0.514 & 0.485\end{bmatrix}$ 0.971	$\begin{bmatrix}0.189 & 0.811\\0.829 & 0.171\end{bmatrix}$ 0.360	$\begin{bmatrix}0.452 & 0.548\\0.660 & 0.339\end{bmatrix}$ 0.791	$\begin{bmatrix}0.492 & 0.508\\0.520 & 0.479\end{bmatrix}$ 0.972	$\begin{bmatrix}0.960 & 0.040\\0.025 & 0.974\end{bmatrix}$ 1.935
0.50	$\begin{bmatrix}0.837 & 0.162\\0.165 & 0.835\end{bmatrix}$ Tr: 1.672	$\begin{bmatrix}0.670 & 0.330\\0.314 & 0.686\end{bmatrix}$ 1.356	$\begin{bmatrix}0.490 & 0.509\\0.530 & 0.469\end{bmatrix}$ 0.959	$\begin{bmatrix}0.193 & 0.807\\0.824 & 0.176\end{bmatrix}$ 0.368	$\begin{bmatrix}0.016 & 0.983\\0.952 & 0.047\end{bmatrix}$ 0.064	$\begin{bmatrix}0.486 & 0.514\\0.506 & 0.494\end{bmatrix}$ 0.979	$\begin{bmatrix}0.940 & 0.059\\0.066 & 0.934\end{bmatrix}$ 1.874
0.60	$\begin{bmatrix}0.820 & 0.180\\0.155 & 0.845\end{bmatrix}$ Tr: 1.665	$\begin{bmatrix}0.672 & 0.327\\0.324 & 0.676\end{bmatrix}$ 1.348	$\begin{bmatrix}0.482 & 0.518\\0.522 & 0.478\end{bmatrix}$ 0.959	$\begin{bmatrix}0.193 & 0.807\\0.824 & 0.176\end{bmatrix}$ 0.368	$\begin{bmatrix}0.040 & 0.959\\0.952 & 0.048\end{bmatrix}$ 0.089	$\begin{bmatrix}0.500 & 0.500\\0.508 & 0.492\end{bmatrix}$ 0.992	$\begin{bmatrix}0.920 & 0.080\\0.038 & 0.962\end{bmatrix}$ 1.882

Fig. 2.   2x2 transition probability matrices, M, and the sum of their diagonal entries, Tr M, representing discrete "gene-like" generators of nucleotide or amino acid sequences as hydrophobic mass energy dynamics, $dU/dt \sim U_i$, from autocovariance transformation (n = 1000, k = 500, $\hat{t}$ = 0.50 Hz^{-1}) of the smooth dynamical systems of Fig. 1. The hydrophobic mass energy frequency parameter λ becomes prepotent at r > 0.40.

dominant 5.0 mode of growth factors BFGF, AFGF, and γ-NGF, at λ = 0.5, r = 0.20, 0.40; the 6.67 mode of the interferons and tumor necrosis factor, at λ = 0.30, r = 0.30, 0.40, 0.50, 0.60; the 8 mode of ubiquitin, at λ = 2.0, r = 0.40; the 10 mode of the growth factors, at λ = 0.3; r = 0.20, 0.25 and λ = 2.0, r = 0.60; and the 13.3 mode of MSH and ACTH, at λ = 0.1, r = 0.15, 0.20, 0.25, 0.30, 0.50, 0.60. Note the nonlinear evolution of dominant broadband spiral mode frequencies over systematic increases in λ. There appears to be a more prominent influence of the polypeptide (or nucleotide) hydrophobic sequence frequency factor λ in determining the location of the poles with the hydrophobic mass energy parameter r more in control of the relative size of the dispersion, Var $\phi^t UU'(\omega)$. It is evident that a wide range of primary sequences of polypeptides and proteins in hydrophobic mass energy dynamics, $dU/dt \sim U_i$, are accessible to dynamical system (11a) across continuous changes in parameters r and λ. Multimodal peptides, such as in the region λ = 0.80, r = 0.20, were also in evidence in many other parameter regions examined but not portrayed here. Attention is drawn to the amorphic, near white

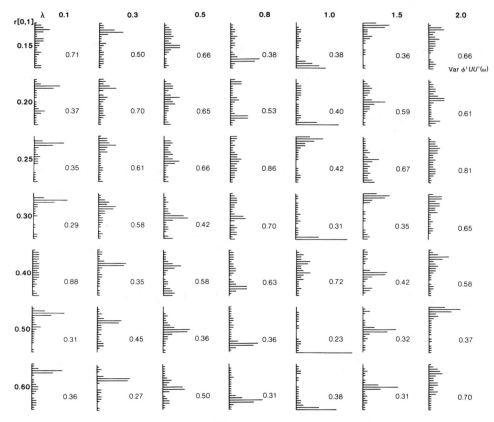

Fig. 3. Fourier transformation of the n = 625 - 675, 50-residue
"peptide" of the dynamical systems of Fig. 1 and Fig. 2 and
their mode dispersions (0.00 ∿ a delta function, 1.0 ∿ white
noise). The full range of hydrophobic mass energy modes
found in naturally occurring polypeptide hormones can be
generated by appropriate values of r and λ. Seen here for
examples: 2.2, CRF; 3.6, calcitonin; 4.0 GRF; 5.0, NGF;
6.6 interferon; 8.0, ubiquitin; 10.0, FGF; 13.3, MSH.
The parameter dependence and maintained mode dominance in
spite of changes in dispersion portray both the sensitivity
and stability of the dynamical systems theory (11a).

noise, power spectrum λ = 0.1, r = 0.40 (Fig. 2, Tr M = 1.160) which
resembles the profile of the ubiquitous intracellular binding protein
chromogranin-A, a polypeptide without a prominent signatory spiral mode.
Tr M (Fig. 2) and the modes of Fig. 3 become almost fixed from r = 0.50
to 0.60 for most λ in spite of their complex orbital geometries (Fig. 1).
This suggests the occurrence of spiral mode phase locking in these geometrically strange, measure-preserving, dissipative dynamical systems.
The λ-dependent stability of the dominant modes across changes in their
dispersion is certainly consistent with the harmlessness of the errors
that do not alter hydrophobic spiral mode dominance in amino acid sequences
as $dU/dt \sim U_i$ seen in the wide range of normal haemoglobin amino acid
sequence variations and those of the α and γ globulins. On the other
hand, dynamical mode loss induced at r = 0.40, λ = 0.1 and 1.0 might
represent a lethal mutation - perhaps not unlike that induced by the
hydrophobic mass energy of, say, cigarette tar upon an oncogene (see
below).

IV. EVIDENCE THAT A SEQUENCE CODE IN dU/dt $\sim$ U_i CAN PREDICT STRUCTURE-
FUNCTION, RECEPTOR BINDING, AND MOLECULAR BIOLOGICAL MECHANISMS OF ACTION
OF POLYPEPTIDE HORMONES

The results that follow are from computations described above.
Transition probability matrices and Fourier transformations of $\phi^t UU'(\hat{t})$ as
discrete time dynamical systems in dU/dt $\sim$ U_i were obtained beginning with
hydrophobic mass energy equivalences from Table I and amino acid sequences
from the Georgetown University, National Biomedical Research Foundation
and the relevant biochemical literature.

Figure 4 lists M's and closely related Tr M for corticotropin
releasing factor, CRF analogues (left) which are equipotent in the
release of adrenocorticotrophic hormone, ACTH, from pituitary cells in a
perfusion system in spite of only $\sim$50% homology with respect to specific
amino acid sequences [19]. On the right are M and Tr M of human GRF and
glucagon, which release growth hormone from some neuroendocrine model cell
systems in spite of far less than 50% amino acid homology. Tr M for real
polypeptides and the proteins (in contrast to the mathematical macromolecules of Figs. 1 through 3) vary between $\sim$0.60 and $\sim$1.35 with values
inverse to the rate of crossings of $\phi^t UU'(\hat{t})$ of its zero set. We would
therefore predict a dominant spiral, broadband mode of lower Fourier
frequency for GRF than for CRF. Figure 5 confirms this expectation, CRF
2.1, a beta strand, GRF $\sim$ 4.0 mode, with an ancillary 4.4 opiate peptide
mode in the former. This is of interest in that CRF releases both ACTH
($\sim$2.2) and enkephalin ($\sim$4.4) from the pituitary.

Human CRF	$\begin{bmatrix} 0.565 & 0.435 \\ 0.688 & 0.313 \end{bmatrix}$		Human GRF	$\begin{bmatrix} 0.652 & 0.348 \\ 0.500 & 0.500 \end{bmatrix}$	
Tr = 0.878			Tr = 1.162		
Ovine CRF	$\begin{bmatrix} 0.368 & 0.631 \\ 0.650 & 0.350 \end{bmatrix}$		Glucagon	$\begin{bmatrix} 0.667 & 0.333 \\ 0.545 & 0.455 \end{bmatrix}$	
Tr = 0.718			Tr = 1.121		
Sucker Urotensin I	$\begin{bmatrix} 0.476 & 0.523 \\ 0.667 & 0.333 \end{bmatrix}$				
Tr = 0.810					
Carp Urotensin I	$\begin{bmatrix} 0.421 & 0.578 \\ 0.600 & 0.400 \end{bmatrix}$				
Tr = 0.821					
Alpha Helical CRF	$\begin{bmatrix} 0.368 & 0.632 \\ 0.600 & 0.400 \end{bmatrix}$				
Tr = 0.768					

Fig. 4. CRF and its analogues (left) release ACTH in pituitary cell
perfusion systems with equal potency although their homology
with respect to specific amino acid composition $\sim$ 50%.
However, M and Tr M calculated from the autocovariance
transformation of their hydrophobic mass energy sequence,
dU/dt $\sim$ U_i, demonstrate their underlying relationship when
viewed as a discrete dynamical system as in Fig. 2. GRF and
glucagon which release growth hormone in model neuroendocrine
cell systems are less than 50% homologous with respect to
specific amino acid sequences but demonstrate similarity in
their discrete "gene-like" dynamical generator.

43

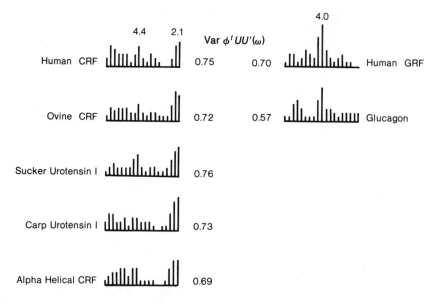

Fig. 5. Fourier transformation of the CRF and GRF families whose
"gene" peptide sequence generators are seen in Fig. 4
demonstrate corresponding hydrophobic mass energy Fourier
mode isomorphisms in their dynamical structures in spite of
a relatively large degree of dispersion.

Figure 6 is the power spectra in $dU/dt \sim U_i$ of the near isomorphic
segments of transferrin and the purified receptor protein to which it
binds with high affinity; in both, the 10.0 mode is dominant. The values
of Var $\phi^t UU'(\omega)$ are relatively close in spite of the apparent differences
in the spectral pattern of dispersion of spiral mode power. One can
imagine the mode "binding" to be analogous to Chothia's intramacro-
molecular helical apposition [3]. Figure 7 contains the spectra of the
relevant peptide segments of the closely related transforming growth
factors (TGF) and epidermal growth factor (EGF) along with the EGF
receptor to which they bind. The 7.2 residue mode of $dU/dt \sim U_i$ is prominent in all four peptides. As more polypeptide receptor sequences

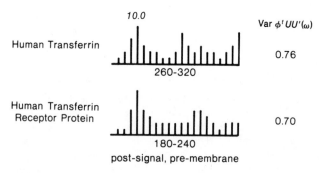

Fig. 6. Among the few polypeptide hormone-receptor pairs that have
been sequenced, transferrin demonstrates a hydrophobic mass
energy dynamical Fourier mode isomorphism at a 10.0 residue
wavelength.

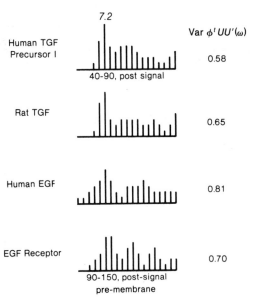

Fig. 7. The transforming growth factors and the related epidermal
growth factor bind to a common receptor and share the
dominance of a 7.2 residue hydrophobic mass energy dynamical
Fourier mode.

become available, the suggested relationship between the hydrophobic
mass energy dynamics of binding and activating ligands and their membrane
proteins will be examined more extensively.

A third line of evidence for the functional importance of the $dU/dt \sim$
U_i sequence in biological information transport comes from our recent
work examining the potential role of the 5 and 10 spiral residue modes in
the regulation of cellular mitogenesis. A key factor in neoplastic transformation appears to involve the regulation of the rate of cell division
rates via unknown polypeptide messengers. The basic and acidic fibroblast
growth factors BFGF and AFGF promote cellular multiplication in mesodermal
and ectodermal tissue as does the gamma chain of nerve growth factor,
γ-NGF (F. Zeytin, personal communication, 1986). The three have <50%
amino acid homology. Power spectral transformations of BFGF, AFGF, and
γ-NGF revealed common 5 and 10 residue modes (and β-strands) (Fig. 8a),
and a search of the protein data bank revealed a similar hydrophobic mass
energy frequency structure in bovine pancreatic ribonuclease, the hydrolytic enzyme that participates in the metabolic turnover of messenger RNA.
Since this enzyme is known to have a low and high activity form (Fig. 8b;
[23]), an order of magnitude apart in K_m, we conjectured that growth
factor BFGF would activate the enzyme and increase the synthesis of RNA
in a neuroendocrine cell system under study [25]. Figure 8c portrays the
experiments confirming this prediction.

Since cancer-related viral and cellular oncogenes in the C-fos family
have protein products which also induce transformations of fibroblasts in
vitro [5], a naturally occurring experiment is available due to the viral-
induced mutation of normal C-fos to the murine osteogenic sarcoma gene
V-fos. C-fos "destablises" messenger RNA causing an increase in messenger
RNA turnover, whereas V-fos "stabilises" messenger RNA such that turnover
is many-fold extended in time. We predicted that the C-fos to V-fos

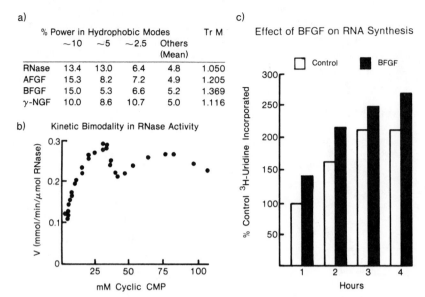

a)

| | % Power in Hydrophobic Modes | | | | Tr M |
	~10	~5	~2.5	Others (Mean)	
RNase	13.4	13.0	6.4	4.8	1.050
AFGF	15.3	8.2	7.2	4.9	1.205
BFGF	15.0	5.3	6.6	5.2	1.369
γ-NGF	10.0	8.6	10.7	5.0	1.116

b) Kinetic Bimodality in RNase Activity

c) Effect of BFGF on RNA Synthesis

Fig. 8. Based on the finding (a) that acid and basic fibroblast
growth factor as well as the gamma chain of nerve growth
factor share a set of modes at 10.0 and 5.0 residues with the
messenger RNA hydrolytic enzyme ribonuclease and (b) that
bovine pancreatic ribonuclease manifests a lower and higher
activity form, the successful prediction was made (c) that
messenger RNA turnover would increase with the administration
of the growth factor BFGF to a model neuroendocrine cell
system (Zeytin and Rusk, unpublished data) [25].

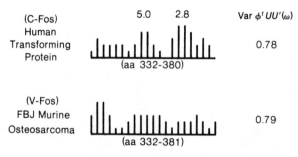

Fig. 9. The protein product of the C-fos oncogene like the growth
factors of Fig. 8 induces fibroblastic transformation and an
increase in messenger RNA turnover. Its shift-mutation product
beginning at a residue 332, called V-fos, generates the murine
osteogenic sarcoma and a marked decrease in the rate of
messenger RNA turnover. From the findings seen in Fig. 8,
the successful prediction was made that some of these characteristic Fourier modes would be lost in the V-fos neoplastic
mutation.

"shift" transformation (known to occur at residue 332 extending to the end, 380-381) would be reflected in a loss of the 5 (or 10) mode which we think facilitates ribonuclease activity and messenger RNA turnover. Figure 9 shows just such loss via the white noise of a 5 mode mutation along with a loss of the β-strand in the malignant transformation of C-fos to V-fos. This appears much like the loss of the dominant mode of Fig. 3: $\lambda = 0.15$, $r = 0.40$ following changes in the parameters r or λ.

V SOME REMARKS

It appears that orbital geometry, transition probability matrices and their traces, and power spectra along with an index of dispersion serve as complementary descriptions of the dynamics of hydrophobic mass energy, dU/dt. No one of them seems sufficient. Theoretical work is still under way concerning the existence of a single and complete invariant measure for dynamical systems in R^3. We note that our choice of $\hat{t}$ was arbitrary due to the absence of a generating partition, $dU/dt \sim U_i$.

We note that Emil Kaiser's group at Rockefeller University has completely substituted amino acid chains with their hydrophobic equivalents in several polypeptides and maintained (or even augmented) their biological activity.

We conjecture that the large amount of "excess" or "unexpressed" nuclear material found in cells composes relatively few patterns of low dimensional hydrophobic mass energy dynamics; a number of individual molecular errors can occur without altering the composite message which is in the language of global motions of hydrophobic mass energy and not specific sequences. Figure 3 demonstrates the persistence of the dominant spiral modes in the face of even a large amount of spectral dispersion. The error-preventing mechanism in such nonlinear dynamical codes is better viewed as the inertia of multidetermined globality than superimpositional redundancy. Like an enormous orchestra playing the same three or four part theme in many ways, several musicians can stop playing or misread the music without the loss of our ability to hear and recognise the musical message.

It follows from this theory of polypeptide-macromolecular coding via the dynamics of hydrophobic mass energy, $dU/dt \sim U_i$, that genetic engineering may be possible using systematic variation in relatively few global parameters. Environmentally induced mutations may proceed via such a mechanism. Barbara McClintock produced the rapid, "nonrandom" reorganisation of the genomes of maize by acute dehydration, certainly a perturbation of and perhaps a bifurcation in the dynamics of their hydrophobic mass energy.

ACKNOWLEDGEMENTS

This work was supported by U.S. Army Research Contract DAAL03-86-K-0095, National Institute of Child and Human Development Grant HD-09690-11, and U.S. Office of Naval Research Contract N00014-86-K-0741.

REFERENCES

[1] G.D. Birkhoff, Mem. Pont. Acad. Sci. Novi Lyncaei 1: 85-126 (1935).
[2] R.G. Bryant and J.E. Jentoft, J. Am. Chem. Soc. 96: 297-289 (1974).
[3] C. Chothia, Nature 248: 338-339 (1974).
[4] C. Chothia, M. Levitt and D. Richardson, Proc. Natl. Acad. Sci. USA 74: 4120-4234 (1977).

[5]   J. Deschamps, R.L. Mitchell, F. Maÿlink, W. Kruyer, D. Schubert and
      I.M. Verma, Cold Spr. Harb. Symp. Quant. Biol. L: 733-745 (1985).
[6]   R.E. Dickerson and I. Geis, "The Structure and Action of Proteins",
      W.A. Benjamin, Reading, Mass., (1969).
[7]   J.T. Edsall, J. Am. Chem. Soc. 57: 1506-1507 (1935).
[8]   D. Farmer, J. Crutchfield, H. Froehling, N. Packard and R. Shaw,
      Ann. N.Y. Acad. Sci. 357: 453-472 (1980).
[9]   H.S. Frank and M.W. Evans, J. Chem. Phys. 13: 507-532 (1945).
[9a] B. Jirgensons, "Optical Activity of Proteins and Other Macromole-
      cules", Springer-Verlag, New York (1973).
[10] J. Leffler and E. Grunwald, "Rates and Equilibria of Organic
      Reactions", Wiley, New York (1963).
[11] R. Lumry and S. Rajender, Biopolymers 9: 1125-1227 (1970).
[12] A.J. Mandell, P.V. Russo and B.W. Blomgren, Ann. N.Y. Acad. Sci.,
      in press (1986).
[13] S. Newhouse, Pub. Math. I.H.E.S. 50: 101-151 (1979).
[14] R.H. Pain, in "Characterization of Protein Conformation and
      Function", F. Franks, ed., 19-36, Symposium Press, London (1977).
[15] J. Palis and F. Takens, IMPA preprint (1985).
[16] R.M. Pashley, P.M. McGuiggan, B.W. Ninham and D. Fennell-Evans,
      Science 229: 1088-1089 (1985).
[17] P.L. Privalov and N.N. Khechinashvili, J. Mol. Biol. 86: 665-684
      (1974).
[18] J.A. Reynolds, D.B. Gilbert and C. Tanford, Proc. Natl. Acad. Sci.
      USA 71: 2925-2927 (1974).
[19] J. Rivier, C. Rivier and W. Vale, Science 224: 889-891 (1984).
[20] G.D. Rose, Nature 272: 586-590 (1978).
[20a] D. Ruelle "Thermodynamic Formalism", Addison-Wesley, Reading, MA
      (1978).
[21] S. Smale, Bull. Math. Soc. 73: 747-817 (1967).
[22] C. Tanford, "The Hydrophobic Effect", Wiley, New York (1980).
[23] E.J. Walker, G.B. Ralston and I.G. Darvey, Biochem. J. 147: 425-
      433 (1975).
[24] C.K. Woodward and B.D. Hilton, Ann. Rev. Biophys. Bioeng. 8: 99-
      127 (1979).
[25] F.N. Zeytin and R. DeLellis, Endocrinology 121: in press (1987).

INTERPLAY BETWEEN TWO PERIODIC ENZYME REACTIONS AS A SOURCE FOR

COMPLEX OSCILLATORY BEHAVIOUR

Olivier Decroly

Faculté des Sciences
Université Libre de Bruxelles
C.P. 231 Boulevard du Triomphe
1050 Brussels, Belgium

ABSTRACT

A single autocatalytic enzyme reaction operating far from equilibrium has been recognized as a major mechanism responsible for instability leading to oscillatory behaviour in biochemistry. We analyse a
biochemical system built as a sequence of two such positive feedback
loops coupled in series, in order to investigate the new types of
dynamical behaviour resulting from the interplay between two 'biochemical oscillators'.

I INTRODUCTION - MODEL AND KINETIC EQUATIONS

Glycolytic oscillations have been studied for long as an example of
biochemical rhythms [19,25]. Models based on the regulatory properties
of the phosphofructokinase, show that an instability-generating mechanism
consisting in a single autocatalytic enzyme reaction operating far from
equilibrium, is sufficient to explain most experimental data concerning
the periodic activity of glycolysis [14]. Other rhythmic phenomena such
as the periodic synthesis of cAMP in the slime mould Dictyostelium
discoideum [10], have been accounted for by a similar positive feedback
process [11], which appears to be well spread as a mechanism for periodic
behaviour in biochemistry [12].

As metabolic regulation can seldom be reduced to a single enzymatic
step, we have investigated the new possibilities of temporal self-
organization resulting from the interaction between two such oscillatory
mechanisms. The periodic forcing of an oscillatory system is known to
produce complex temporal patterns such as chaos, as exemplified by the
glycolytic system [22]. The multiply regulated biochemical system considered here is autonomous. Chaos has also been reported in an autonomous biochemical system, namely, the peroxydase reaction [24].

The model enzymatic system considered [2] is built as a sequence of
two autocatalytic reactions coupled in series, subjected to a constant
input of substrate and a first order degradation of the last product, so
as to maintain non-equilibrium conditions (Fig. 1).

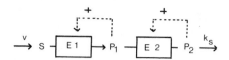

Fig. 1. Sequence of two autocatalytic enzyme reactions coupled in series, analysed as a prototype for complex temporal behaviour in biochemical systems [2]. Substrate S is supplied at a constant rate V, and is transformed into P1 by enzyme E1. P1 is in turn transformed, by enzyme E2, into P2 which is destroyed at a rate characterized by the first order constant k_s. Dashed lines with arrows indicate the positive regulation exerted on enzymes E1 and E2 by their respective products, P1 and P2.

E1 and E2 are considered to be allosteric enzymes activated by their respective products, P1 and P2. The enzymes are supposed to be allo- steric dimers obeying the concerted transition model of Monod, Wyman and Changeux [23], with exclusive binding of effector to the active form of the enzyme. Moreover, the system is considered as spatially homogeneous, i.e. continuous stirring is assumed.

The set of ordinary differential equations (1), which governs the temporal evolution of the normalized (dimensionless) concentrations of S (α), P1 (β) and P2 (γ) in the enzymatic system of Fig. 1 is then obtained as an extension of a two-variables model previously proposed [14] for glycolytic oscillations (see [2] for further details on the construction of the model and for the definition of the different parameters).

$$d\alpha/dt = v-\sigma_1\Phi\ (\alpha,\beta)$$
$$d\beta/dt = q_1\sigma_1\Phi\ (\alpha,\beta)-\sigma_2\eta(\beta,\gamma)$$
$$d\gamma/dt = q_2\sigma_2\eta\ (\beta,\gamma)-k_s\gamma \tag{1}$$

with:
$$\Phi(\alpha,\beta) = \alpha(1+\alpha)(1+\beta)^2/[L_1+(1+\alpha)^2(1+\beta)^2]$$
$$\eta(\beta,\gamma) = \beta(1+\gamma)^2/[L_2+(1+\gamma)^2]$$

Linear stability analysis shows that the system admits only one steady-state, which can be stable or unstable depending on parameter values. The structure of the characteristic equation is such that only a limited number of different types of singularities can be encountered. In particular, for realistic parameter values, it is impossible to approach from a double instability, for which two distinct eigenvalues cross together the imaginary axes. In spite of the apparent simplicity of the phenomenon predicted by linear stability analysis, very complex modes of periodic or aperiodic behaviour can be encountered in numerical simulations, when varying either parameter values or initial conditions.

The non-linear, non-polynomial form of system (1), which results from allosteric enzyme kinetics, prevents a detailed analytical treatment. On the other hand, the results of numerical integration of the differ- ential equations, are so puzzling that they often escape simple straight- forward explanations. Understanding the richness of behaviour of system (1) by numerical simulations, implies a scanning of both spaces of para- meter values and initial conditions and uses a lot of computer time without yielding exhaustive results. However, the use of simpler repre- sentations of the dynamics of the system, such as one-dimensional maps obtained by numerical integration, allows for a qualitative understanding of some of the complex phenomena revealed by direct integration [3-7].

In the following, we present some results of the analysis of system (1). The section II describes the different behavioural modes obtained upon numerical integration of the differential equations. Section III is devoted to the characterization and analysis of the chaotic dynamics obtained in some domains of the parameter space. In section IV, we address the problem of the onset of chaotic behaviour in system (1) and try to make the link between the different behavioural modes, by analysing bifurcation diagrams obtained by numerical simulations and by means of the program AUTO [8]. Finally, section V is devoted to a short discussion oriented towards the possibility of finding other systems displaying similar behaviours in biochemistry or proximate areas.

II NUMERICAL EXPERIMENTS - COMPLEX MODES OF DYNAMICAL BEHAVIOUR

Numerical integration shows that in addition to <u>simple periodic behaviour</u> and <u>excitability</u>, which were already observed for simpler systems [Goldbeter, 1980], the differential system (1) is capable of a large variety of complex behavioural modes (Fig. 2) [2].

Most of complex phenomena can be observed when changing v (0.1 s$^{-1} \leq v \leq 1$ s^{-1} and k_s (0 s$^{-1} < k_s < \infty$ s^{-1}), while the other parameters remain fixed at values given in the legend to Fig. 2a. <u>Simple periodic oscillation of the limit cycle type</u> remain the most common behaviour in the v-k_s parameter space (Fig. 2a). Other patterns of temporal self-organization also appear, including <u>complex periodic behaviour</u> (bursting) (Fig. 2f) and period doubling bifurcations leading to <u>aperiodic (chaotic) oscillations</u> (Fig. 2e).

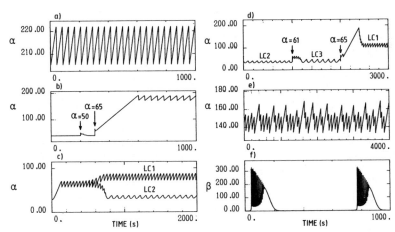

Fig. 2. Time evolution of α(a-e) or β (f) obtained by numerical integration of equations (1) for the parameter values $L_1 = 5.10^8$, $L_2 = 100$, $\sigma_1 = \sigma_2 = 10$ s^{-1}, $q_1 = 50$, $q_2 = 0.02$ (a-e) or $q_2 = 0.1$ (f). For $\bar{v} = 0.45$ s^{-1}, simple periodic oscillations (a), hard excitation (b) and birhythmicity (c) are observed for $k_s = 0.6$, 1.2 and 1.8 s^{-1}, respectively. Trirhythmicity (d) obtains in $v = 0.4$ s^{-1}, $k_s = 1.632$ s^{-1} and chaotic oscillations (e), in $v = 0.25$s^{-1}, $k_s = 1.4$ s^{-1}. Complex periodic oscillations (bursting) (f) are presented here for $v = 0.25$ s^{-1}, $k_s = 7$ s^{-1} (redrawn from ref. [5]).

Another feature of system (1) is that it admits the co-existence between multiple, simultaneously stable attractors: one steady-state and a periodic solution, i.e. hard excitation (Fig. 2b), two limit cycles, or one limit cycle and chaos, i.e. birhythmicity (Fig. 2c), and even three different oscillating regimes either periodic or chaotic [4], i.e. tri-rhythmicity (Fig. 2d). Which of the multiple stable solutions is reached depends on initial conditions (Fig. 2c); the switch from one attractor to the other may also be achieved by perturbing the system while it oscillates (Fig. 2d).

As a consequence of the proximity of the different behavioural domains in the parameter space, the system exhibits an enhanced sensitivity towards changes of parameter values [5]. Moreover, when multiple attractors co-exist, the complex structure of the attraction basins may lead to an increased sensitivity of the asymptotic behaviour to initial conditions, such that it may become very difficult to predict the final state as a function of initial conditions [6], a phenomenon known as final state sensitivity [16].

III CHAOTIC DYNAMICS

In at least three distinct domains of the $v-k_s$ plane, numerical simulations reveal an oscillatory behaviour, which does not show any pattern of repetition - i.e. no periodicity - for times of the order of 3000 oscillations. Trajectories in the phase space never pass twice through the same point, but the object resulting from the union of several trajectories seems to be time independent. Moreover, it remains qualitatively unaffected by a change in initial conditions, as soon as transients have had time to die out. The phase portraits (Fig. 3) share qualitative similarities with the attractors studied by Lorenz [21] and Rössler [26,27], which suggests that chaotic solutions exist in the system. The phase portraits do not permit, however, to distinguish chaos from complete periodic orbits with very long periods, or from a quasi-periodic dynamics on a torus.

An essential feature of chaotic dynamics is the exponential divergence of nearby trajectories. It gives rise to the sensitive dependence to initial conditions (S.I.C.), which prevents to predict the long time behaviour of a chaotic trajectory [Shaw, 1981].

We have tested the S.I.C. of the apparently chaotic dynamics of system (1), by determining two trajectories issued from initial conditions distant by 10^{-8}. Each time these trajectories cross a Poincaré section ($\beta = \beta_s$), we measure the distance, d_n, between them, and plot $\log (d_n)$ versus n, the number of passages through the section (Fig. 4a). The straight line drawn between most points of Fig. 4a indicates the exponential divergence of nearby trajectories, which gives evidence for chaotic dynamics.

A clue for understanding the S.I.C. can be gained by geometrical consideration of the structure of the trajectories in the phase space. We have analysed trajectories issued from initial conditions located on the edges of a parallelogram in a plane $\beta = \beta_s$. After one cycle, when the trajectories return to the plane $\beta = \beta_s$, we see that the parallelogram has been flattened, elongated and bent by the flow, in such a way that it presents the aspect of a horseshoe [29], situated entirely inside the original parallelogram (Fig. 4b).

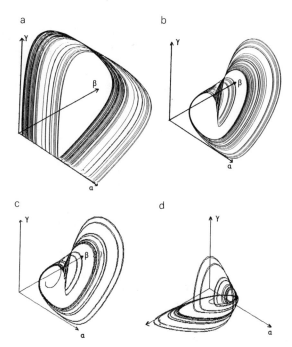

Fig. 3. Aperiodic (chaotic ?) trajectories in the phase space,
         presented in projection. Panels a and b correspond to
         distinct domains of aperiodic behaviour in the parameter
         plane $v-k_s$: $v = 0.45$ s^{-1}, $k_s = 2$ s^{-1}, (a) $v = 0.25$ s^{-1},
         $k_s = 1.4$ s^{-1} (b); other parameter values are as in Fig. 2.
         The origin of these distinct modes of aperiodic behaviour
         is discussed in section IV, as a function of the bifur-
         cations giving rise to chaotic dynamics.

     Thus, the attractivity of the limit set presented in Fig. 3c is
numerically proved, since a vicinity of this set is applied inside itself
by the dynamics. Furthermore, as each tour on the attractor results in
the same process of expansion along one direction, we understand the
origin of exponential separation of trajectories. The bending of the
parallelogram, however, prevents this divergence from being unbounded.
It thus provides the reinjection mechanism which, together with the
separation of trajectories, is supposed to give rise to chaotic
attractors [26,28]. Smale [29] proved on an abstract horseshoe map,
qualitatively similar in shape to that of Fig. 4b, the existence of an
uncountable set of unstable periodic orbits, i.e. the existence of a
chaotic dynamics.

     The fact that the parallelogram of Fig. 4b evolves rapidly to a
nearly one-dimensional curve, suggests that the system is sufficiently
dissipative to justify the study of one-dimensional discrete time return
maps as faithful approximations of the differential system [1,17]. Such
one-dimensional maps, $x_{n+1} = F(x_n)$, have been obtained by numerical
integration of system (1), by taking as discrete variable x_n either the
maximum of a variable ($x_n = \alpha_{max}$, β_{max} or γ_{max}), the amplitude of
successive oscillations ($a_n = \alpha_{max} - \alpha_{min}$,...), or the co-ordinates of the
points of passage of the trajectories through a Poincaré cross-section
($x_n = \alpha_n(\beta_s)$,...) [6,7].

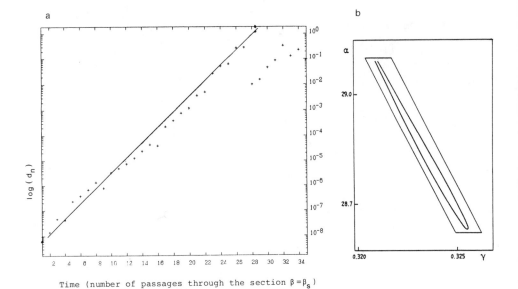

a

b

$\log(d_n)$

Time (number of passages through the section $\beta = \beta_s$)

Fig. 4.   Evidence for sensitive dependence to initial conditions
obtained by numerical analysis of the dynamics of system (1)
in the conditions of Fig. 3c. Panel a shows the logarithm of
the distance d_n between two nearby trajectories, measured as
integration proceeds, for each passage (n = 1, 2,...) through
a Poincaré section $\beta = \beta_s = 184$. Panel b provides a clue
for understanding of this divergence, by showing how a
parallelogram of initial conditions situated on the section
$\beta = \beta_s$ is applied by the flow, on a horseshoe situated inside
the original parallelogram.

In the chaotic domain, the maps obtained by plotting x_{n+1} versus x_n
for successive values of n are smooth curves, generally single humped
[6,7]. This fact results from the shape of the horseshoe map (Fig. 4b)
and supports the evidence that chaos is here deterministic. Aperiodicity
is caused here by a dynamical instability in a deterministic system,
rather than a statistical noise or numerical errors superimposed on a
stable periodic motion: in this latter case, maps would probably yield
in clouds of points instead of continuous curves.

IV BIFURCATIONS LEADING TO CHAOS

Three qualitatively different kinds of transition to chaos have been
characterized in system (1). One of these is the well known scenario of
Feigenbaum [9], by which a periodic orbit doubles its period in success-
ive points of the parameter space (period: $1 \to 2 \to 2^2 \to 2^4 \to ...2^\infty$), towards a
point beyond which the period is infinite (aperiodic motion).
Feigenbaum showed on one-dimensional maps that the parameter values, μ_n,
for which these period-doubling bifurcations occur - i.e. period
$2^n \to 2^{n+1}$ - accumulate at an exponential rate characterized by a universal
constant δ which does not depend on the particular system and has been
numerically computed for discrete dissipative systems [9].

$$\lim_{n \to \infty} \frac{|\mu_{n+1} - \mu_n|}{|\mu_{n+2} - \mu_{n+1}|} = \delta = 4.669... \tag{2}$$

A sequence of period-doubling bifurcations is encountered when entering each of the three domains of chaotic behaviour upon increasing k_s or decreasing v. This result should have been expected from the shape of the maps described earlier. We have attempted to give an estimate for the constant δ obtained in the differential system, so as to compare it with the results obtained on discrete [9] or forced [18] systems. The method used to obtain the parameter values k_{s1}, $k_{s2}\ldots$, at which a periodic orbit of system (1) doubles its period is inspired from the subharmonic stroboscopy described by Hao Bai Lin [18].

Here, as the system is autonomous, we cannot sample at multiples of the driving frequency. An alternative method is to analyse Poincaré sections through the trajectories, and to retain one point of intersection for each 2^n crossings. This allows to identify period-doubling bifurcations up to period 128. For the domains of parameter values corresponding to chaotic dynamics shown in Fig. 3a. and 3c, we have obtained the values $k_{s1},\ldots k_{s7}$ at which the first seven period doubling bifurcations occur, i.e. up to period 128. By plotting $\log|k_{sn+1}-k_{sn}|$ as a function of n, one obtains a straight line of slope $a = -\log \delta$. For the cases analysed the method yields $\delta = 4.5$, which agrees with Feigenbaum's value, in the limits of numerical precision.

On the other side of the chaotic domain, upon increasing k_s, the aperiodic oscillations transform into complex periodic orbits, in a transition which implies extremely complex periodic orbits, arranged in a sequence which can be qualitatively understood by studying a piecewise linear map similar in shape to the maps obtained numerically. This transition has been extensively studied, as well as the origin of complex periodic oscillations, in a recent analysis of bursting [7].

Another way for chaos to disappear (or appear, upon inverse parameter variation) has been described by Grebogi et al. [15] and is referred to as a crisis. During a crisis, a chaotic attractor collides with the boundaries of its attraction basin. Beyond that point, all chaotic trajectories are only transient (metastable chaos); sooner or later, the system escapes the basin of chaos and evolves towards some other attractor in the phase space. It is in such conditions, when metastable chaos co-exists with two stable attractors, that final state sensitivity obtains (as occurs in system (1) for $v = 0.39$ s^{-1} and $k_s = 1.6$ s^{-1} with other parameters fixed as in Fig. 2) [3,6]. The phenomenon is related to the fact that when topological (unstable) chaos occurs between two stable attractors, the basins acquire a fractal nature [30]. Due to processes of self-similarity, their boundaries possess a fine structure at each scale of detail, a situation also described as a fuzzy border [17].

The appearance of chaos upon parameter variation has been investigated by means of bifurcation diagrams obtained by numerical simulations, or with the help of the program AUTO [Doedel, 1981]. This program has the capability to follow singular points or periodic orbits in parameter space, and to display stable as well as unstable solutions as a function of a particular parameter. Bifurcation diagrams yielded the steady-state value α_o, or the maximum value α_m, reached by α in the course of oscillations, as a function of k_s, for different values of v (Fig. 5a-b). Complex asymptotic (stable) behaviours, such as chaos (dark zones in Fig. 5) or complex periodic oscillations (shaded area in Fig. 5a-b, or cross ruled domain in c), were were obtained by numerical simulations, whereas the search for unstable simple periodic orbits is greatly facilitated by the program AUTO.

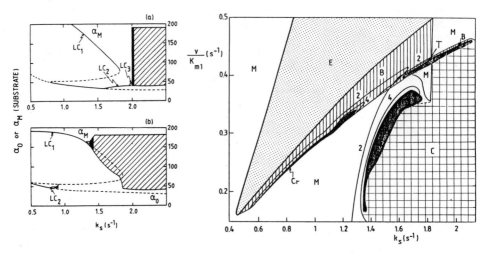

Fig. 5. Bifurcation diagrams obtained as a function of k_s, for v = 0.45 s^{-1} (a) or v = 0.25 s^{-1} (b) [2,7]. Panel (c) summarizes the information of many such diagrams by showing the different behavioural modes of system (1) in the v-k_s parameter plane. Dark zones indicate chaotic behaviour, squared rule area stands for complex periodic orbits (C). M represents simple limit cycle oscillations, E is for hard excitation, B for birhythmicity, and T for trirhythmicity. The lines labelled by 2 and 4 indicates loci of the first two period-doubling bifurcations, whereas Cr indicates a crisis, i.e. bifurcation from stable to unstable chaos.

Figs. 5a-b show bifurcation diagrams obtained in such a way, for high($v\geq$ 0.42 s^{-1}) or low($v\leq$ 0.3 s^{-1}) values of v. The basic structure related to the steady-state and simple periodic orbits remains the same: linear stability analysis shows that the steady state is unstable for low ($\rightarrow$0) and high ($\rightarrow\infty$) values of k_s. A simple limit cycle is obtained in such conditions, for which the instability is due to the first enzyme, i.e. the α-β mechanism, with γ growing unbounded or vanishing, respectively. For intermediate values, however, the two mechanisms of instability co-operate. This gives rise to a small domain of stability of the steady-state, enclosed by two Hopf bifurcations from which two limit cycles, LC1 and LC2, originate.

The structure of the bifurcation diagrams explains the origin of birhythmicity and hard excitation. Another type of birhythmicity is found for values of k_s comprised between 1.98 and 1.99 s^{-1} in Fig. 5b, and concerns the co-existence between LC2 and LC3 [2]. LC3 originates from the same branch as LC2, which is indeed, as revealed by AUTO, the beginning of a branch of stable or unstable periodic orbits winding back and forth in the k_s-α plane, towards the branch of complex periodic orbits. Problems of memory size, however, prevent to follow this branch until stable complex periodic oscillations are established.

As to the occurrence of chaos, bifurcation diagrams reveal that the three zones of chaotic behaviour in Fig. 5c (see also Fig. 3 a-c) result from the doubling - ad infinitum - of limit cycles LC1, LC2 and LC3, respectively. Why a particular limit cycle undergoes such a sequence, as a function of parameter values, and what is the transition implied between Figs. 5a and 5b is currently under investigation.

V CONCLUSION

Fig. 5c shows the different behavioural domains of system (1) in the parameter space. This figure is an enlargement of the domain of the $v-k_s$ parameter plane which has been investigated, and where the dynamics is mainly dominated by periodic oscillations (sometimes complex) and steady behaviour. Chaos and birhythmicity appear to be rare events in the parameter space as compared with simple or complex periodic oscillations; trirhythmicity has only been observed for individual parameter values.

On the other hand, as exemplified in other biochemical systems [13, 20], these complex modes of behaviour seem to appear as soon as two oscillatory mechanisms are coupled, either in series or in parallel. In each case the interaction between two limit cycles seems to be a key-ingredient for the onset of complex periodic oscillations and chaos.

ACKNOWLEDGEMENTS

I wish to thank A. Goldbeter for numerous discussions, and careful reading of the manuscript; O.D. was aspirant du Fond National Belge de la Recherche Scientifique (FNRS) during the completion of this work.

REFERENCES

[1]   P. Collet and J.-P. Eckmann, "Iterated Maps on the Interval as Dynamical Systems", Birkäuser Verlag, Basel and Boston (1980).
[2]   O. Decroly and A. Goldbeter, Proc. Natl. Acad. Sci. USA 79:6917 (1982).
[3]   O. Decroly and A. Goldbeter, Phys. Lett. 105A:259 (1984).
[4]   O. Decroly and A. Goldbeter, C.R. Acad. Sc. Paris Ser. II, 298:779 (1984).
[5]   O. Decroly and A. Goldbeter, in "Fluctuations and Sensitivity in Nonequilibrium Systems", eds. W. Horsthemke and D.K. Kondepudi, Springer, Berlin-Heidelberg-New York-Tokyo, Proc. in Phys. 1:214 (1984).
[6]   O. Decroly and A. Goldbeter, J. theor. Biol. 113:649 (1985).
[7]   O. Decroly and A. Goldbeter, J. theor. Biol. (in press) (1986).
[8]   E.J. Doedel, "Proc. 10th Manitoba Conf. on Num. Maths and Comput. Winnipeg, Canada", Cong. No. 30:265.
[9]   M.J. Feigenbaum, J. Stat. Phys. 19:25 (1978).
[10] G. Gerisch and U. Wick, Biochem. Biophys. Res. Commun. 65:364 (1975].
[11] A. Goldbeter, in "Mathematical Models in Molecular and Cellular Biology", p 248, ed. L.A. Segel, Cambridge University Press (1980).
[12] A. Goldbeter and S.R. Caplan, Ann. Rev. Biophys. Bioeng. 5:449 (1976).
[13] A. Goldbeter, J.-L. Martiel and O. Decroly, in "Dynamics of Biochemical System" p 173, ed. J. Ricard at A. Cornish-Bowden, Plenum Press, New York and London (1984).
[14] A. Goldbeter and G. Nichols, Prog. Theor. Biol. 4:56 (1976).
[15] C. Grebogi, E. Ott and J.A. Yorke, Physica 7D:181 (1983).
[16] C. Grebogi, S.W. McDonald, E. Ott and J.A. Yorke, Phys. Lett. 99A: 415.

[17]   I. Gumowski and C. Mira, "Dynamique Chaotique", ed.   Cepadues,
       Toulouse (1980).
[18]   B.-L. Hao, Phys. Lett. 87A:267 (1982).
[19]   B. Hess and A. Boiteux, in "Regulatory Function of Biological
       Membranes", p 148, ed. J. Jarnefelt, Elsevier, Amsterdam (1968).
[20]   Y.-X. Li, D.-F. Ding and J.-H. Xu, Comm. Theor. Phys. 3:629 (1985).
[21]   E.N. Lorenz, J. Atm. Sc. 20:130 (1963).
[22]   M. Markus, D. Kuschmitz and B. Hess, FEBS Lett. 172:235 (1984).
[23]   J. Monod, J. Wyman and J.-P. Changeux, J. Mol. Biol. 12:88-118
       (1965).
[24]   L.F. Olsen and H. Degn, Nature (London) 267:177 (1977).
[25]   E.K. Pye, Can. J. Bot. 47:271 (1969).
[26]   O.E. Rössler, Z. Naturforsch. 31a:259 (1976).
[27]   O.E. Rössler, Bull. Math. Biol. 39:275 (1977).
[28]   R. Shaw, Z. Naturforsch. 36a:80 (1981).
[29]   S. Smale, Bull. Amer. Math. Soc. 73:747 (1967).
[30]   S. Takesue and K. Kaneko, Prog. Theor. Phys. 71:35 (1984).

DYNAMICS OF CONTROLLED METABOLIC NETWORK AND CELLULAR BEHAVIOUR

Somdatta Sinha and [*]R. Ramaswamy

Centre for Cellular and   [*]Tata Institute of
Molecular Biology   Fundamental Research
Hyderbad, India   Bombay, India

ABSTRACT

The existence of elaborate control mechanisms for the various biochemical processes inside and within living cells is responsible for the
coherent behaviour observed in its spatio-temporal organisation.
Stability and sensitivity are both necessary properties of living systems
and these are achieved through negative and positive feedback loops. We
have studied a three-step reaction scheme (i) with endproduct inhibition,
and, (ii) with allosteric activation coupled to endproduct inhibition, to
observe the variety of behaviour exhibited by the systems under different conditions. The more complex system showed a wider variety of behaviour comprising of steady state, simple limit cycle, complex oscillations,
and period bifurcations leading to chaos. This system also shows the
existence of two distinct chaotic regimes under the variation of a single
parameter. In comparison, the single loop system showed only steady,
bistable and periodic behaviour. The variety of functions observed in
living systems may be controlled by the interplay of few basic processes -
the higher the level of complexity of the network, more diverse is the
behaviour.

I

The biggest triumph of the reductionist approach of molecular
biology has been the elucidation of almost all the biochemical reactions
or pathways controlling different functions in the bacteria Eschericia-
coli. Hence, as a fair approximation, a living cell can be considered as
a collection of interacting biochemical pathways integrated into an
overall reaction network [6] through both enzymatic and genetic control
elements (which are little more than molecules with conformational
flexibility). It is known that when elements of a certain degree of
complexity become organised into a totality of an entity belonging to a
higher level of organisation, the coherence of the higher level depends
on properties which the isolated elements could not exhibit until they
entered into certain relations with one another [9]. If this is true
for living systems, this then tells us that the coherent behaviour (or
the property of co-ordinated activities) observed in the spatio-temporal
organisation seen in more complex systems (i.e. higher organisms) which
have evolved from the simpler cells (which in turn are composed of basic

biochemical elements) cannot always be understood by reducing it to its
rudimentary elements; and it is helpful to study different levels of
organisations to obtain valuable information about their functions and
their interplay.

Most of the molecular processes controlling different cellular functions are regulated by a few common phenomena, such as, sequential reactions with rate limiting steps, competition for common sites, allosteric
changes, endproduct inhibition or repression, etc. Hence it is useful to
formulate general models of different levels of complexity incorporating
the above mentioned common features and study their dynamical properties.
In this paper we present the results of study of two generalised biochemical control systems with (a) a single negative feedback describing endproduct repression, and, (b) coupled negative and positive feedback representing endproduct inhibition and allosteric activation. We first
describe the two general models and then compare their behaviour to sub-
stantiate our view.

II

(a) The process considered here (see figure 1) is a three step reaction
with the endproduct (C) repressing the production of A after combining
with another substance R. This process is similar to the repressor mediated repression of bacterial operons where the endproduct can repress the
operon only after combining with the repressor molecule and these two
interactions have very different specificities and dissociation constants.
The time evolution of such a process can be written as, in non-dimension-
al form

$$\frac{dx}{dt} = (\frac{\gamma}{1+\gamma}) \frac{1}{1+(1+\gamma)z}n + \frac{1}{1+\gamma} -k_1 x$$

$$\frac{dy}{dt} = x-k_2 y \ ; \quad \frac{dz}{dt} = y-k_3 z-g$$

where k_1, k_2, k_3 are degradation rates, n is the cooperativity of repression, g is related to the rate of utilisation of the endproduct (say, use
of amino acids in protein synthesis in a cell), and γ is a parameter
describing the strength of repression. To study the above process, the
tryptophan operon was taken as the model system for which the parameter
values are known from experiments [7].

Linear stability analysis of the steady states and numerical simulation of the equations near the steady states shows the following behaviour of the system when one of the parameter (γ) was varied

   STABLE → BISTABLE → PERIODIC.

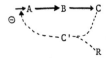

Fig. 1. Three step reaction process with endproduct repression of
         the production of A. C is the complex formed of C and R.

60

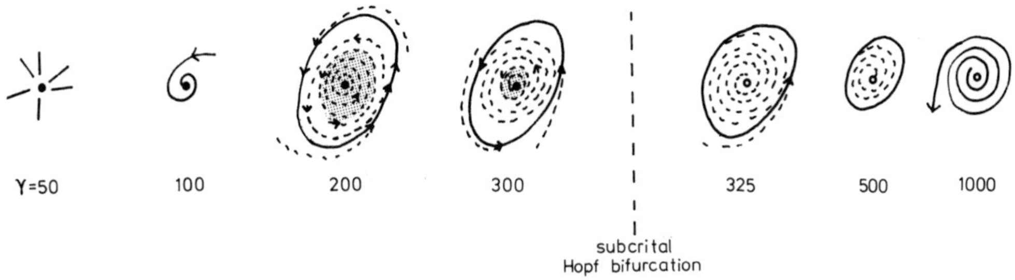
γ=50
100
200
300
325
500
1000
subcrital
Hopf bifurcation

Fig. 2. The behaviour of the system (z shown here) near to the bifurcation value of γ. It shows the stable and unstable limit cycle schematically. The values of the parameters are n=2, g=1, α_1=1, α_2=.01, α_3=.01. o denotes stable and o denotes unstable steady states. The shaded region is the unstable limit cycle (not drawn in scale).

At bistability the system showed the coexistence of an unstable and stable limit cycle surrounding the stable steady state. Figure 2 is a schematic diagram of the above behaviours of the system on changing γ. Bifurcation analysis shows that the steady state loses its stability through a subcritical Hopf bifurcation between γ values of 55 and 60. Figure 3 shows the bistable pattern of the endproduct synthesis for another set of parameter values.

(b) To describe a higher level of organisation, we consider the coupling of a negative and a positive feedback describing endproduct inhibition and allosteric activation in a three step reaction process (see figure 4). Here the B to C reaction is catalysed by an allosteric enzyme E which is assumed to be a dimer obeying the concerted transition model (4). The dimensionless form of the equations describing the above process is

$$\frac{dx_1}{dt} = \frac{a_1}{a_2+x_3^h} - kx$$

$$\frac{dx_2}{dt} = a_3x_1-\phi(x_2,x_3); \quad \frac{dx_3}{dt} = a_4\phi(x_2,x_3)-qx_3;$$

$$\phi(x_2,x_3) = \frac{Tx_2(1+x_2)(1+x_3)^2}{L+(1+x_2)^2(1+x_3)^2}$$

where a_1, a_2, a_3 and a_4 are parameters which are functions of the pseudo-Michaelis constants and dissociation constants of the reactions, k and q are the degradation rates and L and T are the allosteric constant and maximum activity of the enzyme E. n denotes the cooperativity of repression. The values of the parameters were chosen from existing literature on both hypothetical metabolic reactions and specific processes [1,2,7]. The system was studied using linear stability analysis and numerical simulation of the equations.

To reduce the dimensions of the parameter space to be explored we fixed the values of a_1, a_2, a_3, a_4, L, T and n; and k and q were varied around their basal values 1 and .01. The following behaviours were observed in the system on the variation of one parameter (k):

PERIODIC → PERIOD DOUBLING → CHAOS → REVERSE BIFURCATIONS
↓
COMPLEX BIFURCATIONS
↓
PERIODIC ← REVERSE BIFURCATIONS ← CHAOS ← PERIOD DOUBLINGS

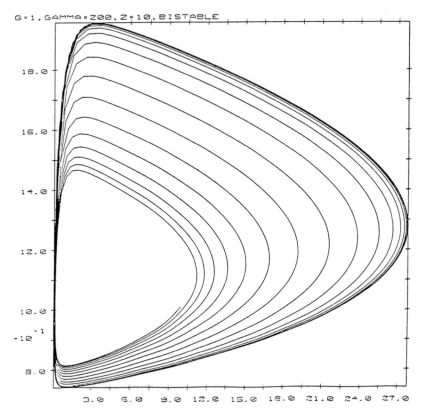

Fig. 3.   The bistable pattern of synthesis of z, (a) evolves to steady
state, and (b) evolves to periodic pattern.

Figure 5 shows the variety of patterns of the endproduct synthesis in
some of the above mentioned cases. Table 1 describes the behaviour of the
system (x_3 here) on changing k for increasing values of q. It shows that
most of the variety of behaviour vanishes as q increases.

III

The existence of elaborate control mechanisms for various biochemical
processes inside a living cell as well as at the supra-cellular level is
responsible for the coordinated behaviour observed in the spatio-temporal
organisation of the living systems. But the complexity involved in even
the simplest living system is bewildering enough to make us look for
basic regulatory structures and then study their interactions in a general
form.

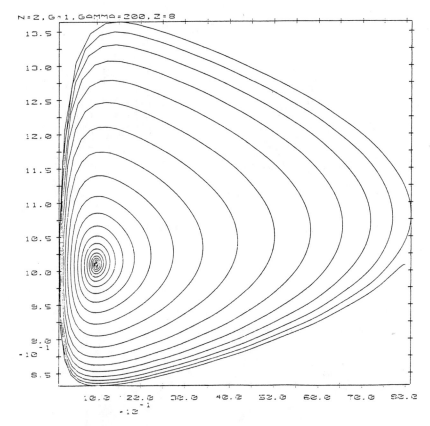

Fig. 3 (continued)
The parameter values are n=2, g=1, $\gamma=200$, $\alpha_1=1$, $\alpha_2=\alpha_3=.01$
and the initial conditions are z=8 in (a) and z=10 in (b).
x and y are kept at the steady state value.

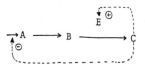

Fig. 4. Three step reaction process involving a positive and a
negative feedback loop.

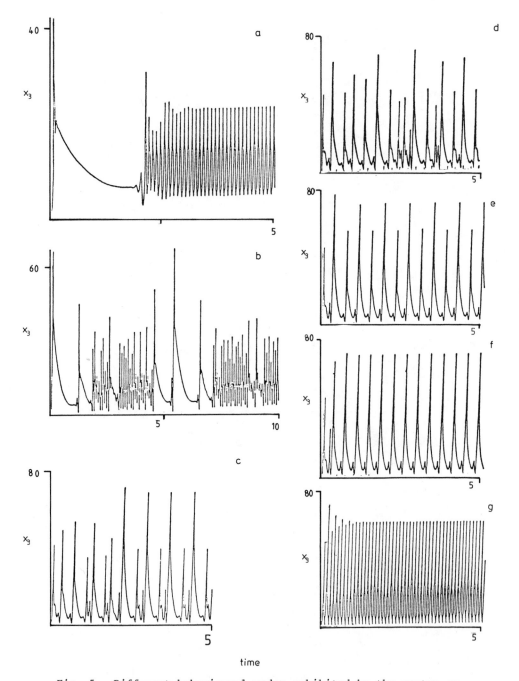

Fig. 5.   Different behavioural modes exhibited by the system on
changing the parameter q for k = 0.1.
(a) limit cycle, (b) chaos, (c) complex oscillations,
(d) chaos, (e) period 4 oscillations, (f) period 2
oscillations, (g) limit cycle.

Table 1. Complex Behavioural Modes of the System (z, shown here) as a function of q, and k.

q	k	Behaviour
.01	.001	stable
	.005–10.0	limit cycle
.05	.001–.002	limit cycle
	.003–.004	chaos
	.005–.012	complex oscillations
	.02–1.0	limit cycle
	.02–1.0	

q	k	Behaviour
0.1	.001	limit cycle
	.002–.003	chaos
	.004–.008	complex oscillations period doublings
	.0082–.00835	chaos
	.03–1.0	limit cycle
1.0	.001–<.002	stable
	>.002–<.009	limit cycle
	.009–.02	one period doubling
	>.02 –.5	limit cycle
	>.5 –10	stable

q	k	Behaviour
0.5	.001	limit cycle
	.003	complex oscillations
	.004	chaos
	.005–.02	complex oscillations period doublings
	.03–.5	limit cycle
	>1.0	stable

Earlier work on biochemical control systems involving a single negative feedback loop showed that under variation of conditions it can either evolve to the stable steady state or show limit cycle oscillations [7,8]; and systems involving one positive feedback loop showed more variety, such as, multistability and excitability along with stable and periodic behaviours [1,2,5]. This means that the decision-making ability to perform under different conditions and yet maintain stability is present even at this elementary level of organisation. In our study with one negative feedback loop we have shown that the system can also show bistable behaviour, i.e. can function either stably or in a periodic manner depending on the concentrations of the molecules involved. Our study also describes how an altered system (with changed strength of repression) tends towards destabilisation. It also shows that even at this rudimentary level of complexity the same system can behave different- ly under different initial conditions - a property presumably needed by cells in a tissue for choosing different differentiation pathways even when other parameters are the same.

Increase in complexity in biochemical networks is necessary for the coordination of different activities in the cell. There have been very few studies of general nature involving more complex control loops. Two studies [1,3] on a higher level of organisation, i.e. coupling of two positive feedback loops in one reaction sequence, and coupling of one positive feedback with recycling of the product into the substrate, showed a wider variety of behaviour of temporal self-organisation.

In living systems, the property of stability is necessary to main- tain and conserve energy in cellular economy. This property of desensit- isation to external perturbation is achieved through negative feedback. Feedback activation plays an important role in amplification for the switching and rapid - response processes. Hence, in living systems, an optimal combination of sensitivity and stability is achieved through the coupling of positive and negative feedback processes. This level of organisation in the biochemical reaction network can involve both genetic (endproduct repression) and metabolic (enzyme activation) processes since both are interdependent in cellular functions. Our study on such a system shows that increase of complexity induces newer, emergent behav- iour in the system on variation of a parameter, which were not seen in single control loops. In real biological systems, the degree of complex- ity is many-fold higher and naturally one observes a variety of struc- tures and functions. Our study points towards the fact that even at this low level of complexity newer behaviour is exhibited which is not observ- ed if the system is reduced to more elementary functional subunits.

REFERENCES

[1] O. Decroly and A. Goldbeter, Proc. Natl. Acad. Sci. USA 79:6917 (1982).
[2] A. Goldbeter and G. Nicolis, Progr. Theor. Biol. 4:65 (1976).
[3] F. Moran and A. Goldbeter, Biophysical Chem. 20:149 (1984).
[4] J. Monod, J. Wyman and J.P. Changeux, J. Mol. Biol. 12:88 (1965).
[5] B.O. Palsson and E.N. Lightfoot, J. Theor. Biol. 113:231 (1985 a, b, c).Co., Reading (1976).
[6] M.A. Savageau, in 'Biochemical System Analysis', Addision Wesley Publ. Co., Reading (1976).
[7] J.J. Tyson, J. Theor. Biol. 103:313 (1983).
[8] J.J. Tyson and H.G. Othmer, Progr. Theor. Biol. 5:1 (1978).
[9] C.H. Waddington, Organisers and Genes, Cambridge University Press, Cambridge (1947).

PERIODIC FORCING OF A BIOCHEMICAL SYSTEM WITH MULTIPLE MODES OF

OSCILLATORY BEHAVIOUR

Federico Moran[1] and Albert Goldbeter[2]

[1]Department of Biochemistry   [2]Faculté des Sciences
Faculty of Chemistry   Université Libre de Bruxelles
Universidad Complutense   Campus Plaine, C.P. 231
28040 Madrid, Spain   B-1050 Brussels, Belgium

ABSTRACT

A two variable system that represents two enzymic reactions shows birhythmicity and multi-threshold excitability. Sinusoidal forcing of this system in the region of birhythmicity can give complex birhythmicity.

I   INTRODUCTION

A forced nonlinear oscillator is probably the simplest system (i.e. of lowest dimension) capable of exhibiting complex behaviour, including chaos [20]. In thise sense, it is well known that practically all non-linear oscillators exposed to periodic perturbation display chaotic oscillations, for certain values of the controlling parameters [18]. Such complex dynamics has been demonstrated both in experimental systems [1,9, 13,15] and models [3,4,10-12,14,19]. The interest and biological relevance of this phenomenon stem from the numerous biological and biochemical systems that are exposed to an external periodic perturbation in physiological conditions.

On the other hand we have previously studied the occurrence of complex oscillatory phenomena in autonomous conditions. Thus, we developed a two-variable biochemical model capable of presenting a wide variety of complex dynamic behaviour, mainly birhythmicity (i.e. the coexistence between two stable limit cycles) [16] and excitability with multiple thresholds [17].

The aim of the present study is to determine the consequences of imposing an external periodic perturbation in a system with multiple modes of oscillatory behaviour. As shown below, this situation may account for new modes of complex dynamic behaviour, including chaos and the coexistence between two stable tori. The model considered is particularly suitable as it comprises only two variables. This property allows us to study the forced system with the usual techniques employed for 3-dimensional systems, such as stroboscopic plots or phase space representation [14,19].

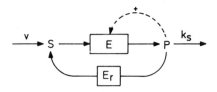

Fig. 1. Model of a product-activated enzyme (E) reaction with recycling
of product into substrate, catalysed by enzyme Er.

III MODEL AND KINETIC EQUATIONS

The model considered is closely related to those previously proposed
for the best known examples of metabolic oscillations [2,6-8]. This model
consists basically of two enzymatic reactions, namely, that catalysed by
an allosteric enzyme activated by its reaction product, and a cooperative
recycling of the product into the substrate (Fig. 1). As pointed out
above, this system has been previously studied for the emergence of birhythmicity and for multi-threshold excitability [16,17].

The model comprises two variables whose time evolution is governed by
the kinetic equations (for details see [16,17]):

$$\frac{d\alpha}{d} = v - \sigma_M\phi(\alpha,\gamma) + \frac{\sigma_1 \gamma^n}{k^n+\gamma^n}$$

$$\frac{d\gamma}{dt} = q\sigma_M\phi(\alpha,\gamma) - \frac{\sigma_1 \gamma^n}{k^n+\gamma^n} - k_s\gamma \tag{1}$$

with $\phi(\alpha,\gamma) = \dfrac{\alpha(1+\alpha)(1+\gamma)^2}{L + (1+\alpha)^2 + (1+\gamma)^2}$

where the variables and parameters are:

α,γ : normalised concentration of substrate and product;
v : normalised input rate of substrate;
σ_M,σ_i: normalised maximum rate of enzymes E and Er;
L : allosteric constant of enzyme E;
q : ratio of the Michaelis constant of enzyme E, divided by dissociation constant of the product;
K : normalised Michaelis constant of enzyme Er;
k_s : apparent first-order rate constant for product removal;
n : Hill coefficient of recycling process.

System (1) admits a single steady state with $\gamma_0 = qv/k_s$. Linear
stability analysis shows that the steady state is unstable whenever the
slope of the product nullcline on the steady state is sufficiently negative:

$$(d\alpha/d\gamma)_0 < -(1/q) \tag{2}$$

This property will be used, in the following studies, to characterise
the instability regions on the product nullcline in the phase plane.

The equation used for the periodic variation of v is:

$$v = V_0 + V_a \sin(2\pi t/T_1) \tag{3}$$

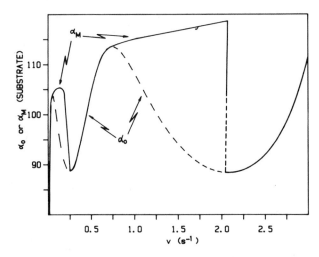

Fig. 2.   Bifurcation diagram as a function of parameter v. The diagram
shows the steady state concentration of substrate α_O (lower
curve), and the maximum substrate concentration in the course
of oscillations, α_M (upper curves). Stable and unstable
regimes are represented by solid and dashed lines, respectively.
Parameter values are L=5 10^6, σ_M=5.8 s^{-1}, K=12, σ_i=2.2 s^{-1},
n=4, q=20, k_s=1 s^{-1}.

where V_O is the mean value of v, whereas V_a and T_1 denote the amplitude
and the period of the perturbation, respectively.

We shall consider in turn the situation where the periodic perturbation extends over two distinct oscillatory domains, and over a single
instability domain where two variable limit cycles coexist.

III   FORCING THE SYSTEM OVER TWO OSCILLATORY DOMAINS

In the absence of recycling the system presents only one region of
instability [5,6,8,21]. In these conditions, an appropriate variation in
the substrate rate input can shift the steady state into the instability
region and a single type of sustained oscillation is obtained.

The inclusion of the recycling reaction induces, for the appropriate
values of the recycling-controlling parameters σ_i and K, a bump in the
product nullcline. According to equation (2) this may produce two distinct instability regions; adequate changes in parameter v can then
switch the steady state from one to the other region of instability [16,
17]. Figure 2 shows the bifurcation diagram as a function of parameter v.
In this figure both the steady state level of substrate (α_o) and the maximum amplitude of substrate oscillations (α_M) are represented. The two
domains of oscillations are separated by a region of stable steady-states
where oscillations do not occur.

When the substrate input is periodically varied so that the steady
state passes through the two oscillatory domains, the system alternatively
passes from one to the other type of oscillatory behaviour. As shown in
Fig. 3, a series of high-frequency and large-amplitude bursts in γ occur
when v is above the mean; on the other hand, a few small-amplitude peaks
are obtained when v goes under the mean.

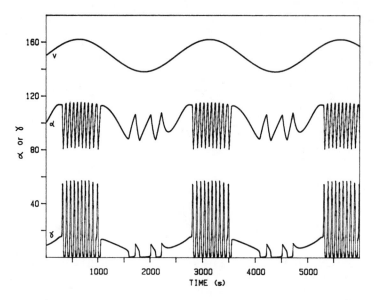

Fig. 3. Periodic forcing through multiple oscillatory domains. Shown are the time course of substrate, product, and substrate input (upper trace, not scaled). Parameter values are those of Fig. 2 with V_o=0.51 s^{-1}, V_a=0.4 s^{-1}, T_1=2500 s.

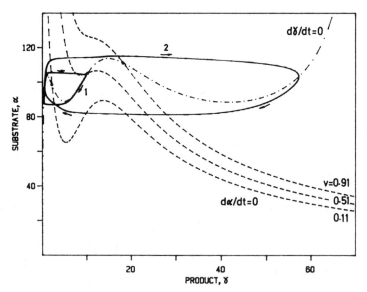

Fig. 4. Phase plane portrait of the autonomous system for parameter values of Fig. 3. The product nullcline is shown in the phase plane, together with the substrate nullclines for the three values of v. The two stable limite cycles plotted correspond to the extreme values of v reached in the periodic variation of this parameter: 0.11 (1) and 0.91 (2) s^{-1}.

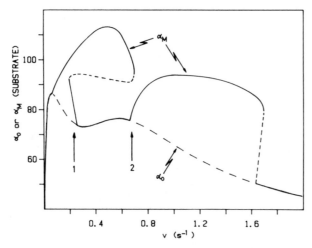

Fig. 5. Bifurcation diagram as a function of v. The diagram is drawn
as in Fig. 2, for parameter values $L=5 \cdot 10^6$, $\sigma_M=10 \text{ s}^{-1}$, $K=10$,
$\sigma_i=1.3 \text{ s}^{-1}$, $n=4$, $q=1$, $k_s=0.06 \text{ s}^{-1}$. Arrows indicate the
domains of birhythmicity.

This behaviour is made clear by the phase plane representation of
this dynamics in Fig. 4, where the two limit cycles corresponding to the
extreme values of v are represented. The substrate nullclines corresponding to the extreme and mean values of v are also plotted. The two
limit cycles are characteristic of each oscillatory region. The first
one, obtained for small v values, yields oscillations of small amplitude
and long period (345 s), whereas the second one yields the opposite, i.e.
a rhythm of large amplitude and short period (64 s). The periodic alternance between these two very different oscillatory regimes is responsible
for the bursting-like behaviour shown in Fig. 3.

IV FORCING THE SYSTEM IN THE DOMAIN OF BIRHYTHMICITY

It is interesting to determine the behaviour of the forced system
when the autonomous system presents birhythmicity. In this situation the
two oscillatory regimes coexist for the same values of the controlling
parameters. Therefore, the perturbation needed to switch the system
between the two periodic regimes should, obviously, be much smaller.

Figure 5 represents the bifurcation diagram as a function of v for
values of the parameters yielding birhythmicity. As previously pointed
out [16], in this situation the phenomenon of birhythmicity appears for
two narrow ranges of v values (indicated by the arrows in Fig. 5). The
larger domain (arrow 1) has been selected in order to perform the period
forcing studies. For the average value of $v=0.255 \text{ s}^{-1}$ the autonomous
system presents two stable limit cycles (Fig. 6, curves 2a and 2c) whose
periods are 458 and 354 s for the larger- and smaller-amplitude oscillations respectively.

The phase plane portrait for the two extreme values of v (0.055 and
0.455 s^{-1}) is also represented in Fig. 6. During the continuous variation
of v between these values the system evolves, in a cyclical manner, from
stable steady-state (1) to single oscillations, then to birhythmicity (2a -
2c) and, finally, to hard excitation (3a,3b).

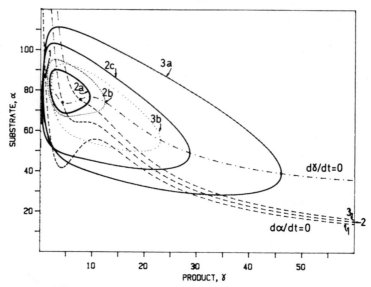

Fig. 6. Phase-plane portrait of the autonomous system for the mean and
the two extreme values of v, reached in the periodic variation
of this parameter: stable steady-state for v=0.055 s^{-1}(1);
birhythmicity for v=0.255 s^{-1} (2a-2c); and hard excitation
for v=0.455 s^{-1} (3a,3b). Stable limit cycles (———) and
unstable periodic trajectories (....) are plotted together
with substrate (-----) and product (-.-.) nullclines for the
three values of parameter v.

Holding the perturbation amplitude V_a equal to 0.2 s^{-1}, the system
responds in very different manners to relatively small variations in the
forcing period T_1. As it occurs in other forced systems [10,14],
different values of T_1 may give rise to different types of entrainment,
quasiperiodic oscillations, or chaos. In the present system the existence
of two autonomous oscillators favours the occurrence of the latter, more
complex, mode of behaviour.

In Fig. 7, both the time evolution of the forced system and the phase
space portrait are shown for two close values of T_1. For T_1=155 s (Fig.
7a,b) a complex oscillation of period 9 times T_1 is obtained. A small
increment in the perturbation period, up to 157 s, drives the system to
an aperiodic oscillation (Fig. 7c,d). The plot of Fig. 7d has been
obtained by drawing only several dozens of cycles (drawing more cycles
would yield a black stain). The stroboscopic plot, when taking points at
v phase values of 0° (v=0.255 s^{-1}), is shown in Fig. 8a. Successive
enlargements of this map (Fig. 8b,c) corroborate the existence of a fine
structure of the attractor at successive levels of precision.

A region of chaos with period doubling sequence is obtained when T_1
reaches values close to 242 s. Figure 9 shows the route to chaos by means
of stroboscopic representation. Starting from an initial cycle of period
$3T_1$ (Fig. 9a) the orbit bifurcates successively (Fig. 9b,c,d) until chaos
is reached (Fig. 9e,f). Whereas in Fig. 9e there are six islands, they
evolve to three (Fig. 9f) which correspond to the initial three points.

Of particular interest is the behaviour obtained for T_1=200 s (Fig.
10 a,b). Then, two stable complex oscillations coexist, yielding a
phenomenon of "complex birhythmicity". As in the case of simple birhythmicity [16] the system can switch between the two stable periodic regimes
by means of adequate perturbation at the appropriate phase of each

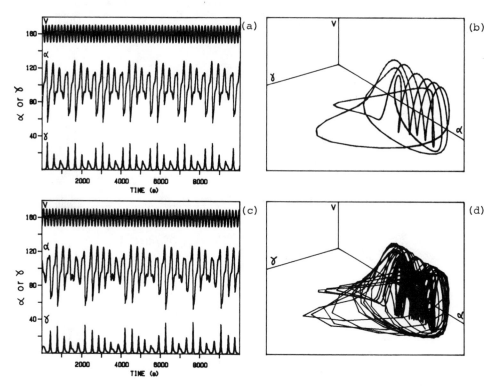

Fig. 7. Time course and phase-space portrait of the forced system for
T_1=155 (a,b) and 157 (c,d) s. The time-course curve of the
variable α has been shifted 20 units to avoid its intersection
with the one corresponding to γ. The time course of parameter
v is also represented by the upper curve (not scaled). The
trajectory plotted in (d) corresponds to only several cycles.
Other parameter values are those of Fig. 5 with V_o=0.255 s^{-1}
and V_a=0.2 s^{-1}.

oscillation. This is indicated in Fig. 10a by the arrows. The two limit
cycles are shown in the phase space, in Fig. 10b. Further increment in
the value of T_1 up to 250 s, allows us to obtain the two types of
oscillations in a unique periodic trajectory. Figure 10c shows indeed
the alternance of large- and small-amplitude oscillations over a cycle.
The complex limit cycle corresponding to this behaviour is shown in
Fig. 10d; it displays a good resemblance with the two cycles of Fig. 10b.

Upon reducing the amplitude V_a from 0.2 to 0.1 s^{-1}, the two limit
cycles observed in Fig. 10b transform into two coexistent quasiperiodic
orbits. The stroboscopic plot of these orbits is shown in Fig. 11. In
both cases, closed curves without any fine structure can be observed.

V DISCUSSION

We have analysed, in a simple two-variable biochemical model, the
effect of period forcing on a system with multiple oscillatory domains.
As a result of the behavioural complexity of the autonomous system, the
forced one presents some new modes of oscillatory behaviour, which should
be added to the classical ones described for periodically forced nonlinear
oscillators [20].

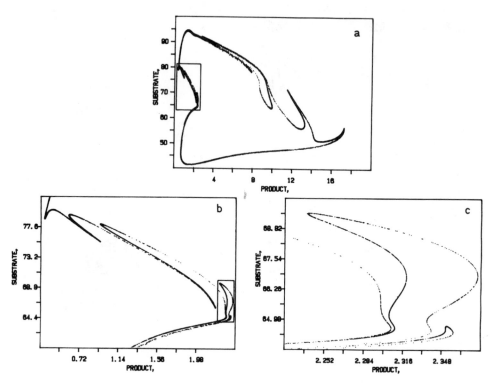

Fig. 8 (a) Stroboscopic phase portrait of the forced system for the motion described in Fig. 7d. Points are taken at mean value $v=0.255$ s^{-1}. (b) Enlargement of section indicated by box in (a). (c) Enlargement of section indicated by box in (b).

One of the goals of the present analysis was to show how a system presenting autonomous birhythmicity responds to an external periodic perturbation. In such a simulation, two internal, nonlinear oscillators can be driven by the periodic input. This interaction yields a wide variety of complex behaviour. Owing to the presence of two "internal" frequencies, it is not possible to calculate properly the fundamental harmonic entrainment. On the other hand, the existence of these two internal oscillations gives rise to complex behaviour including cycles of period 17 to 23 times T_1 (data not shown), period doubling bifurcations, and chaos. Of particular interest is the coexistence of two complex limit cycles which bifurcate to two simultaneously stable, quasiperiodic oscillations, corresponding to two distinct tori. The presence of multiple attractors in the system allows it to respond in a very different manner to various perturbations or different initial conditions. Whether the two tori may transform into two coexisting strange attractors is currently under investigation.

ACKNOWLEDGMENTS

This work has been supported in part by grant 1472-82 from the Comision Asesora de Investigacion Cientifica y Tecnica (Spain). We thank Drs. J.L. Martiel and F. Montero for fruitful discussions.

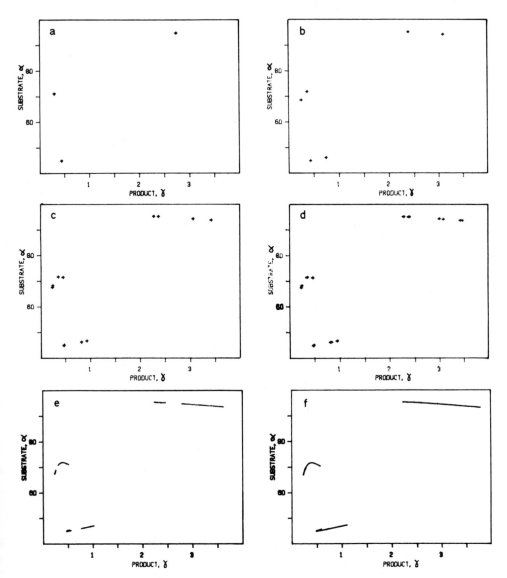

Fig. 9. Successive stroboscopic phase portraits indicating regions of period doubling bifurcation and chaos, as a function of T_1.
(a) $T_1=246$ s, period 3×2^0, (b) $T_1=244$ s, period 3×2^1,
(c) $T_1=243$ s, period 3×2^2, (d) $T_1=242.9$ s, period 3×2^3,
(e) $T_1=242.6$ s, chaos with 6 branches, (f) $T_1=242.3$ s,
chaos with 3 branches. Other parameter values are those of
Fig. 7.

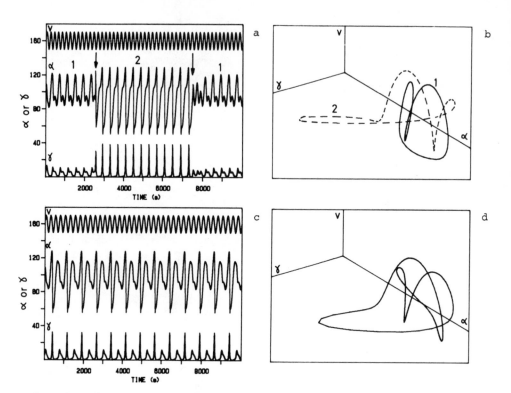

Fig. 10.   Time course and phase-space portrait of the forced system for
T_1=200 (a,b) and 250 (c,d) s.   Other parameter values and
plotter conditions are those of Fig. 7.   The arrows indicate
the perturbation that switches the system between the two
stable complex oscillations: α=109, γ=3, t=2515 s (from 1 to
2), and α=89, γ=3, t=7480 s (from 2 to 1) (initial conditions
are α=89, γ=3, t=0 s).

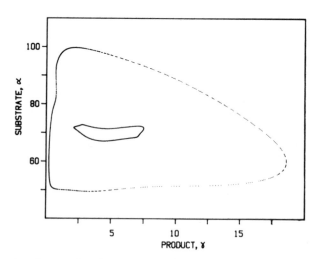

Fig. 11.   Stroboscopic phase portrait of the forced system for V_a =
0.1 s^{-1}.   Other parameter values are those of Fig. 10a.

76

REFERENCES

[1]   A. Boiteux, A. Goldbeter and B. Hess, Proc. Natl. Acad. Sci. USA
      72: 3829-3833 (1975).
[2]   O. Decroly and A. Goldbeter, Proc. Natl. Acad. Sci. USA 79: 6917-
      6921 (1982).
[3]   H. Degn, in "Chemical Applications of Topology and Graph Theory",
      ed. R.B. King, 364-370, Elsevier, Amsterdam (1983).
[4]   M.R. Guevara and L. Glass, J. Math. Biol. 14: 1-23 (1982).
[5]   A. Goldbeter and T. Erneux, C.R. Acad Sci. (Paris) Ser. C. 286:
      63-66 (1978).
[6]   A. Goldbeter and R. Lefever, Biophys. J. 12: 1302-1315 (1972).
[7]   A. Goldbeter, J.L. Martiel and O. Decroly, in "Dynamics of Bio-
      chemical Systems, eds. J. Ricard and A. Cornish-Bowden, 173-212
      (1984).
[8]   A. Goldbeter and G. Nicolis, in "Progress in Theoretical Biology",
      eds. F. Snell and R. Rosen, Vol. 4, 65-160, Academic Press, New
      York (1976).
[9]   H. Hayashi, S. Ishizuka, M. Ohta and K. Hirakawa, Phys. Lett. 88A:
      435-438 (1982).
[10] T. Kai and K. Tomita, Progr. Theor. Phys. 61: 54-73 (1979).
[11] M. Markus and B. Hess, Proc. Natl. Acad. Sci. USA 81: 4394-4398
      (1984).
[12] M. Markus and B. Hess, Arch. Biol. Med. Exp. 18: 261-271 (1985).
[13] M. Markus, D. Kuschmitz and B. Hess, FEBS Lett. 172: 235-238 (1984).
[14] M. Markus, D. Kuschmitz and B. Hess, Biophys. Chem. 22: 95-105
      (1985).
[15] G. Matsumoto, K. Aihara, M. Ichikawa and A. Tasaki, J. theoret.
      Neurobiol. 3: 1-14 (1984).
[16] F. Moran and A. Goldbeter, Biophys. Chem. 20: 149-156 (1984).
[17] F. Moran and A. Goldbeter, Biophys. Chem. 23: 71-77 (1985).
[18] L.F. Olsen and H. Degn, Quart. Rev. Biophys. 18: 165-225 (1985).
[19] K. Tomita, Phys. Rep. 86: 114-167 (1982).
[20] K. Tomita, in "Chaos", ed. A.V. Holden, 211-236, Manchester Univ.
      Press (1986).
[21] D. Venieratos and A. Goldbeter, Biochimie 61: 1247-1256 (1979).

PERIODIC BEHAVIOUR AND CHAOS IN THE MECHANISM OF INTERCELLULAR

COMMUNICATION GOVERNING AGGREGATION OF <u>DICTYOSTELIUM</u> AMOEBAE

A. Goldbeter and J.L. Martiel[*]

Faculté de Sciences   [*]Faculté de Médicine
Université Libre de Bruxelles Université de Grenoble
Campus Plaine   F-38700 La Tronche
C.P. 231   France
B-1050 Brussels, Belgium

ABSTRACT

     Upon starvation, <u>Dictyostelium discoideum</u> amoebae aggregate in a
wave-like manner around centers, by a chemotactic response to cAMP
signals. These signals are emitted by the centers in the form of pulses,
with a periodicity of several minutes. Relay of the cAMP signals by
chemotactically responding cells allows one center to control the
aggregation of up to 10^5 amoebae. We analyse a model for the cAMP signalling system, based on the activation of cAMP synthesis upon binding of
cAMP to a cell surface receptor, and on desensitisation of this receptor
upon incubation with cAMP. The analysis shows that the self-amplifying
properties of the signalling system may give rise to an instability leading to autonomous oscillations of cAMP. The model also accounts for the
relay of suprathreshold cAMP pulses; such behaviour reflects the excitability of <u>D. discoideum</u> cells. Besides relay and periodic oscillations,
the model also shows evidence for bursting and for aperiodic oscillations
of cAMP, i.e. chaos (J.L. Martiel and A. Goldbeter, Nature 313:590-592,
1985). The results on chaotic behaviour may be related to experimental
observations by Durston (Devel. Biol. 38:308-319, 1974) who found that in
contrast to the wild type which aggregates in a periodic manner, the
<u>D. discoideum</u> mutant Fr17 is characterised by the property of aperiodic
aggregation. Ways to test the possibility that Fr17 behaviour represents
an example of chaos in intercellular communication will be discussed, in
the light of the modelling study.

I INTRODUCTION

     One of the simplest systems of intercellular communication is that
which governs aggregation of cellular slime moulds after starvation.
When deprived of nutrients, these amoebae acquire a chemotactic response
to signals emitted by aggregation centers. Thereby they form a multicellular mass which further develops into a fruiting body composed of a
stalk surmounted by a mass of spores [1].

     In some species such as <u>Dictyostelium discoideum</u>, the aggregation
possesses a wave-like nature. Concentric or spiral patterns of amoebae

develop on agar, owing to the periodic release of the chemotactic factor by the centres and to the relay of the signal towards the periphery of the aggregation field [6,10,29]. The chemoattractant has been identified in D. discoideum as cyclic AMP (cAMP) [23]. Periodic secretion and relay (i.e. amplification) of cAMP pulses have been confirmed in cell suspension experiments [11]. The period of cAMP oscillations in cell suspensions is close to 9 min; such period is also that of wave-like aggregation on agar.

We review here the mechanism of cAMP oscillations in Dictyostelium and briefly present a model for this phenomenon, based on receptor desensitisation [15,27]. Besides periodic behaviour, we obtain evidence for the occurrence of aperiodic oscillations (i.e. chaos) in this realistic model for the signalling system [28]. The domain of complex oscillatory phenomena such as chaos is found in parameter space by means of an empirical method of general applicability to autonomous systems. (The conditions for the occurrence of chaos in a three-variable version of the model are explored analytically by Martiel [26]. We relate the theoretical prediction of chaos to the behaviour of the D. discoideum mutant Fr17, which is known to aggregate in an aperiodic manner [8], and discuss the physiological significance of aperiodic oscillations in intercellular communication.

II MECHANISM OF cAMP OSCILLATIONS IN DICTYOSTELIUM

The source of oscillatory behaviour in chemotactic signalling is the positive feedback loop which governs cAMP synthesis in D. discoideum (Fig. 1). Binding of extracellular cAMP to a membrane receptor triggers the synthesis of cAMP, through activation of adenylate cyclase; the intracellular cAMP thus formed is transported into the extracellular medium [6,10,29]. The ensuing self-amplification would lead to a runaway phenomenon, were it not for the limiting factors that keep the signalling system from a biochemical 'explosion'.

Initial attempts to model the signalling system [12,17] suggested that substrate consumption might be the process limiting autocatalysis: if ATP were synthesised at a constant rate, activation of adenylate cyclase could lead to a transient depletion of ATP near the adenylate cyclase reaction site. The subsequent decrease in cAMP synthesis, coupled

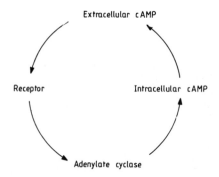

Fig. 1. The positive feedback loop governing cAMP synthesis in the slime mould Dictyostelium discoideum.

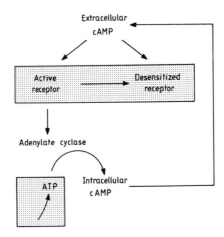

Fig. 2. Receptor desensitisation and substrate consumption limit the
autocatalytic synthesis of cAMP in D. discoideum. When coupled
to the positive feedback exerted by cAMP, each of these two
limiting processes (dashed areas) may produce sustained
oscillatory behaviour.

to cAMP hydrolysis by phosphodiesterase, would allow for replenishment of
substrate until a new burst occurs in cAMP production. In this mechanism,
cAMP oscillations are accompanied by a significant variation in ATP [12,
17].

Two types of experiments have since indicated that ATP consumption is
not the primary factor that limits autocatalytic synthesis of cAMP in
Dictyostelium. First, the intracellular level of ATP does not vary significantly in the course of cAMP oscillations [34]. While such observation does not rule out the possibility of ATP compartmentation, subsequent experiments showed more directly that another process is involved.
Theibert and Devreotes [35] used caffeine to uncouple the cAMP receptor
from adenylate cyclase and found that cells incubated with caffeine and
cAMP for a few minutes do not respond to cAMP signals upon removal of the
uncoupling agent. This suggests that incubation with extracellular cAMP
sets on a process which prevents cAMP synthesis, somewhere on the path
between the receptor and adenylate cyclase. The current view is that
phosphorylation of the cAMP receptor [7,21] is the process that mediates
the decrease in cAMP synthesis after the initial amplification of a cAMP
signal.

This finding supports our previous assumption [27] that explosive
cAMP synthesis is prevented by the passage of the cAMP receptor into a
desensitised state upon incubation with cAMP. Our model for cAMP signalling in Dictyostelium [15,27] is thus based on self-amplification and self-
limitation through receptor desensitisation, as schematised in Fig. 2.
(An alternative model has been presented [32,33], on the basis of a
putative inhibition of adenylate cyclase by calcium ions).

When the hypothesis of a constant input of substrate is retained,
the model of Fig. 2 corresponds to the detailed reaction sequence (1):

$$R \underset{k_{-1}}{\overset{k_1}{\rightleftharpoons}} D, \tag{1a}$$

$$R+2P \underset{d_1}{\overset{a_1}{\rightleftharpoons}} RP_2 \; ; \quad D+2P \underset{d_2}{\overset{a_2}{\rightleftharpoons}} DP_2, \tag{1b,c}$$

81

$$RP_2 \underset{k_{-2}}{\overset{k_2}{\rightleftharpoons}} DP_2 \quad ; \quad RP_2 + C \underset{d_3}{\overset{a_3}{\rightarrow}} E,$$ (1d,e)

$$E + S \underset{d_4}{\overset{a_4}{\rightleftharpoons}} ES \overset{k_4}{\rightarrow} E + P_i,$$ (1f)

$$C + S \underset{d_5}{\overset{a_5}{\rightleftharpoons}} CS \overset{k_5}{\rightarrow} C + P_i,$$ (1g)

$$P_i \overset{k_i}{\rightarrow} ; \quad P_i \overset{k_t}{\rightarrow} P; \quad P \overset{k_e}{\rightarrow},$$ (1h,i,j)

$$\overset{v}{\rightarrow} S$$ (1k)

In the above steps, R and D represent the active and desensitised forms of the cAMP receptor; P and P_i represent extracellular and intracellular cAMP, respectively; S is the substrate ATP; C is the free form of adenylate cyclase, which is taken as less active than the complex E formed with RP_2. The rate constants k_t, k_i, and k_e relate to the transport of intracellular cAMP into the extracellular medium and to cAMP hydrolysis by the intracellular and extracellular forms of phosphodiesterase. Finally, v denotes the normalised rate of ATP synthesis.

Steps (1b,c) are based on the assumption of cooperative binding of cAMP to the receptor. Such nonlinearity is justified in view of some experimental observations [2]. Similar results obtain when the the nonlinearity is transferred to the activation step (1e), i.e. when two cAMP-receptor complexes are assumed to bind to the free enzyme C in a cooperative manner, as observed for some hormone receptors (J.L. Martiel and A. Goldbeter, submitted for publication).

When modelling cAMP oscillations in well stirred cellular suspensions [11], the dynamics of the signalling system is governed by a set of seven ordinary differential equations:

$$\frac{d\rho}{dt} = k_1(-\rho + L_1\delta) + d_1(-\rho\gamma^2 + x)$$

$$\frac{d\delta}{dt} = k_1(\rho - L_1\delta) + d_2[1 - \delta(c^2\gamma^2 + 1) - \rho - x - \mu(1 - \bar{c}(1 + \alpha\theta))]$$

$$\frac{dx}{dt} = k_2[L_2(1 - \rho - \delta) - (1 + L_2)x - L_2\mu(1 - \bar{c}(1 + \alpha\theta))] + d_1(\rho\gamma^2 - x) + $$
$$+ \mu d_3[\phi - \bar{c}(\epsilon x + \phi(1 + \alpha\theta))]$$

$$\frac{d\bar{c}}{dt} = d_3[\phi - \bar{c}(\epsilon x + \phi(1 + \alpha\theta))]$$

$$\frac{d\alpha}{dt} = v - \sigma\alpha\phi[1 - \bar{c}(1 + \alpha\theta(1 - \lambda) - \lambda\theta)]$$

$$\frac{d\beta}{dt} = q\sigma\alpha\phi[1 - \bar{c}(1 + \alpha\theta(1 - \lambda) - \lambda\theta)] - (k_i + k_t)\beta$$

$$\frac{d\gamma}{dt} = (k_t\beta/h) - k_e\gamma + 2\eta[d_1(-\rho\gamma^2 + x) + d_2(1 - \rho - \delta(1 + c^2\gamma^2) - x - \mu(1 - \bar{c}(1 + \alpha\theta)))]$$ (2)

In these equations, α, β and γ denote the normalised concentrations of ATP, intracellular cAMP, and extracellular cAMP; ρ, and x denote, respect-

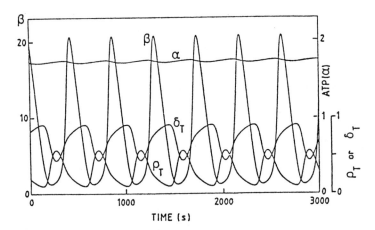

Fig. 3. Periodic synthesis of cAMP (β) in the model for the signalling system based on receptor modification. Shown are the normalised concentrations of intracellular cAMP (β), ATP (α), and the inverse relationship between the fractions of receptor in active, R (p_T) and inactive, D(δ_T) states. The curves are obtained by numerical integration of the seven kinetic equations that govern the time evolution of the model. Parameter values are $\lambda=\theta=0.01, L_1=300, c=10^2, L_2=0.03, k_1=1.5\times10^{-2}s^{-1}, k_2=6\times10^{-3}s^{-1}, d_1=d_2=d_3=1s^{-1}, h=5, k_e=0.08s^{-1}, k_i=0.06s^{-1}, k_t=0.04s^{-1}, \varepsilon=0.2, \eta=0.1, q=4000, \mu=0.1, \sigma=0.01s^{-1}, v=2\times10^3s^{-1}$.

ively, the fraction of receptor in the states R, D, and RP_2, while $\bar{c}$ is the fraction of adenylate cyclase in the free form C (see references 16 and 27 for further details and definition of the parameters; in these references, c should be defined as $(K_R/K_D)^{1/2}$).

System (2) has been analysed for oscillations, relay of suprathreshold cAMP pulses (i.e. excitability), and adaptation to constant stimuli [15,16,27]. The analysis of reductions of this system to three or four variables yields results similar to those obtained with the complete set of equations (2) (J.L. Martiel and A. Goldbeter, submitted for publication; see also [26]).

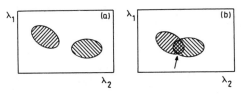

Fig. 4. Strategy for search of complex oscillations and chaos in parameter space. In (a), two distinct oscillatory domains (dashed areas) are schematised in the space of two parameters, λ_1 and λ_2. In (b), by appropriate change in the value of one or more of the remaining parameters $\lambda_3, \ldots, \lambda_n$, the two instability domains are brought together until they intersect. Complex oscillatory phenomena (such as bursting and multiple limit cycles) as well as chaos often occur in (or near) the region where the two domains overlap.

Oscillations obtained by integration of eqs. (2) show that the periodic synthesis of cAMP pulses is accompanied by only slight variation in ATP, and by a periodic alternance of the receptor between the active and desensitised states (Fig. 3). Such result agrees with the periodic evolution from the dephosphorylated to the phosphorylated state of the receptor observed experimentally in the course of cAMP oscillations [22].

For parameter values slightly different from those producing oscill-ations, eqs. (2) admit a stable steady state which is excitable, since suprathreshold perturbations in extracellular cAMP lead to the synthesis of a cAMP pulse before the system returns to the stable steady state [16]. Such behaviour accounts for the relay of suprathreshold cAMP pulses observed during aggregation on agar [6,10,29] and in cell suspension experiments [11].

III PREDICTION OF CHAOS IN cAMP SIGNALLING

Linear stability analysis around the steady-state solution of eqs. (2) shows that two distinct domains of instability may sometimes coexist in parameter space. The similar existence of two distinct instability domains in a model for two autocatalytic enzyme reactions coupled in series [4,5] was clearly ascribable to the fact that each of the two positive feedback loops could produce sustained oscillations. In the present model, the coexistence of two instability domains is at first view surprising, given the presence of a single positive feedback loop in the cAMP signalling process (see Figs. 1 and 2).

The multiplicity of instability domains originates here from the fact that two mechanisms may nevertheless underlie cAMP oscillations. These mechanisms, coupled in parallel, share the feedback exerted on adenylate cyclase by extracellular cAMP, but differ by the process limiting cAMP autocatalysis. In the first mechanism, the limiting effect is due to substrate consumption, whereas in the second it arises from receptor desensitisation (see Fig. 2).

The coexistence of two distinct instability domains in parameter space is schematised in Fig. 4(a) where parameters λ_1 and λ_2 may represent, for example, k_{-1} and v in the reaction sequence (1). Previous experience with a similar situation in a model for a multiply regulated biochemical system [4,5] suggested to look for the possible occurrence of complex oscillations and chaos by means of the empirical method schematised in Fig. 4(b). This method consists in changing the value of some other parameter(s) of the system so as to move the two instability domains toward each other in the λ_1 and λ_2 plane, until they overlap. It is in this region - or just before junction of the two domains - that complex oscillatory phenomena occur.

The rationale for this approach rests on the observation that the oscillations which arise in each of the two domains of instability differ, since they originate from distinct oscillatory mechanisms which coexist within the same system. In the region where the two instability domains overlap, the two oscillatory mechanisms are simultaneously active. Their interaction gives rise to various kinds of complex oscillatory phenomena.

In the present model, the strategy outlined above led to the observ-ation of complex periodic oscillations of the bursting type [28]. Aperiodic oscillations in cAMP were also observed in this manner, by integration of eqs. (2) [28]. An example of such chaotic oscillations is shown in Fig. 5. Here, as in Fig. 3, the oscillations in cAMP are

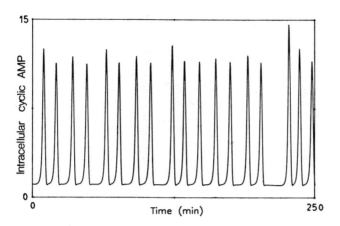

Fig. 5. Aperiodic oscillations (chaos) in the model for cAMP synthesis in D. discoideum [28]. The curve is obtained by integration of eqs. (2) for $v=7.545 \times 10^{-4} s^{-1}$; other parameter values are as in Fig. 3.

accompanied by a slight variation in ATP and by an alternation of the receptor between the active and desensitised states.

The aperiodic oscillations of Fig. 5 correspond in the phase space to the evolution towards a strange attractor. Shown in Fig. 6 is the projection of this attractor in a three-variable space. In contrast to periodic behaviour which corresponds to a closed curve such as a limit cycle, the trajectory here never passes twice through the same point. Moreover, the distance between two points initially close to each other on the attractor increases exponentially with time.

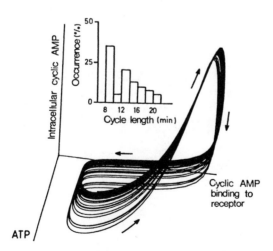

Fig. 6. Strange attractor corresponding to the chaotic behaviour of Fig. 5 [28]. The curve is obtained by projection of the trajectory on a three-variable space. The inset shows the histogram of cycle lengths for some 500 successive cycles on the attractor.

The numerical integration of eqs. (2) for more than 500 cycles on the strange attractor yields the histogram of cycle lengths shown in the inset of Fig. 6. For the parameter values considered, most cycles last from 8 to 10 min, but some cycles extend to 22 min.

Further evidence that the irregular oscillatory behaviour of Fig. 5 represents chaos is provided by the return map in Fig. 7. There, the maximum in intracellular cAMP is plotted as a function of the preceding maximum, yielding a map of $(\beta_M)_{n+1}$ vs $(\beta_M)_n$. The curve obtained is continuous but not closed; such property is characteristic of chaos, whereas a closed, continuous curve would correspond to quasi-periodicity.

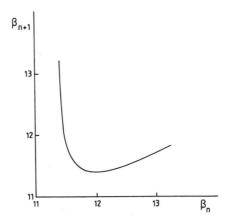

Fig. 7. Return map for the aperiodic oscillatory behaviour of Fig. 5, showing evidence for chaos. The maximum intracellular cAMP concentration in the course of oscillations is plotted as a function of the maximum reached during the preceding cycle.

As in many chaotic systems, aperiodic oscillations appear in the model through a series of period-doubling bifurcations. Succcessive values of parameter v corresponding to these bifurcations obey the universal pattern observed by Feigenbaum [9]. Indeed, the values v_i (in s^{-1}) corresponding to the first three period-doubling bifurcations, from period-1 up to period-8, are found to be:

$$v_1 = 7.804 \times 10^{-4},$$

$$v_2 = 7.63 \times 10^{-4},$$

$$v_3 = 7.594 \times 10^{-4}$$

These values yield the ratio

$$\frac{v_1 - v_2}{v_2 - v_3} = 4.83$$

which is close to Feigenbaum's value of 4.669... [9]. Analytical results on bifurcations leading to chaos are given in [26] for a three-variable version of this model.

IV APERIODIC SIGNALLING IN THE MUTANT FR17: AN INSTANCE OF CHAOTIC
    BEHAVIOUR?

    Whereas periodic oscillations in cAMP synthesis account for the
behaviour of wild type D. discoideum amoebae on agar and in cell suspensions, the question arises as to whether the theoretical prediction of
chaos corresponds to any experimental observation. It seems that a set of
experimental results obtained by Durston on a pacemaker mutant of
D. discoideum might be interpreted in terms of chaos.

    Durston [8] measured the time interval separating successive waves of
amoebae aggregating on agar. In the wild type, this time interval remains
in a narrow band, as expected for periodic aggregation when measurements
are performed for hundreds of waves. In contrast, the time interval
between successive waves in the mutant Fr17 is highly irregular: it has
a mean of 14 min, but varies by a factor of up to three, thus yielding a
histogram of intervals much broader than in the wild type reminiscent of
the one shown in Fig. 6. Before the significance of chaotic phenomena
was clearly recognised, Durston referred to the behaviour of Fr17 as
'aperiodic signalling', and stressed that the analysis of such behaviour
would throw much light on the dynamics of the signalling process.

    That the aperiodic behaviour of Fr17 truly represents chaos should
be further investigated, a most promising approach being that of cell
suspension experiments. Preliminary observations in such conditions [3]
showed the occurrence of 'erratic' oscillations of cAMP; the reduced
number of cycles recorded does not permit, however, to characterise this
behaviour as chaotic. More prolonged observations of cAMP synthesis in
Fr17 suspensions are needed to reach such conclusion.

    Interestingly, a spontaneous transition from irregular to regular
oscillations is sometimes observed during development of Fr17 after starvation [8]. From a theoretical viewpoint, such transition may correspond
to the bifurcation of a strange attractor into a limit cycle upon continuous variation of some control parameter. This transition would be
analogous to the passage from excitable to oscillatory behaviour which
occurs in the wild type after starvation [11]. It has previously been
suggested that a continous increase in the activity of adenylate cyclase
and phosphodiesterase underlies this developmental transition. In a
similar manner, the conjecture that Fr17 behaves chaotically could further
be tested on agar or in cell suspensions by addition of exogenous phosphodiesterase: the model predicts that the addition of a sufficient amount of
phosphodiesterase should bring the intercellular communication system of
Fr17 from the chaotic domain into the much larger domain of periodic
signalling where the wild type operates.

V CONCLUDING REMARKS

    The cAMP signalling system of D. discoideum provides the opportunity
of studying periodic as well as chaotic modes of intercellular communication. That aperiodic oscillations occur in a model for the cAMP signalling system comprising a single positive feedback loop indicates that
multiple feedback processes are not a prerequisite for chaos. Here, the
phenomenon originates from the interplay between two oscillatory mechanisms coupled in parallel within the signalling system; both mechanisms
share the same positive feedback which governs cAMP synthesis in
D. discoideum.

    The importance of the interplay between the two oscillatory mechan-
isms in the origin of complex oscillations is corroborated by the empiri-

ical method successfully used for finding the chaotic domain in parameter space. This method, previously applied to another autonomous system [5], consists in bringing together two instability domains in parameter space (see Fig. 4). Complex oscillatory phenomena, including bursting and chaos, occur in the region where the two domains of instability overlap or in the region separating the two domains, just before their junction.

In the autonomous systems, chaos generally results from the coupling in series or in parallel between two endogenous oscillatory mechanisms. A close link can thus be established with non-autonomous systems in which chaos often results from the interplay between an endogenous oscillator and an external periodic force. Examples of non-autonomous chaos in biological systems are numerous [30] and include, at the biochemical level, the periodically force glycolytic system in yeast [24,25].

Besides the cAMP signalling system of the slime mould, other instances of autonomous chaos in biochemical or cellular processes have been described and analysed, e.g. the peroxydase reaction [30,31], and the electrical activity of molluscan neurones subjected to drugs [19]. If confirmed to represent chaos, the aperiodic signalling in Fr17 would provide a first example, at the cellular level, of a transition from periodicity in the wild type to chaotic behaviour, solely through mutation (the mutant Fr17 is characterised by a precocious activation of adenylate cyclas [3,20].

What is the physiological significance of chaos versus periodic behaviour in cell to cell signalling processes. Resorting to Fr17 for deciding this issue is difficult, in view of the poor performance of this mutant during aggregation and subsequent development; such troubles could indeed be due to factors other than the mode of intercellular communication.

In Dictyostelium, it is nevertheless unlikely that aperiodic signalling provides any advantage to aggregating cells, given the fine tuning that exists in the wild type between the frequency of pulse generation and the time required by the receptor to recover from desensitisation upon removal of the stimulus. The delivery of signals at regular intervals appears to provide maximum responsiveness in hormonal and intercellular communication systems subjected to receptor desensitisation [13,14]. Such view accounts for the effectiveness of cAMP pulses delivered at the physiological frequency on the development of Dictyostelium cells [36]. Therefore, any signalling with intervals between successive pulses now below and then above the physiological (normal) value - as in the chaotic regime - would prove less effective than the periodic mode of intercellular communication.

REFERENCES

[1]   J.T. Bonner, 'The Cellular Slime Molds', Princeton University Press, Princeton, N.J. (1967).
[2]   B. Coukell, Differentiation, 20:29-35 (1981).
[3]   M.B. Coukell & F.K. Chan, FEBS Lett. 110:39-42 (1980).
[4]   O. Decrouly, This volume pp   (1987).
[5]   O. Decrouly & A. Goldbeter, Proc. Nat. Acad. Sci. USA, 79:6917-6921 (1982).
[6]   P.N. Devreotes, in 'The Development of Dictyostelium discoideum', ed. W.F. Loomis, pp 117-168, Academic Press, New York (1982).
[7]   P.N. Devreotes & J.A. Sherring, J. Biol. Chem. 260:6378-6384 (1985).
[8]   A.J. Durston, Devel. Biol. 38:308-319.

[9]   M.J. Feigenbaum, J. Stat. Phys. 19:25-52 (1978).
[10] G. Gerisch, Ann. Rev. Physiol. 44:535-552 (1982).
[11] G. Gerisch, D. Malchow, W. Roos & U. Wick, J. Exp. Biol. 81:33-47
      (1979).
[12] A. Goldbeter, Nature 252:540-542 (1975).
[13] A. Goldbeter, in 'Molecular Mechanisms of Desensitisation', eds.
      T. Konijin, P.J.M. Van Haastert, H. Van der Wel & M.D. Houslay,
      pp 43-62, Springer, Berlin (1986).
[14] A. Goldbeter, in 'Temporal Disorder in Human Oscillatory Systems',
      eds. L. Rensing, U. an der Heiden & M.C. Mackey, in press,
      Springer, Berlin (1987).
[15] A. Goldbeter & J.L. Martiel, in 'Sensing and Response in Micro-
      organisms', eds. M. Eisenbach & M. Balaban, pp 185-198, Elsevier,
      Amsterdam (1985).
[16] A. Goldbeter, J.L. Martield & O. Decroly, in 'Dynamics of Bio-
      chemical Systems', eds. J. Ricard & A. Cornish-Bowden, pp 173-212
      Plenum, New York (1984).
[17] A. Goldbeter & L.A. Segel, Proc. Nat. Acad. Sci. USA 74:1543-1547
      (1977).
[18] A. Goldbeter & L.A. Segel, Differentiation 17:127-135 (1980).
[19] A.V. Holden, W. Winlow & P.G. Haydon, Biol. Cybern. 43:169-173.
[20] R.H. Kessin, Cell 10:703-708 (1977).
[21] C. Klein, J. Lubs-Haukeness & S. Simons, J. Cell. Biol. 100:715-720
      (1985).
[22] P. Klein, A. Theibert, D. Fontana & P.N. Devreotes, J. Biol. Chem.
      260:1757-1764 (1985).
[23] T.M. Konijin, J.G.C. Van de Meene, J.T. Bonner & D.S. Barkley,
      Proc. Nat. Acad. Sci. USA 58:1152-1154.
[24] M. Markus & B. Hess, Proc. Nat. Acad. Sci. USA 81:4394-4398 (1984).
[25] M. Markus, D. Kuschmitz & B. Hess, FEBS Lett. 172:235-238 (1984).
[26] J.L. Martiel, In preparation (1987).
[27] J.L. Martiel & A. Goldbeter, C.R. Acad. Sci. (Paris) Ser. III 298:
      549-552 (1984).
[28] J.L. Martiel & A. Goldbeter, Nature 313:590-592.
[29] P.C. Newell, in 'Microbial Interactions (Receptors and Recognition),
      Ser. B. Vol. 3, ed. J.L. Reissig, pp 3-57, Chapman and Hall,
      London (1977).
[30] L.F. Olsen & H. Degn, Nature 267:177-178 (1977).
[31] L.F. Olsen & H. Degn, Quart. Rev. Biophys. 18:165-225 (1985).
[32] H.G. Othmer, P.B. Monk & P.E. Rapp, Math. Biosci. 77:79-139 (1985).
[33] P.E. Rapp, P.B. Monk & H.G. Othmer, Math. Biosci. 77:35-78 (1985).
[34] W. Roos, C. Scheidegger & G. Gerisch, Nature 266:259-261 (1977).
[35] A. Theibert & P.N. Devreotes, J. Cell Biol. 97:173-177 (1983).
[36] B. Wurster, Biophys. Struct. Mech. 9:137-143 (1982).

TURBULENT MORPHOGENESIS OF A PROTOTYPE MODEL REACTION-DIFFUSION SYSTEM

Jürgen Parisi[1], Brigitte Röhricht[1], Joachim Peinke[1] and
Otto E. Rössler[2]

[1]Physikalisches Institut II   [2]Institut für Physikalische
Universität Tübingen   und Theoretische Chemie
Morgenstelle 14   Universität Tübingen
D-7400 Tübingen   Morgenstelle 8
F.R.G.   D-7400 Tübingen
                                 F.R.G.

ABSTRACT

Based on the well-established Rashevsky-Turing theory of morphogenesis, we report on a simple two-cellular symmetrical reaction-
diffusion model capable of eliciting symmetry-breaking phase transitions
and boiling-type turbulence. Such self-organising cooperative processes
are experimentally demonstrated with spatio-temporal nonlinear transport
phenomena in semiconductors. The present reaction-diffusion model may
acquire a rather general significance, as it represents the most
convenient prototype model of many different synergetic systems in nature.

I INTRODUCTION

The well-known Rashevsky-Turing (RT) theory [10,19] was originally
designed to account for a qualitative understanding of the spontaneous
occurrence of differentiation in multicellular systems based on identical,
individually stable cells. Assuming a symmetrically built, homogeneous
reaction system, symmetry-breaking RT behaviour implies evocation of
instability of the symmetrical steady state under a parametric change.
Following the original Turing reaction-diffusion-transport equations [19]
as the first concrete dynamical model of biological morphogenesis, several
more examples have been described in the literature. The well-known
Prigogine-Nicolis Brusselator [4,9] as well as the Gierer-Meinhardt
equations [1,3] are of the same type, providing concrete models of chemical reaction kinetics and biological hydra morphogenesis, respectively.
Based on a version of the original RT system, Rössler [14,15] discussed a
two-compartment chemical reaction-diffusion model. Detailed numerical
analysis of the underlying bifurcation behaviour towards turbulent flow was
given by Kennedy and Aris [2]. Very recently, the same system has further
been introduced into physics [12] and biochemistry [18]. Turing-type
equations are of interest, therefore, for mathematical, chemical, physical,
biological, and biochemical systems.

Representing a general dynamical model for many different synergetic systems in nature, we concentrate in the following on the simple prototype version of a two-cellular symmetrical reaction-diffusion system. We report on numerical evidence of a symmetry-breaking nonequilibrium phase transition from phase-locked coherent to phase-lagged differentiated behaviour of the two subsystems. We further indicate the structural change of the system flow that is evocated, from stable morphogenesis to boiling-type turbulence. Finally, such self-organising cooperative processes are experimentally demonstrated with spatio-temporal nonlinear transport phenomena in semiconductors.

II THE MODEL

As the essential behavioural characteristic of RT systems, breakdown of symmetry can in the simplest case be realised by a two-cellular symmetrical morphogenetic system consisting of cross-inhibitorily coupled, potentially oscillating two-variable subsystems (4-D flow) [14]. The reaction scheme of such a simple RT system is sketched in Fig. 1. The two morphogens A are self-inhibiting via B and, to a lesser extent, cross-inhibiting each other via the symmetrical coupling between the two morphogens B. The excess of self-inhibition over cross-inhibition is compensated by a path of self-activation (autocatalysis of A) which is not mediated to the other side. The effects of constant pools and reaction partners are comprised in the effective rate constants $K_1, \ldots, K_5$. Following the argumentation introduced by Rössler [14], the system of Fig. 1 can be described under the assumption of a Michaelis-Menten-type kinetics by the following set of simultaneous ordinary differential equations:

$$
\begin{aligned}
\dot{a} &= (K_1 - K_3)\, a - K_2\, b\, \frac{a}{a + K} + K_5 \\
\dot{b} &= K_3\, a - K_4\, b + D\,(b' - b) \\
\dot{b}' &= K_3\, a' - K_4\, b' + D\,(b - b') \\
\dot{a}' &= (K_1 - K_3)\, a' - K_2\, b'\, \frac{a'}{a' + K} + K_5
\end{aligned}
\tag{1}
$$

where a, b, a', b' denote the concentration of the two morphogens A and B in compartment 1 (unprimed) and 2 (primed), respectively. D is the

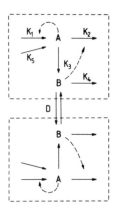

Fig. 1. Reaction scheme of a simple Rashevsky-Turing system (constant pools omitted from the scheme, catalytic rate control indicated by dashed arrows).

92

diffusion coefficient for the morphogen B, and K represents the phenomenological Michaelis-Menten constant. We note that the above equations
presuppose isothermy, homogeneity in either compartment, and fast relaxation of intermediate products as usual.

The above two-compartment structure of the 4-D flow implies the two
essential characteristics of any RT system, namely the spontaneous
occurrence of evocation behaviour in the symmetrical homogeneous reaction
system and the existence of a certain delay in the inner dynamics of each
subsystem. Based on these main phenomena, the RT reaction-diffusion model
described in Eq. (1) is capable of eliciting symmetry-breaking non-
equilibrium phase transitions from phase-locking of the two subsystems to
a phase-reversal. Moreover, the system flow that is evocated may further
bifurcate from stable morphogenesis to boiling-type turbulence.

Numerical evidence on the spatio-temporal correlation of the two
subsystems during the above transition from the symmetrical state of
phase-locked periodic oscillations to symmetry-breaking structures of
phase-lagged periodic and chaotic oscillations under variation of the
control parameter K_5 (termed an evocation parameter by Turing [19]) is
presented in Fig. 2. The projection of the trajectorial flow in part (a)
corresponds to the symmetrical oscillatory behaviour of the two-cellular
reaction-diffusion system (control parameter $K_5 = 2$). Subfigures (b) -
(d) indicate the scenario from the fully differentiated state (b) towards
the irregularly recurring differentiation (d) between the two main
variables a and a', obtained by increasing the parameter K_5 from 0.4 up
to 0.8 (all other parameter values are kept constant; see figure legend).
It is emphasised that the apparently nonperiodic flow of Fig. 2d ends up
with a sort of screw-type bichaotic flow - two screw-type chaotic regimes
separated by a symmetrical saddle-limit cycle - as described by Rössler
[14]. Such kind of turbulent morphogenesis in 4-D state space admits 3-D
cross-sections of the generalised horseshoe type [2,14,17].

III COMPARISON WITH A REAL PHYSICAL SYSTEM

Since the above self-organising cooperative processes appear to
reflect a general property of real synergetic systems, let us now see
whether recent experimental investigations on spatio-temporal nonlinear
transport phenomena in the avalanche breakdown regime of current-carrying
semi-conductors [6,7,8,11,13] can be interpreted in terms of the main
behavioural characteristics of the RT reaction-diffusion model. Our
experimental semiconductor system consists of a single homogeneously
p-doped germanium crystal, electrically driven into the post-breakdown
regime due to impurity impact ionisation at liquid-helium temperatures.
Spatially separated and coupled sample subsystems are realised by means
of an appropriate arrangement of ohmic contact probes, dividing the
semiconductor crystal into different parts which by themselves show non-
linear dynamical behaviour.

As discussed elsewhere [5] in detail, the current transport phenomena
in semiconductors include spontaneous formation of a rich variety of both
spatial and temporal dissipative structures. Experimental evidence of
analogous symmetry-breaking nonequilibrium phase transitions and various
kinds of boiling-type turbulence suggests a reaction scheme for the
semiconductor system qualitatively similar to that of the phenomenological
RT model indicated in Fig. 1. Specifically, the two morphogens A (activator) and B (inhibitor) of the RT model can be interpreted as the number
density of moving charge carriers (reflected in the electric current) and
the mean energy per carrier, respectively. Diffusive coupling between the

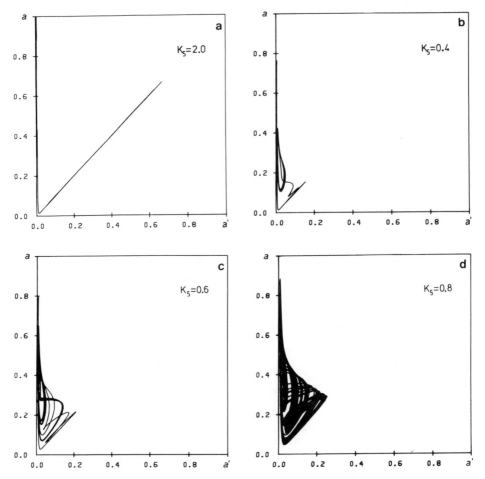

Fig. 2. Symmetry-breaking transition from phase-locked periodic flow
(a) towards phase-lagged periodic (b) and chaotic (d) flow of
the 4-D Rashevsky-Turing system. The phase portraits of the
dominant variables a vs. a' were obtained at different values
of the evocation parameter K_5 (as indicated in the figure)
and for the following set of constant parameter values:
D = 12, K = 0.03, K_1 = 10.8, K_2 = K_3 = 6, and K_4 = 3. The
initial conditions of the morphogen concentrations were
chosen as follows: a(0) = 0.75, b(0) = b'(0) = 0.6, and
a'(0) = 0.005. The phase portraits are displayed from t_o = 0
up to t_{end} = 100 time units. Numerical simulation was
performed on a DEC VAX 750 computer with peripherals, using
a standard Runge-Kutta-Merson integration routine.

two electronic subsystems is effected through energy exchange via
acoustical phonons emitted by the hot carriers. The long-range phonon
propagation benefits from the large lattice heat conductivity of
germanium. However, the rather complicated nonlinear behaviour of the
autocatalytic impact ionisation process of the charge carriers can be
only roughly approximated with the simple kinetics of the present
activator variable [16].

IV CONCLUSIONS

Our experiments deal with a challenging example of a convenient synergetic system, the dynamic possibilities of which can qualitatively be described by the above 4-D reaction-diffusion model. Reflecting general behavioural characteristics of many different systems in nature, the well-known Turing equation from chemical systems theory (Eq. (1)) turned out to possess a certain dynamical universality. Nevertheless, the specific significance of turbulent morphogenesis for self-differentiating biological processes is lately under controversial discussion. The still open question arises whether chaos is a pathological or even a necessary element in the formation and functioning of living organisms.

ACKNOWLEDGEMENTS

The authors are grateful to R.P. Huebener, U. Rau and E. Schöll for stimulating discussions and the Stiftung Volkswagenwerk for financial support.

REFERENCES

[1] A. Gierer and H. Meinhardt, Kybernetik 12: 30 (1972).
[2] C.R. Kennedy and R. Aris, in "New Approaches to Nonlinear Problems in Dynamics", P.J. Holmes, ed., SIAM, Philadelphia, p. 211 (1980).
[3] H. Meinhardt and A. Gierer, J. Cell. Sci. 15: 321 (1974).
[4] G. Nicolis and I. Prigogine, Self-organisation in Nonequilibrium Systems, Wiley, New York (1977).
[5] J. Parisi, J. Peinke, B. Röhricht and R.P. Huebener, in "Proceedings of the 18th International Conference on the Physics of Semiconductors", World Scientific, Singapore, p. 1571 (1986).
[6] J. Peinke, A. Mühlbach, R.P. Huebener and J. Parisi, Phys. Lett. 108A: 407 (1985).
[7] J. Peinke, B. Röhricht, A. Mühlbach, J. Parisi, Ch. Nöldeke, R.P. Huebener and O.E. Rössler, Z. Naturforsch. 40a: 562 (1985).
[8] J. Peinke, J. Parisi, B. Röhricht, B. Wessely and K.M. Mayer, Phys. Rev. Lett. (1987, to be published).
[9] I. Prigogine and G. Nicolis, J. Chem. Phys. 46: 3542 (1967).
[10] N. Rashevsky, Bull Math. Biophys. 2: 15, 65, 109 (1940).
[11] B. Röhricht, B. Wessely, J. Peinke, A. Mühlbach, J. Parisi and R.P. Huebener, Physica 134B: 281 (1985).
[12] B. Röhricht, J. Parisi, J. Peinke and O.E. Rössler, Z. Phys. B - Condensed Matter 65: 259 (1986).
[13] B. Röhricht, J. Parisi, J. Peinke and R.R. Huebener, Z. Phys. B - Condensed Matter (1987, to be published).
[14] O.E. Rössler, Z. Naturforsch. 31a: 1168 (1976).
[15] O.E. Rössler and F.F. Seelig, Z. Naturforsch. 27b: 1444 (1972).
[16] E. Schöll, J. Parisi, B. Röhricht, J. Peinke and R.P. Huebener, Phys. Lett. A 119A: 419 (1987).
[17] R. Shaw, The Dripping Faucet as a Model Chaotic System, Aerial Press, Santa Cruz (1985).
[18] O. Sporns, S. Roth and F.F. Seelig, Physica D (1986, to be published).
[19] A.M. Turing, Philos. Trans. R. Soc. London B 237: 37 (1952).

PERIODIC SOLUTIONS AND GLOBAL BIFURCATIONS FOR NERVE IMPULSE EQUATIONS

M. Rosário L.M.S. Álvares-Ribeiro

Grupo de Matemática Aplicada
Universidade do Porto
R. das Taipas, 135
4000 Porto, Portugal

ABSTRACT

Software for the Apple Macintosh microcomputer has been developed using the harmonic balance method for the detection of periodic solutions of feedback systems. This software, based on graphical criteria, also provides a good notion of the system's dynamics and the way bifurcations occur, as well as the stability characteristics of the limit cycles. Here we report the results of its application to the FitzHugh equations for the nerve impulse. We numerically detect periodic solutions and global bifurcation points that, although theoretically predicted, had never been located. We describe amplitude, frequency and stability characteristics for these solutions, as well as the type and location of the bifurcation points.

I INTRODUCTION

A simplified model for the nerve impulse is provided by FitzHugh equations [1]

$$\dot{v} = \lambda+u+v-v^3/3 \qquad 0<\gamma<1$$

$$\dot{u} = -\rho(v+\gamma u) \qquad 0<\rho<1 \tag{1}$$

where v is the difference of electric potential across the nerve cell's membrane, u represents the permeability of the cell membrane to certain ions, and λ represents the current applied to the nerve cell as stimulus.

A description of (1) is given by the feedback system in figure 1. G and h are given by

$$G(s) = \frac{s+\gamma\rho}{s^2+(\gamma\rho+v_0^2-1)s+\gamma\rho v_0-\gamma\rho+\rho} \qquad s\in C \tag{2}$$

$$h(x) = v_0 x^2+x^3/3 \qquad\qquad x\in R$$

$$\lambda = v_0^3/3+(1-\gamma)v_0/\gamma \tag{3}$$

$$x = v-v_0$$

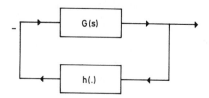

Fig. 1.   G(s) is the Transfer Function and so represents the system's
linear part and the function h describes its nonlinearity.

Since (1) is symmetric in λ, it follows from (3) that we can take
v_0 (first component of the critical point of (1)) to be non-negative.

II   METHOD

We seek periodic solutions, $x(t)$, of the system in figure 1 that,
using the harmonic balance [4], are the solutions of the system

$$G(ikw)c_k(a_0,a_1)+a_k = 0 \qquad\qquad k\in N_o \qquad\qquad (4)$$

where $a_0,c_0\in R,a_k,c_k(a_0,a_1,\ldots)\in C$, $k\in N, x(t) = \Sigma a_k e^{-ikwt}$ and $h(xt)) =$
$\Sigma c_k(a_0,a_1,\ldots)e^{-ikwt}$;   with a positivity condition for some coefficient
a_q, with $q\in N$ (as we are working with an autonomous system the time origin
is arbitrary).

Under certain conditions [4], the solutions of (4) with $G(ikw) = 0$
0 $\forall k>1$ and $a_1\in R^+$, are near the periodic solutions of the system in figure
1;   the describing function method is to look for solutions of the
equations

$$G(0)c_0(a_0,a_1)+a_0 = 0 \qquad a_0,c_0(a_0,a_1)\in R \qquad\qquad (5)$$

$$G(iw)c_1(a_0,a_1)+a_1 = 0 \qquad a_1\in R^+,c_1(a_0,a_1)\in C \qquad\qquad (6)$$

The describing function is defined by:

$$N(a_0,a_1) = c_1(a_0,a_1)/a_1$$

Solving (5) for a_1 we are reduced to the equation

$$G(iw) = -1/N(a_0,a_1(a_0))$$

i.e.:   we look for intersections of $\{z = G(iw):w\in R_0^+\}$ and $\{z = -1/N(a_0,a_1$
$(a_0)):a_0\in R,a_1(a_0)\in R\}$.

III   SOFTWARE

The software developed for the Apple Macintosh microcomputer plots
graphics of $-1/N(a_0,a_1(a_0))$ and the Nyquist locus of $G(iw)$, and defines
'regions' on the screen corresponding to them;   the intersection points
(w,a_0,a_1) of $\{z = G(iw):w\in R_0^+\}$ and $\{z = -1/N(a_0,a_1(a_0)):a_0\in R,a_1(a_0)\in R^+\}$ are
obtained using the Macintosh routines dealing with 'regions'.

In this way, we avoid numeric procedures in a time efficient program
that outputs the essential characteristics (period, center and amplitude)
of the periodic solution instead of an amount of values to be treated
later.   Moreover, the stability of the limit cycles is indifferent for
its detection.

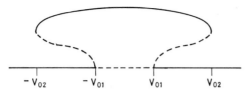

$$- V_{02} \qquad - V_{01} \qquad V_{01} \qquad V_{02}$$

Fig. 2.   Qualitative prediction [3] for the bifurcation diagram with the parameter v_0 for $\gamma = 0.95$ and $\rho = 0.1$. The dashed line represents unstable solutions.

Working on a phase gain diagram, we can determine the stability characteristics of the corresponding limit cycles from the way the curves $z = G(iw)$ and $z = -1/N(a_0, a_1(a_0))$ cross each other [2]:   in figure 6 we see that they do so at the two intersection points in a different way. According to the graphic criteria described in [2], the stable periodic solution is the one which has center $a_0 = -0.824$, and amplitude $2a_1 = 2.06$;   the other is unstable.

IV   RESULTS

For the system in figure 1 we have that

$$-1/N(a_0, a_1(a_0)) = -1/\{a_0(2v_0 + a_0) + [a_1(a_0)]^2\} \qquad (7)$$

where

$$[a_1(a_0)]^2 = [-a_0^3 - v_0 a_0^2 (1 - v_0 - 1/\gamma] / [2(v_0 + a_0)].$$

For $\rho = 0.1$ and $\gamma = 0.95$ the qualitative prediction for the bifurcation diagram [3] with the parameter v_0 is presented in figure 2. The points $v_0 = v_{01}$ and $v_0 = v_{01}$ corresponds to Hopf bifurcations while for $v_0 = -v_{02}$ and $v_0 = v_{02}$ we see that the two limit cycles coalesce and disappear.

In figure 3 we can see that, for given values of ρ and γ, the Hopf bifurcation points are detected by a change on the transfer function behaviour as we would expect since it represents the system's linear part.

The global bifurcation points correspond to a change on both transfer function and describing function behaviour as figures 4 and 5 show.

From our numerical results (figure 3) we conclude that a Hopf bifurcation occurs for $v_0 \in]0.950, 0.955[$, and in figures 3 and 5 we see that after this bifurcation there are two limit cycles with the same period.   Over this range, the describing function graph is nearly the same, and the change on the transfer function behaviour implies that before bifurcation they cross each other when the describing function is negative (so we have one periodic solution) and afterwards when the describing function is positive (and then we have two periodic solutions).

In figure 6 we can see the period w, the center a_0, and the amplitude $2a_1$, of these limit cycles for a particular value of v_0. As previously noted, using the graphic criteria described in [2], we obtain

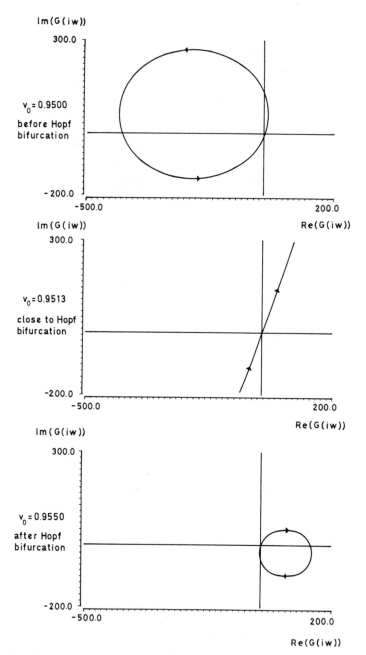

Fig. 3. Nyquist locus of the Transfer Function G for $\gamma = 0.95$ and $\rho = 0.1$.

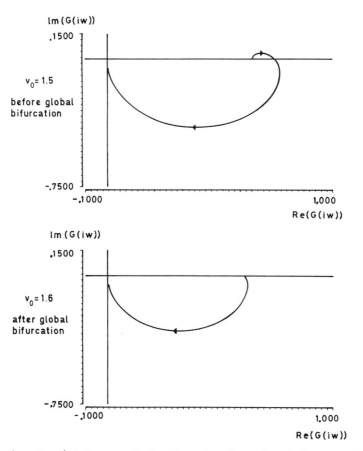

Fig. 4. Nyquist locus of the Transfer Function G for γ = 0.95
and ρ = 0.1.

the stability characteristics predicted in [3]: the stable solution
is the one corresponding to a positive third component of the vectorial
product of the tangent vectors to $-1/N(a_0, a_1(a_0))$ and to $G(iw)$ at the
intersection point taken by this order; the other one is unstable. The
same criteria implies that, before Hopf bifurcation, the periodic
solution (is only one) is stable.

We have found that as we increase v_0 the centers and amplitudes
of the two limit cycles become closer and disappear for $v_0 \in [1.5, 1.6]$
(figures 4 and 5); after this there are no periodic solutions i.e.
$v_{02} \in]1.5, 1.6[$.

V CONCLUSIONS

This method of detecting periodic solutions allows an easy determination of global bifurcations without recourse to numerical
integration, thus avoiding convergence and stability difficulties.

We compute the global bifurcation points at $v_0 = v_{02}$ and $v_0 = -v_{02}$
that, although already predicted [3], had never been numerically
detected, as well as the periods, centers, and amplitudes of the

101

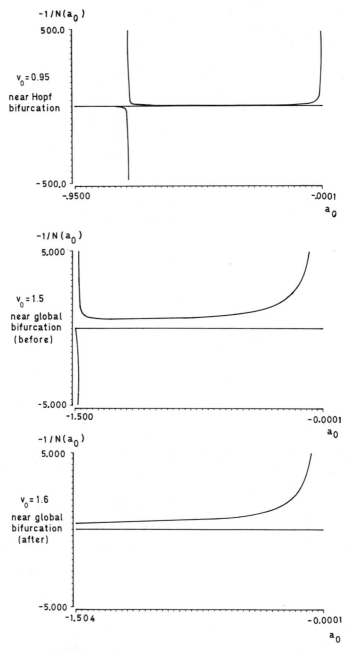

Fig. 5. Graph of $-1/N(a_0)$ where N is the Describing Function for $\gamma = 0.95$.

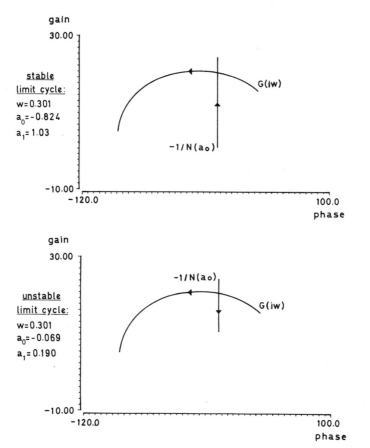

Fig. 6. Detection of the limit cycles for $\gamma = 0.95$ and $\rho = -0.1$.

periodic solutions. Moreover, even far from the global bifurcation points, when we find two limit cycles they always have the same period.

The bifurcation diagram of figure 2 shows the existence of hysteresis - sudden change from the rest state to repetitive activity, that occurs at different values of λ, as the stimulus is increased or decreased. This behaviour is predicted in [3] and here we calculate the corresponding range for λ.

This procedure can be extended to higher order, n, by considering, in (4), $G(ikw) = 0$ $\forall k > n$ and so we may obtain a better precision on the periodic solutions detection.

Here we present the first order case. The location of the other types of bifurcation for the FitzHugh equations predicted in [3] are currently under consideration using this approach.

ACKNOWLEDGEMENT

I wish to thank Dr. J. Basto Goncalves and Dr. Isabel S. Labouriau for numerous fruitful discussions, valuable hints, and careful reading of and commenting on the manuscript. I also thank Dr. P. Lago for his useful advice and for help on the software feature. I am grateful for financial support from research grant 27/85/UP. and INIC.

REFERENCES

[1] R. FitzHugh, Impulses and physiological states in theoretical
     models of nerve membrane, Biophys. J. 1:445-466 (1961).
[2] A. Gelb and W. van der Velde, Multiple-input Describing Functions
     and Nonlinear System Design, McGraw-Hill (1968).
[3] Isabel S. Labouriau and Nelma R.A. Moreira, Solucões Periódicas das
     equacões de FitzHugh para o impulso nervoso, VII Congresso dos
     Matematicos de Expressão Latina (1985).
[4] Alistair Mees, Dynamics of Feedback Systems. John Wiley (1981).

HOMOCLINIC AND PERIODIC SOLUTIONS OF NERVE IMPULSE EQUATIONS

Isabel Salgado Labouriau

Grupo de Matemática Aplicada
Faculdade de Ciencias
Universidade do Porto
4000 Porto, Portugal

ABSTRACT

We study the clamped Hodgkin and Huxley equations for the nerve impulse of the squid giant axon (HH).

Results on generalized Hopf bifurcation are used to describe the periodic solutions of HH in a region of the parameter space where there is a single equilibrium solution. The equations are shown to have a two dimensional attracting centre manifold, and a parameter value is found for which HH are equivalent to a Hopf-Takens bifurcation of codimension 2. In this way we obtain a description of the periodic solution branch and of its stability when a special parameter, the stimulus intensity, is varied. Other codimension 2 singularities present in HH are the cusp catastrophe and the Bogdanov-Takens cusp. The study of these singularities provides a description of the way a nerve cell may switch from repetitive activity (periodic solutions) to action potentials (homoclinic solutions).

I INTRODUCTION

In this paper we use singularity theory techniques to study the interaction of periodic and equilibrium solutions of the Hodgkin and Huxley equations (HH) for the propagation of nerve impulse on the squid giant axon [4]. These equations, stated explicitly in section II, relate the difference of electric potential (V) across an axon's membrane to the membrane's permeability to Na^+ and K^+ ions (described by the variables m, n and h) as a response to an externally applied current stimulus (I).

We treat the space-clamped HH as a family of ordinary differential equations on R^4 (V-m-n-h space). We study here the bifurcation of periodic and homoclinic solutions to these equations regarding the stimulus intensity I and the potassium equilibrium voltage V_K as bifurcation parameters.

In experiments, the value of V_K can be made to vary by changing the ionic concentrations around the axon. Application of a sufficiently large stimulus may induce a periodic fluctuation of V - the cell is said to fire repetitively [2]. Under most experimental conditions, however, nerve cells respond to stimulation with a single voltage pulse, followed by a return to equilibrium, called an action potential. It has a characteristic amplitude independent of stimulus intensity and is initiated at a sharply defined threshold.

This paper presents an attempt to describe the way in which an axon changes its type of response from repetitive firing to single action potentials. We shall indentify the later to homoclinic solutions of the differential equations - solutions approaching the same equilibrium point as $t \rightarrow \pm \infty$. The stability of equilibria of HH is described in section III, where it is explained that in the region of parameter space we work with the system is essentially two dimensional, and therefore, homoclinic solutions may only appear for parameter values corresponding to multiple equilibria. We also describe the results of [7] on the way a stable periodic solution branch appears at a Hopf bifurcation.

As the value of V_K is changed, a pair of complex eigenvalues of the linearization of HH tends to zero. The simplest description of what takes place around this point is given by the Bogdanov-Takens cusp [1, 8], a bifurcation occuring generically in two parameter families of vector fields. In section IV we discuss homoclinic and periodic orbits near this point, and their biological interpretation is given in section V.

II NERVE IMPULSE EQUATIONS

The Hodgkin and Huxley equations [4] relate the difference of electric potential across the cell membrane (V) and the ionic conductances (m, n, and h), to the stimulus intensity (I), and temperature (T) as follows

$$
\text{HH} \begin{cases}
\dfrac{dV}{dt} = -G(V,m,n,h)-I \\[2mm]
\dfrac{dm}{dt} = \Phi(T)[(1-m)\alpha_m(V)-m\beta_m(V)] \\[2mm]
\dfrac{dn}{dt} = \Phi(T)[(1-n)\alpha_n(V)-n\beta_n(V)] \\[2mm]
\dfrac{dh}{dt} = \Phi(T)[(1-h)\alpha_h(V)-h\beta_h(V)]
\end{cases}
$$

where $\Phi(T) = 3^{\frac{T-6.3}{10}}$

$G(V,m,n,h) = \bar{g}_{Na}m^3h(V-V_{Na})+\bar{g}_K n^4(V-V_K)+g_L(V-V_L).$

$\alpha_m(V) = \Psi(0.1V+2.5)$ $\beta_m(V) = 4e^{(V/18)}$

$\alpha_n(V) = \Psi(0.1V+1)$ $\beta_n(V) = 0.125e^{(V/80)}$

$\alpha_h(V) = 0.07e^{(V/20)}$ $\beta(V) = (2+e^{(V+30)/10})$

with $\Psi(0) = 1$, and $\Psi(x) = \dfrac{x}{e^x-1}$ for $x \neq 0$. Notice that $\alpha_J(V)+\beta_J(V) \neq 0$

for all V and for J = m,n or h. The constants g_{ion}, V_{ion} were obtained from experimental data by Hodgkin and Huxley, with the values given below

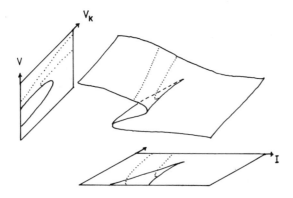

Fig. 1. Schematic representation of equilibrium voltages V_* for the Hodgkin and Huxley equations. The stability of equilibria changes at Hopf bifurcation points lying on the dotted curves, where periodic solutions are generated. The surface of equilibria is folded along saddle-node bifurcations on the two solid curves joining at a cusp point. The projection on to the $I \times V_K$ plane is analogous to the findings of [5].

$$\bar{g}_{Na} = 120 \text{ mS/cm}^2 \qquad \bar{g}_K = 36 \text{ mS/cm}^2 \qquad g_L = 0.3 \text{ mS/cm}^2$$

$$V_{Na} = -115 \text{ mV} \qquad V_K = 12 \text{ mV} \qquad V_L = 10.599 \text{ mV}.$$

III EQUILIBRIA AND STABILITY

For all the results in this paper we use the temperature $T = 6.3°C$. For the value of V_K of +12 mV quoted above (which will be called here normal value) the HH equations have a unique temperature-independent steady-state for each value of I. The linearization of HH around one of these points has two real negative eigenvalues and a pair of complex eigenvalues whose real part changes sign at two Hopf bifurcation points, where periodic solutions are generated [3]. Therefore HH has a I-parametrized family of two-dimensional attracting centre manifolds - invariant manifolds where small amplitude periodic and homoclinic orbits lie.

When V_K is decreased the equations acquire two new steady-states at a cusp catastrophe. This is shown schematically in Figure 1, where the stability on the centre manifold is also indicated. The linearization of HH still has two real negative eigenvalues and the real part of the other two changes sign at Hopf bifurcation and saddle-node points.

Any equilibrium solution of HH, (V_*, m_*, n_*, h_*) must satisfy

$$J_* = \frac{\alpha_J(V_*)}{\alpha_J(V_*) + \beta_J(V_*)} = J_\infty(V_*) \qquad \text{for } J = m, n, h$$

as well as $G(V_*, m_\infty(V_*), n_\infty(V_*), h_\infty(V_*)) = f(V_*) = -I$. Therefore we can change coordinates in HH so as to have the origin of R^4 as a steady-state, using the equilibrium voltage as a new bifurcation parameter $\lambda = V_* = f^{-1}(-I)$. The new set-up has the advantage of eliminating one error factor in numerical computations, as it is no longer necessary to compute $f^{-1}(-I)$. Moreover, we are working on the $\lambda \times V_K$ plane, instead of $I \times V_K$ (see Figure 1) and this provides a better representation of the surface of equilibria - we are interested in equilibrium points of the form

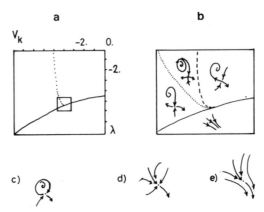

Fig. 2.   Qualitative phase portraits for HH near a Bogdanov-Takens
          bifurcation. (a) Numerical plot of Hopf bifurcation points
          (dotted line) and saddle-node points (solid line) for HH, on
          the $\lambda \times V_K$ plane. This exact picture can be seen as a detail
          of the vertical plane of Figure 1. (b) The small square in
          (a) enlarged, showing persistent qualitative phase portraits
          near the Bogdanov-Takens point where the two curves in (a)
          meet tangentially. The closed orbit created at the Hopf
          bifurcation disappears as its period tends to infinity when
          the dashed line is crossed. (c) (d) and (e) Transition
          qualitative phase portraits for HH on the centre manifold:
          (c) Homoclinic orbit (dashed line on b); (d) saddle-node
          (solid line on b); (e) Bogdanov-Takens cusp (intersection
          point).

$(\lambda, m_\infty(\lambda), n_\infty(\lambda), h_\infty(\lambda)) = f(\lambda, V_K)$, with $I = -f(\lambda, V_K)$. Saddle-node points,
where two equilibria come together and disappear, are characterized by
$\partial f / \partial \lambda = 0, \partial^2 f / \partial \lambda^2 \neq 0$, and the cusp point by $\partial f / \partial \lambda = 0, \partial^2 f / \partial \lambda^2 = 0, \partial^3 f /$
$\partial \lambda^3 \neq 0$.

IV HOMOCLINIC AND PERIODIC SOLUTIONS

     As λ and V_k are varied the Hopf bifurcation points for HH follow the
two curves indicated schematically on the vertical plane of Figure 1.   The
projection of these curves into the $\lambda \times V_K$ plane near one of the two points
of real intersection is shown in Figure 2a. Comparing the two planes of
Figure 1 shows another advantage of working on the $\lambda \times V_K$ plane: the
projection on to the $I \times V_K$ plane creates two artificial intersections of
the Hopf bifurcation curves with the curve of saddle-node points.

     At the point where the curve of Hopf bifurcations joins the saddle-
node curve tangentially, the linearization of HH has a double zero eigenvalue and HH is locally equivalent to a Bogdanov-Takens bifurcation [1,
8]. Periodic solutions, created at the Hopf bifurcation, disappear at
homoclinic orbits when the curve of homoclinic points is crossed. This
last global behaviour cannot be detected by a linear analysis around the
equilibrium. The local structure of the Bogdanov-Takens bifurcation is
represented on Figure 2b, along with the local trajectories for HH on
the centre manifold.

     We had already established elsewhere [6,7] that the two periodic
orbits created at Hopf bifurcation points coalesce and disappear for the

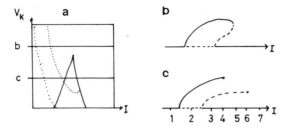

Fig. 3. Amplitude diagrams for periodic solutions of HH at two
different values of V_K. (a) V_Kx diagram corresponding to
horizontal plane of Figure 1, showing the two choices of V_K
used in (b) and (c). Dotted curves - Hopf bifurcations;
solid curves - saddle-node bifurcations. (b) The two periodic
orbits created at Hopf bifurcation points coalesce and disappear for values of V_K where the equilibrium is unique for
all [6,7]. (c) Near the Bogdanov-Takens bifurcation point
the both periodic solution branches terminate at homoclinic
orbits and their periods tend to ∞. Phase portraits for this
case are shown in Figure 4. In both cases b and c the dashed
line represents unstable solutions.

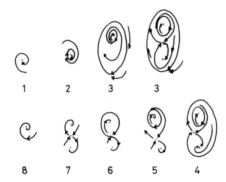

Fig. 4. Qualitative phase portraits for HH on the two dimensional
attracting centre manifold, near the Bogdanov-Takens bifurcation point. The choice of V_K and the numbering of phase
portraits are indicated in Figure 3c.

normal value of V_K, when the equilibrium is unique for all I. The
amplitude of the periodic solutions as a function of I, as well as their
stability are shown schematically in Figure 3. The qualitative behaviour
must remain the same for other values of V_K unless some global bifurcation takes place: this is what must happen near the Bogdanov-Takens
point, where the unstable periodic solution terminates at a homoclinic
orbit and its period tends to infinity. By the Poincaré-Bendixson
Theorem, and recalling that the orbits under discussion lie on a two-
dimensional invariant manifold, it can be shown that at this point the

stable cycle has either disappeared, since it is not possible to have two stable equilibria and a saddle point inside a stable closed orbit, or its stability has changed. In Figure 3c we show the amplitude and period diagrams for HH supposing the stable periodic solution branch has a homoclinic termination. This hypothetical termination is currently being verified using numerical estimates of the periods of stable periodic solutions as a function of λ. Qualitative phase portraits on the centre manifold in this case, are shown in Figure 4.

V DISCUSSION OF EXPERIMENTAL RESULTS

In an experiment, a nerve cell would be stimulated by a decrease in the applied current intensity, I. In the situation of Figure 3b, a sufficiently large stimulus initiates repetitive activity as the equilibrium solution looses stability at the Hopf bifurcation point on the right (Figure 5a). The onset of repetitive firing is observed with a well defined positive period, and a positive amplitude. When stimulation is decreased (i.e. I increases) repetitive activity stops in the same way, but with different values of amplitude, period and stimulus intensity (Figure 5b). This behaviour has been described in experiments with squid giant axons under low Ca^{++} concentrations [2]. If stimulation is increased sufficiently, the repetitive activity stops at the Hopf bifurcation point on the left, where the amplitude tends to zero and the period approaches a positive limit value (Figure 5c).

For values of V_K corresponding to multiple equilibria there is always a value I_0 of I for which HH has a homoclinic orbit. Under experimental conditions corresponding to parameter values in this region, a reduction in I still leads to repetitive activity starting at finite period and amplitude values (Figure 6a). For values of I close to I_0 this activity may be seen as very widely spaced bursts, as the periodic orbit slows down near the saddle point. When stimulus intensity is decreased in this setting, the activity terminates on a 'periodic orbit of infinite period' with positive amplitude – an action potential (Figure 6b). This is the qualitative behaviour of squid axons under normal Ca^{++} concentration.

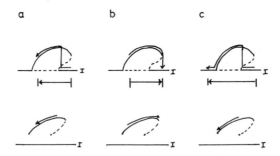

Fig. 5.   Amplitude (top) and period (bottom) diagrams for HH under
          an experiment for V_K in the region where the equilibrium is
          unique for all I , dashed lines standing for unstable solutions.
          Nerve cells are stimulated by a decrease in I . Supra-
          threshold stimulation initiates repetitive activity with a
          positive period as the equilibrium solution looses stability
          at a Hopf bifurcation (a). When stimulation is decreased
          repetitive activity stops in the same way, with different
          values of amplitude, period and stimulus intensity (b). If
          stimulation is sufficiently increased the activity stops at
          the second Hopf bifurcation (c).

a   b

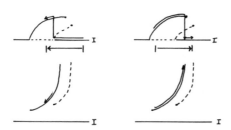

Fig. 6. Amplitude (top) and period (bottom) diagrams for HH under an experiment for I and V_K near the Bogdanov-Takens point, with conventions as in Figure 5. Suprathreshold stimulation initiates repetitive activity with a positive period (a) as in Figure 5a, but a decrease in stimulation (b) terminates the activity at a stable 'periodic orbit of infinite period' - an action potential.

REFERENCES

[1] R.I. Bogdanov, Trudy Sem. Petrovsk. 2:73-65 (1976), English translation Versal Deformation of a singularity of a vector field on the plane in the case of zero eigenvalue, Sel. Math Sov. 1:389-421 (1981).

[2] R. Guttman, S. Lewis and J. Rinzel, Control of repetitive firing in squid axon membrane as a model for a neuroneoscillator, J. Physiol. 305:377-395 (1980).

[3] B. Hassard, Bifurcation of periodic solutions of the Hodgkin-Huxley model for the squid giant axon, J. Theoret. Biol. 7:401-420 (1978).

[4] A.L. Hodgkin and A.F. Huxley, A quantitative description of membrane current and its application to conduction and excitation in nerve, J. Physiol. 117:500-544.

[5] A.V. Holden, M.A. Muhamad and A.K. Schierwagen, Repolarizing currents and periodic activity in nerve membrane, J. Theoret. Neurobiol. 4:61-71 (1985).

[6] I.S. Labouriau, Degenerate Hopf bifurcation and nerve impulse, SIAM J. Math. Anal. 16:1121-1133 (1985).

[7] I.S. Labouriau, Degenerate Hopf bifurcation and nerve impulse - Part II, preprint, Grupo de Matematica Aplicada, FCUP (1986).

[8] F. Takens, Singularities of Vectorfields, Publ. Math. IHES 43:47-100 (1975).

HIGH SENSITIVITY CHAOTIC BEHAVIOUR IN SINUSOIDALLY DRIVEN

HODGKIN-HUXLEY EQUATIONS

P. Arrigo, L. Marconi, G. Morgavi, S. Ridella, C. Rolando
and F. Scalia.

Consiglio Nazionale delle Ricerche
Istituto per i Circuiti Elettronici
Casella Postale 191
16121 Genova, Italy

ABSTRACT

In this paper the interaction between low frequency, low amplitude
electric field and the Hodgkin-Huxley (HH) membrane model is considered
in order to study non-thermal biological effects due to small fields
applied for long time.

For the situation presented a small change in amplitude can induce
a chaotic behaviour. The results are presented in terms of membrane
voltage and maximal Lyapunov exponents.

The goal of this paper is to investigate the Hodgkin-Huxley
equations (HH) [1] with small sinusoidal current stimulation to see if
chaotic effects can be detected. Our interest in this topic comes from
the study of non-thermal effects due to the interaction between low
amplitude, low frequency electromagnetic field and biological material.

We solved the Hodgkin-Huxley equations shown in Appendix 1 using the
Runge-Kutta method. We chose it because it was available and well tested
on our computer even if it is not the best with regard to the accuracy
and performance.

We computed then the maximal Lyapunov exponent [2] in order to
measure the chaotic behaviour of the solution.

The method we used is described in Appendix 2.

First of all, to test our solution, we tried to repeat the results
presented by Jensens et al. in [3] and by Holden and Muhamad [4].

In Figures 1, 2, 3 and 4 some results we obtained are shown: they
are congruent with [3,4].

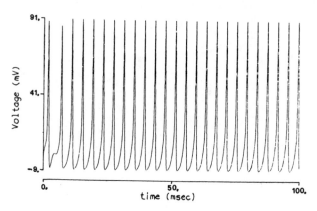

Fig. 1. Resulting membrane voltage, V(t) in response to an applied
current density of −15+9*cos(2*π*250*t) [μA/cm²] at
temperature T = 18.5°C.

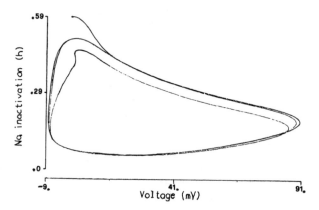

Fig. 2. Resulting projection of the trajectory on to the V-h plane
in response to an applied current density of
−15+9*cos(2*π*250*t) [μA/cm²] at temperature T = 18.5°C.

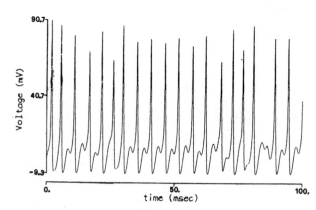

Fig. 3. Resulting membrane voltage, V(t) in response to an applied
current density of −15+9*cos(2*π*370*t) [μA/cm²] at
temperature T = 18.5°C.

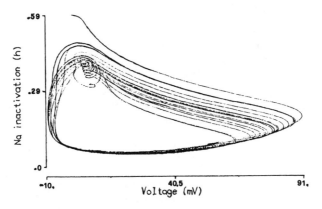

Fig. 4. Resulting projection of the trajectory on to the V-h plane in response to an applied current density of -15+9*cos(2*π*370*t) [μA/cm²] at temperature T = 18.5°C.

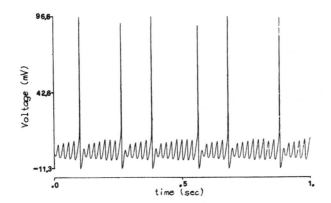

Fig. 5. Resulting membrane voltage, V(t) in response to an applied current density of 1.515*cos(2*π*50*t) [μA/cm²] at temperature T = 6.3°C.

After an extensive analysis of the Hodgkin-Huxley equations we obtained for

$I = A * \cos 2 * \pi *f*t+B$

different behaviour for very small changes in B. We chose the value A = 1.515 μA/cm², f = 50 Hz.

In Figures 5 and 6 the results for V(t) and H(V) are shown for B = 0. The behaviour is chaotic.

In Figures 7 and 8 the results for V(t) and H(V) are shown for B = -0.007 μA/cm²: in that case the behaviour is periodic.

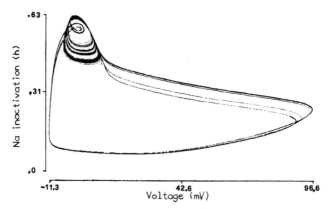

Fig. 6. Resulting projection of the trajectory on to the V-h plane
in response to an applied current density of
1.515*cos(2*π*50*t) [μA/cm²] at temperature T = 6.3°C.

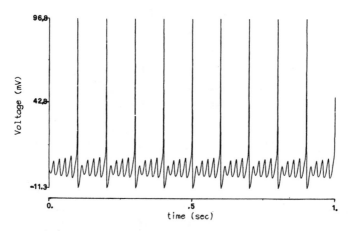

Fig. 7. Resulting membrane voltage, V(t) in response to an applied
current density of -0.007+1.515*cos(2*π*50*t) [μA/cm²] at
temperature T = 6.3°C.

In Figure 9 the behaviour of the maximal Lyapunov exponent λ_1 for
each applied signal is plotted. It shows that in the chaotic
situation λ_1 is positive, in the periodic one λ_1 is negative.

116

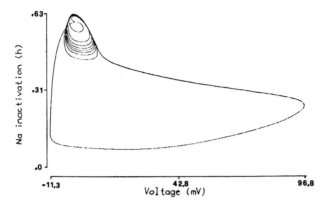

Fig. 8. Resulting projection of the trajectory on to the V–h plane in response to an applied current density of −0.007+1.515*cos (2*π*50*t) [μA/cm²] at temperature T = 6.3°C.

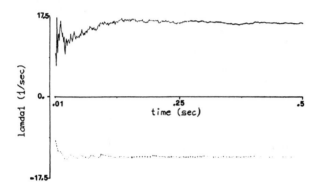

Fig. 9. Maximal Lyapunov exponents:
_____ chaotic situation,
.... periodic situation.

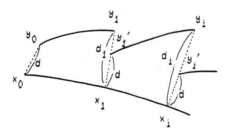

Fig. 10. Trajectories in four dimensional space used for definition of the maximal Lyapunov exponent.

117

APPENDIX 1

The HH equations describing the membrane behaviour are the following:

$$\frac{dV}{dt} = \frac{1}{C} \{I(t) - \bar{g}_{Na} \, h \, m^3 \, (E-E_{Na}) - \bar{g}_k \, n^4 (E-E_K) - g_1 (E-E_1)\}$$

$$\frac{dm}{dt} = T_f\{\alpha_m(V) \, (1-m) - \beta_m(V)m\}$$

$$\frac{dh}{dt} = T_f\{\alpha_h(V) \, (1-h) - \beta_h(V)h\}$$

$$\frac{dn}{dt} = T_f\{\alpha_n(V) \, (1-n) - \beta_n(V)n\}$$

where

$$T_f = 3^{(T-6.3)/10}$$

V = membrane voltage in mV

m, h and n are the sodium activation, the sodium inactivation and the potassium inactivation respectively

$$\alpha_m(V) = 0.1(25-V)/(\exp(0.1 \, (25-V)-1)$$

$$\beta_m(V) = 4 \, \exp(-V/18)$$

$$\alpha_h(V) = 0.07 \, \exp(-V/20)$$

$$\beta_h(V) = 1/(\exp(0.1(30-v)+1)$$

$$\alpha_n(V) = 0.01(10-V)/(\exp(0.1(10-V)-1)$$

$$\beta_n(V) = 0.125 \, \exp(-V/80)$$

The values of the parameters are:

$C = 0.01 \, F/m^2$

$\bar{g}_{Na} = 120.0 \, mS \, cm^{-2}$ sodium conductance

$\bar{g}_K = 36.0 \, mS \, cm^{-2}$ potassium conductance

$g_1 = 0.3 \, mS \, cm^{-2}$ leakage conductance

$E_{na} = -55 \, mV$

$E_K = 72 \, mV$

$E_1 = 49.387 \, mV$

T is the temperature in centigrade

$E_R = 60 \, mV$ at temperature T = 6.3°C

$V = E-E_R$

APPENDIX 2

In order to define the maximal Lyapunov exponent [2], we introduced the four dimensional space (V,m,h,n).

Let x_o and y_o be two adjacent points in this space at distance d (see Fig. 10). On the trajectory originating at the point x_o, the point x_1 represents the solution of the Hodgkin-Huxley equations at the time $t = \Delta t$. A trajectory beginning at y_o reaches the point y_1 after the time step Δt. This point is at distance d_1 from the corresponding point x_1 on the trajectory through x_o. The next piece of the y-trajectory originates at the point y_1', on the connection line between x_1 and y_1 and at distance d from x_1. This construction is continued for $i = 2$ and so on as shown in Fig. 10.

The maximal Lyapunov exponent is then defined as

$$\lambda_1 = \lim_{n \to \infty} \frac{1}{n\Delta t} \sum_i^n \ln \frac{|d_i|}{|d|}$$

REFERENCES

[1] A.L. Hodgkin and A.F. Huxley, A quantitative description of membrane current and its application to conduction and excitation in nerve, J. Physiol. 117:500-544 (1952).

[2] G. Benettin, L. Galgani and J.M. Strelcyn, Kolmogorov entropy and numerical experiments, Phys. Rev. A14:2338-2345 (1976).

[3] J.H. Jensen, P.L. Christiansen, A.C. Scott and O. Skovgaard, Chaos in nerve, ACI 83,Vol. 2:15.6-15.9 (19).

[4] A.V. Holden and M.A. Muhamad, The identification of deterministic chaos in the activity of single neurones, J. Electrophysiol. Tech. 11:135-147 (1984).

FORCED OSCILLATIONS AND ROUTES TO CHAOS IN THE HODGKIN-HUXLEY AXONS
AND SQUID GIANT AXONS

K. Aihara[1] and G. Matsumoto[2]

[1]Department of Electronic   [2]Electrotechnical Laboratory
Engineering   Tsukuba Science City
Faculty of Engineering   Niihari-gun
Tokyo Denki University   Ibaraki 305
2-2 Nishiki-cho   Japan
Kanda
Chiyoda-ku
Tokyo 101, Japan

ABSTRACT

Nonlinear responses of a neural oscillator to sinusoidal force are analysed theoretically with the Hodgkin-Huxley equations and experimentally with squid giant axons.

First, the periodically forced oscillations in the nerve membranes are qualitatively classified into (1) synchronised oscillations, (2) quasi-periodic oscillations and (3) chaotic oscillations by examining the Poincaré sections and the return maps.

Second, it is confirmed that there exist the three types of routes to the chaos, namely, (1) the successive period-doubling bifurcations, (2) the intermittency and (3) the collapse of the quasi-periodicity. The global structure of the three routes and many Arnold's tongues of the synchronised oscillations in the parameter space is also examined.

Last, simple mapping models of the neural responses are discussed.

I INTRODUCTION

Neurons communicate with each other by trains of action potentials. The nonlinear and nonequilibrium dynamics of the nerve membranes produce not only the generation and propagation of action potentials with an all-or-nothing law, but also self-sustained oscillations [3,23,26,34], bi-stability [4,24]42] and chaos [5-9,20,21,25,36]. Information processing in the brain seems to be supported by the abundant dynamical behaviours of the neurons and by interactions between the neurons. In this chapter responses of a self-sustained neural oscillator to sinusoidal forcing are analysed, both numerically and experimentally, as a simple example of the interation between the neurons.

II METHODS

    Intact giant axons of squid (Doryteuthis bleekeri) were used in the
experimental analysis. A self-sustained oscillation was induced by
immersing the axon in a 1:9 mixture of natural sea water and 550 mM NaCl
[34]. The self-sustained oscillation is in the soft-oscillation mode in
this condition; the limit cycle representing the oscillation is a unique
attractor [3]. The self-oscillatory membrane was stimulated by a sinusoidal current Asin(2πFt). The amplitude A and the frequency F of the
sinusoidal force were changed as the bifurcation parameters.

    The sinusoidally forced oscillations in squid giant axons were
analysed by stroboscopic observation [9,36]. Namely, the membrane potential V(t) and its time differential dV(t)/dt were stroboscopically sampled
at a fixed phase of the sinusoidal force. The sampled time-series V_i and
dV_i/dt (i=1,...,1280) were displayed on two-dimensional planes in the
following two ways; one is (V_i, dV_i/dt) and the other is (V_i, V_{i+1}).
The former and the latter are a Poincaré section projected on to the
V-dV/dt plane and a successive transfer function, or a return map of the
membrane potential, respectively.

    The Hodgkin-Huxley ordinary differential equations [22] were used in
the numerical analysis (see ref. [8] for details of the numerical
calculations on the forced Hodgkin-Huxley oscillator). In the experiments, squid giant axons were spatially clamped by an internal current
electrode with the conducting length of 5 mm. This space-clamp condition
makes it possible to compare the numerical solutions of the Hodgkin-
Huxley ordinary differential equations with the experimental results.

III FORCED OSCILLATIONS IN NERVE MEMBRANES

    The sinusoidally forced oscillations in both the Hodgkin-Huxley axons
and squid giant axons are qualitatively classified into (1) synchronised,
or phase-locked oscillations, (2) quasi-periodic oscillations, and (3)
chaotic oscillations by examining the Poincaré sections, the return maps
and the power spectra [5-9,36].

    Figure 1 shows a 4/5-synchronised oscillation in a squid giant axon.
Figures 1 (a),(b) and (c) correspond to the waveform, the Poincaré section
and the return map at the phase 180°, respectively. In general, a n/m-
synchronised oscillation (m,n:positive integers) is a periodic oscillation
such that the fundamental period of the oscillation equals exactly m times
the period 1/F of the force and that n action potentials are generated at
locked phases during the fundamental period m/F. It should be noted that
m and n are not always relatively prime; for example, a q/p-synchronised
oscillation changes into a 2q/2p-synchronised oscillation through a
period-doubling bifurcation. The Poincaré sections of a n/m-synchronised
oscillation are composed of m distinct points; the trajectory is a limit
cycle in the phase space V x dV/dt x S^1.

    An index of the forced oscillation, or the excitation number is
defined as the ratio of the number of action potentials to that of stimuli
used for their generation [6,20,33,38]. The excitation number of a n/m-
synchronised oscillation is exactly equal to n/m.

    An invariant closed curve asymptotically emerges in every Poincaré
section of a quasi-periodic oscillation [5-6,8-9,36]. The trajectory of a
quasi-periodic oscillation is in the form of a two-dimensional torus in
V x dV/dt x S^1.

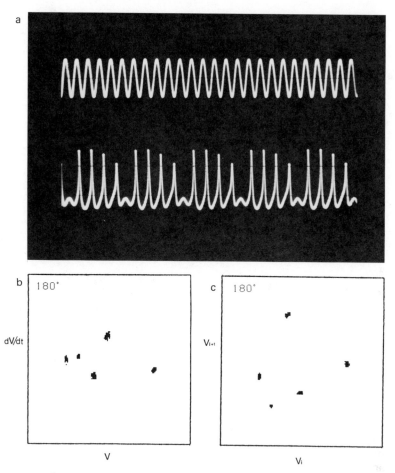

Fig. 1. A 4/5-synchronised oscillation in a squid giant axon (A = 2.58 μA, F = 250 Hz and the natural frequency F_N of the self-sustained oscillation = 175 Hz). (a) The waveforms (above: the stimulation, below: the response of the membrane potential). (b) The Poincaré section on the V-dV/dt plane at the phase 180° of the sinusoidal force. (c) The return map of the membrane potential at the phase 180° of the sinusoidal force.

Figures 2 (a) and (b) show the Poincaré sections of a chaotic oscillation in the Hodgkin-Huxley axon and those in a squid giant axon, respectively. Folding and mixing dynamics peculiar to chaotic attractors in forced oscillators [43] is clearly seen in Fig. 2. The strange attractor of Fig. 2(a) is also chaotic in the meaning that the maximum Lyapunov exponent λ_M [27,41] is positive; the numerically calculated value of λ_M is 0.04/ms, or 0.24 bits per one forcing period.

IV   ROUTES TO CHAOS IN NERVE MEMBRANES

Different synchronised oscillations can be produced both in the Hodgkin-Huxley axons and in squid giant axons with changing the amplitude A and the frequency F of the sinusoidal force. The region of each n/m-

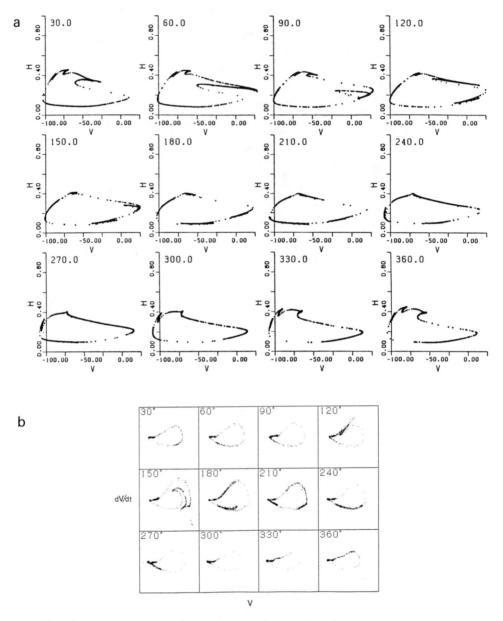

Fig. 2. Poincaré sections of chaotic oscillations. The number in
each section designates the phase (°) of the sinusoidal
force at which the corresponding section is plotted. (a)
In the Hodgkin-Huxley axon (A = 40.0 μA/cm^2, F = 242.1 Hz
and F_N = 174.6 Hz). (b) In a squid giant axon (A = 2 μA,
F = 270 Hz and F_N = 184 Hz).

synchronised oscillation forms an Arnold tongue in the parameter space
A x F. The Arnold tongue has a cusp-contact with the F-axis at the point
of A = 0 and F = mF_N/n where F_N is the natural frequency of the self-
sustained oscillator. When A is sufficiently large or sufficiently small,
almost all the responses are synchronised oscillations or quasi-periodic
oscillations, respectively. Regions of chaotic oscillations exist in the
intermediate range of A.

The routes from synchronised oscillations to chaotic oscillations [5, 8,10] are successive period-doubling bifurcations [37] or intermittency [40]. Figure 3 demonstrates a route with period-doubling bifurcations in a squid giant axon [10].

Collapse of a two-dimensional torus [29,39] can also be observed as routes from quasi-periodic oscillations to chaotic oscillations by carefully adjusting A and F [2,10]. The routes are usually interrupted by phase-locking in the experiments and reduced to the routes with period-doubling or intermittency.

When the bifurcation parameters A and F are changed in global ranges, alternating period-chaotic sequences are generated both in the Hodgkin-Huxley axons and in squid giant axons [7]. Figure 4 shows waveforms in the chaotic region which exists between the Arnold tongue of 2/3-synchronisation and that of 3/4-synchronisation in the parameter space A x F. Since the waveforms in Fig. 4 are mixed patterns of those of the two neighbouring synchronised oscillations, the excitation numbers of the chaotic oscillations are between 2/3 and 3/4. Figure 5 shows the transition characteristics of the excitation number ρ in the Hodgkin-Huxley axon [10]. "PD" and "I" in Fig. 5 denote the route with period-doubling bifurcations and that with intermittency, respectively. In the route with period-doubling bifurcations of Fig. 5, the excitation numbers are in the form of 2^k x $2/2^k$ x 3 (k: positive integers) and constantly equal to 2/3. On the other hand, chaotic phases appear intermittently between nearly periodic phases of 3/4-synchronisation in the route with intermittency of Fig. 5. These bifurcation structures produce the smooth transitions of the excitation number from the synchronised oscillations to the chaotic oscillations by both routes "PD" and "I" in Fig. 5.

V DISCUSSION

The excitation number of a chaotic oscillation has an intermediate value of the adjacent synchronised oscillations in the alternating periodic-doubling sequences, as shown in Figs. 4 and 5. Moreover, both successive period-doubling bifurcations and intermittency produce smooth transitions of the excitation number. In other words, the neural

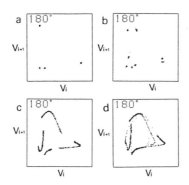

Fig. 3. Return maps in a route to chaos with period-doubling
bifurcations in a squid giant axon [10] (F = 270 Hz and
F_N = 192 Hz). (a) a 3/4-synchronised oscillation (A =
1.5 µA), (b) a 6/8-synchronised oscillation (A = 2.0 µA),
(c) an oscillation with four bands (A = 2.2 µA) and (d) a
chaotic oscillation (A = 2.32 µA).

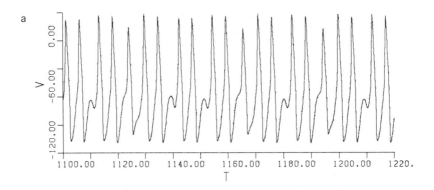

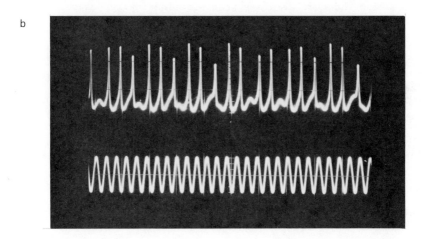

Fig. 4. Waveforms in the chaotic region which exists between the
Arnold tongue of 2/3-synchronisation and that of 3/4-
synchronisation in the parameter space A x F. (a) The
response of the membrane potential in the Hodgkin-Huxley
axon (A = 40.0 μA/cm^2, F = 242.7 Hz and F_N = 174.6 Hz).
(b) The response of the membrane potential (above) and the
stimulation (below) in a squid giant axon (A = 2.41 μA,
F = 270 Hz and F_N = 183 Hz).

oscillator can respond smoothly to changes of A and F by using these
chaotic modes [6]. The soft responses may be important in sensory
neurons such as mechanoreceptors and auditory neurons.

On the other hand, the excitation number of a chaotic oscillation
definitely decreases less than that of a neighbouring synchronised
oscillation through type III intermittency when squid giant axons in the
resting state are stimulated by a train of pulses [35]. The similar
decrease of the excitation number can also occur with periodic forcing
of the Hodgkin-Huxley oscillator in the mode of hard-oscillation, or
co-existence of a stable limit cycle and a stable equilibrium point
(Aihara, unpublished). The characteristics of the excitation number are
probably determined by global dynamical flows in the phase space after
local bifurcations.

There are many researches to describe responses of neurons by simpler
models of difference equations and maps. Responses of a self-oscillatory

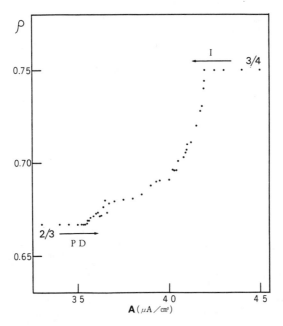

Fig. 5. The transition characteristic of the excitation number ρ in
the Hodgkin-Huxley axon [10]. "PD" and "I" denote the route
with period-doubling bifurcations and that with intermittency,
respectively.

neuron stimulated by a train of pulses can be analysed by circle maps of
the phase of the oscillator immediately before each stimulation [12,17,19,
30,44]. The circle maps from S^1 to S^1 are also effective for examining
responses of a self-sustained oscillator stimulated by sinusoidal force
because the attractor is a two-dimensional torus under weak forcing. It
has been demonstrated that simple classes of circle maps have abundant
dynamical structures related to responses of forced neural oscillators
[12,16-19,28-32,39,44].

We found that responses to periodic force of a resting membrane are
also described by one-dimensional maps on an interval. Figure 6(a) is a
successive transfer function (V_i,V_{i+1}) of the membrane potential which was
obtained by stroboscopic observation of the Hodgkin-Huxley resting axon
stimulated by a train of pulses [11]. The value of the membrane potential
V_i was stroboscopically sampled at each trailing edge of the stimulating
pulses. Figure 6(a) clearly shows that the transfer function can be
approximated by a one-dimensional map on an interval. Figures 6 (b) and
(c) are, respectively, the bifurcation diagram and the excitation number
calculated by the one-dimensional map $V_{i+1}=F(V_i)+a$, where the function
$F(V_i)$ and the parameter a are a fitted function of Fig. 6(a) and an
introduced additive bifurcation-parameter, respectively [11]. The
characteristics of Figs. 6 (b) and (c) are similar to those experimentally
observed in squid giant axons [35]. A detailed analysis of the one-
dimensional maps is now in progress [11].

Responses of a resting neuron have been studied qualitatively by
simpler models [31,38,47]. For example, Nagumo and Sato [38] constructed
a simple difference equation based upon the Caianiello's neuronic equation
[13] in order to analyse responses of a resting neuron to a train of
pulses. It has been clarified that the difference equation has the

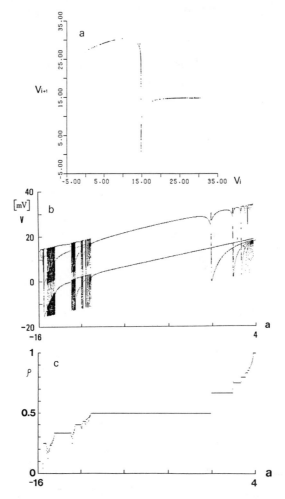

Fig. 6. Responses of a one-dimensional map derived from the
Hodgkin-Huxley equations [11]. (a) A return map in the
Hodgkin-Huxley resting axon stimulated by a train of pulses.
(b) The bifurcation diagram of a one-dimensional map
$V_{i+1} = F(V_i) + a$, where $F(V) = c_1 V^2 + c_2 V + c_3 + 1/(V-c_4)$ (for
$V \leqq 14.8$ mV), $= c_5 V + c_6 + c_7 \exp(-(V-c_8)/c_9)$ (for $V > 14.8$ mV).
(c) The property of the excitation number, corresponding
to Fig. 6(b).

response characteristics of a complete devil's staircase [38,45,46];
that is, the equation has chaotic solutions only at a Cantor set of the
parameter values with zero Lebesque measure. Recent experiments on
squid giant axons, however, have elucidated that resting membranes respond to a train of pulses not only periodically but also chaotically
according to the values of the amplitude and the interval of pulses [35].
These alternating periodic-chaotic sequences experimentally observable
in the forced resting axons can be qualitatively described by the following difference equation which is a modified Nagumo-Sato model [2,11]:

$$y(t+1) = y(t)/b + a - g(f(y(t))) \tag{1}$$

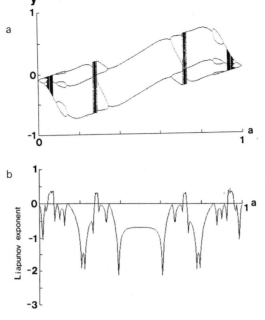

Fig. 7.   Responses of eq. (1) where f(y) = 1/(1/exp(-yε)), g(z) = z,
b = 2.0, ε = 0.03 and y(0) = 0.5 [11].   (a)   The bifurcation
diagram.   (b)   The Lyapunov exponent.

where y(t) is the internal state of the neuron, or roughly speaking the
depolarisation caused by each stimulus minus the variable threshold;
b(>1) is the decay of the refractory memory;   a is the magnitude of the
stimulus;   f is an output function representing the relationship between
the internal state and the output of a graded action potential;   g is a
function representing the relationship between the graded output and the
magnitude of the refractory for the next stimulation [2,38].   When f(y) =
1 (for y ≧ 0), = 0 (for y < 0) and f(z) = z, eq. (1) is reduced to the
Nagumo-Sato model.   The function f is generally an increasing function
reflecting smoothly continuous stimulus-response properties of nerve
membranes [14], or a continuous type of threshold in neural dynamics [15].
Orbital instability of the chaos in nerve membranes is a result of
sensitive separation of the orbits near this continuous type of threshold
separatrix [6,36].   Therefore, the continuity of the function f in eq. (1)
is crucial for the description of the chaos in the nerve membranes.   When,
for example, f(y) = 1/(1+exp(-y/ε)) and g(z) = z, eq. (1) has alternating
periodic-chaotic responses as shown in Fig. 7 [11].   The above function
f(y) was determined to be the same with the stochastic output function of
"Boltzmann machines" [1].   It is an important future problem to clarify
dynamics of neural networks composed of neurons of eq. (1).

ACKNOWLEDGEMENT

The authors wish to thank T. Takabe for his help in the numerical
calculations of Figs. 6 and 7.

129

REFERENCES

[1]   D.H. Ackley, G.E. Hinton and T.J. Sejnowski, Cognitive Sci. 9: 147
       (1985).
[2]   K. Aihara, M. Kotani and G. Matsumoto, in "Structure, Coherence and
       Chaos",ed. P.L. Christiansen, Manchester Univ. Press,
       Manchester (1987).
[3]   K. Aihara and G. Matsumoto, J. Theor. Biol. 95: 697 (1982).
[4]   K. Aihara and G. Matsumoto, Biophys. J. 41: 87 (1983).
[5]   K. Aihara and G. Matsumoto, in "Chaos", ed. A.V. Holden, 257,
       Manchester Univ. Press, Manchester and Princeton Univ. Press,
       Princeton, N.J. (1986).
[6]   K. Aihara and G. Matsumoto, in "Dynamical Systems and Nonlinear
       Oscillations", ed. G. Ikegami, 254, World Scientific, Singapore
       (1986).
[7]   K. Aihara, G. Matsumoto and M. Ichikawa, Phys. Lett. A111: 251
       (1985).
[8]   K. Aihara, G. Matsumoto and Y. Ikegaya, J. Theor. Biol. 109: 249
       (1984).
[9]   K. Aihara, T. Numajiri, G. Matsumoto and M. Kotani, Phys. Lett.
       A116: 313 (1986).
[10]   K. Aihara, T. Numajiri et al., in preparation.
[11]   K. Aihara, T. Takabe et al., in preparation.
[12]   J. Belair, J. Math. Biol. 24: 217 (1986).
[13]   E.R. Caianiello, J. Theor. Biol. 2: 204 (1961).
[14]   K.S. Cole, R. Guttman and F. Bezanilla, Proc. Nat. Acad. Sci. 65:
       884 (1970).
[15]   R. Fitzhugh, in "Biological Engineering",ed. H.P. Schwan, 1,
       McGraw-Hill, New York (1969).
[16]   J.M. Gambaudo, P. Glendenning and C. Tresser, Phys. Lett. 105A:
       97 (1984).
[17]   L. Glass, M.R. Guevara, A. Shrier and R. Perez, Physica 7D: 89
       (1983).
[18]   J. Guckenheimer, Physica 1D: 227 (1980).
[19]   M.R. Guevara, L. Glass, M.C. Mackey and A. Shrier, IEEE, SMC-13:
       790 (1983).
[20]   R. Guttman, L. Feldman and E. Jakobsson, J. Memb. Biol. 56: 9
       (1980).
[21]   H. Hayashi, S. Ishizuka, M. Ohta and K. Hirakawa, Phys. Lett. A88:
       435 (1982).
[22]   A.L. Hodgkin and A.F. Huxley, J. Physiol. (London) 117: 500
       (1952).
[23]   A.V. Holden, Biol. Cybern. 38: 1 (1980).
[24]   A.V. Holden, P.G. Haydon and W. Winlow, Biol. Cybern. 46: 167
       (1983).
[25]   A.V. Holden, W. Winlow and P.G. Haydon, Biol. Cybern. 43: 169
       (1982).
[26]   A.F. Huxley, Ann. N.Y. Acad. Sci. 81: 221 (1959).
[27]   J.H. Jensen, P.L. Christiansen, A.C. Scott and O. Skovgaard,
       Physica 13D: 269 (1984).
[28]   K. Kaneko, Prog. Theor Phys. 72: 1089 (1984).
[29]   K. Kaneko, in "Chaos and Statistical Methods", ed. Y. Kuramoto
       83, Springer, Berlin (1984).
[30]   J.P. Keener and L. Glass, J. Math. Biol. 21: 175 (1984).
[31]   J.P. Keener, F.C. Hoppensteadt and J. Rinzel, SIAM. J. Appl. Math.
       41: 503 (1981).
[32]   R.S. MacKay and C. Tresser, Physica 19D: 206 (1986).
[33]   M. Marek, L. Vroblova and I. Schreiber, Lecture at MIDIT Workshop
       on Structure, Coherence and Chaos in Dynamical Systems (1986).
[34]   G. Matsumoto, in "Nerve Membrane, Biochemistry and Function of

Channel Proteins", eds. G. Matsumoto and M. Kotani, 203, University of Tokyo Press, Tokyo (1981).

[35] G. Matsumoto, K. Aihara, Y. Hanyu, N. Takahashi, S. Yoshizawa and J. Nagumo, in preparation. See also the paper by G. Matsumoto in this volume.

[36] G. Matsumoto, K. Aihara, M. Ichikawa and A. Tasaki, J. Theoret. Neurobiol. 3: 1 (1984).

[37] R.M. May, Nature 261: 459 (1976).

[38] J. Nagumo and S. Sato, Kybernetik 10: 155 (1972).

[39] S. Ostlund, D. Rand, J. Sethna and E. Siggia, Physica 8D: 303 (1983).

[40] Y. Pomeau and P. Manneville, Commun. Math. Phys. 74: 189 (1979).

[41] I. Shimada and T. Nagashima, Prog. Theor. Phys. 61: 1605 (1979).

[42] I. Tasaki, J. Physiol. (London) 148: 306 (1959).

[43] J.M.T. Thompson and H.B. Stewart, "Nonlinear dynamics and Chaos", John Wiley and Sons, Chichester (1986).

[44] C. Torras i Genís, J. Math. Biol. 24: 291 (1986).

[45] I. Tsuda, Phys. Lett. 85A: 4 (1981).

[46] M. Yamaguchi and M. Hata, in "Competition and Cooperation in Nerve Nets", ed. S. Amari and M.A. Arbib, 171, Springer, Berlin (1982).

[47] S. Yoshizawa, H. Osada and J. Nagumo, Biol. Cybern. 45: 23 (1982).

QUANTIFICATION OF CHAOS FROM PERIODICALLY FORCED SQUID AXONS

R.M. Everson

Department of Applied Mathematical Studies
The School of Mathematics and the Centre for Nonlinear Studies
The University of Leeds
Leeds LS2 9JT

ABSTRACT

We analyse time series collected by measuring the membrane potential of a squid giant axon subject to sinusoidal forcing. 1/1 phase-locked oscillations, quasiperiodic oscillations and chaos are observed. Phase space portraits are reconstructed using the method of delays and singular systems analysis, allowing Lyapunov exponents and entropies to be calculated. Comparisons with numerically integrated Hodgkin-Huxley equations show good correspondence with experiment.

I INTRODUCTION

Recent years have brought widespread recognition that periodically forced oscillators may produce chaotic responses. Chaotic oscillations have been induced in biological membranes such as cardiac cells, molluscan neurons, pacemaker neurons and squid giant axons. In this chapter we analyse the response of the membrane of squid giant axons to periodic forcing. In the normal physiological state, an unstimulated axon remains at rest, with constant membrane potential. A small impulsive current produces an action potential, a voltage spike of roughly fixed duration, before the axon returns to equilibrium. Immersing the axon in calcium-deficient seawater induces repetitive firing of the axon, producing a train of action potentials. The axon behaves as a relaxation oscillator, which may be forced with a small current applied through an internal electrode. We exhibit here three types of response: 1/1-entrainment of the membrane potential to the stimulation, quasiperiodic oscillations and chaos. First the experimental method and analytical techniques are discussed, after which we examine each response in turn.

II DATA COLLECTION AND ANALYSIS

Aihara et al. [1] have collected data from giant axons of squid (Doryteuthis bleekeri). Self-sustained oscillations were induced by immersing the axons in a 1:9 mixture of natural seawater and 550 mM NaCl. The space clamped neural oscillator was forced by a sinusoidal current $I \sin (2\pi f_s t)$ through an internal current electrode and the membrane potential was recorded on an analogue tape recorder.

The data was tranferred to computer and digitised to 12 bits precision with a sampling time of 88 μs. The stimulating and response frequencies are about 200 Hz, yielding roughly 50 samples per cycle.

Vector phase portraits were constructed from the scalar time series, $x(t_i)$, using the "method of delays" developed by Crutchfield et al. [8] and Takens [10]. Attractors were initially embedded in 30 dimensional Euclidean space before projecting down on to a few significant directions using singular value decomposition (SVD), pioneered in this context by Broomhead and King [3]. This technique results in considerable smoothing of the data and allows maximum information to be extracted.

As we shall show, the attractors can be visualised in 3 dimensional space, but algorithms for calculating Lyapunov exponents performed better with embeddings in higher dimensions. Measurements of the marginal redundancy [5], which we discuss below, suggest that only three or four coordinates are independent. Also, the Hodgkin-Huxley equations [6] which are widely accepted to describe the axon's dynamics, have solutions moving on a four dimensional manifold. Thus the Whitney embedding theorem [11] leads one to expect that, at worst, the attractor can be embedded in 2 x 4 + 1 = 9 dimensions. Indeed, it was never found necessary to embed in more than seven dimensions.

Dynamical systems are easily characterised by behaviour of nearby trajectories. Periodic behaviour is associated with nearby trajectories converging and loss of information: after long times it is difficult to distinguish between trajectories as they converge to a periodic orbit. Conversely, nearby trajectories diverge exponentially fast for chaotic attractors and information about the initial conditions is revealed as they take increasingly different paths. Lyapunov exponents measure the average logarithmic rate of separation of nearby points on the attractor. An attractor is chaotic if at least one exponent is positive, implying that nearby points on the attractor diverge.

For a trajectory $\{x_i\}, x_i = f^{(i)}(x_0)$ on an m-dimensional manifold, and a vector $1 \in T_x$, the tangent space at x, the following limit exists under very general conditions:

$$\lambda(x,1) = \lim_{n \to \infty} \frac{1}{n} |Df^{(n)}(1)|$$

The tangent vector 1 may be thought to join $\{x_i\}$ to an initially close trajectory, $\{y_i\}, y - x = 1$. Thus the limit measures the rate of separation in the direction 1. $\lambda(x,1)$ is the Lyapunov exponent and there are m, not necessarily distinct, exponents.

If the Jacobian matrix, Df, is available, it is a simple matter to evaluate the growth of vectors in the tangent space as the equations evolve. However, in an experimental situation Df is certainly not available. Wolf [12] has developed an algorithm that avoids this difficulty by choosing a fiducial point $x(t_0)$ and its nearest neighbour $y(t_0)$. The vector $y(t_0) - x(t_0)$ approximates the tangent vector and the initial separation $|y(t_0) - x(t_0)|$ is denoted L_0'. One then follows the pair of trajectories for a time t_0 until the distance, L_0 between them exceeds some value ε.

The evolved first data point, $x_0(t_1)$, is retained and a new neighbour, $y(t_1)$, sought so that the distance $L_1 = |y(t_1) - x(t_1)|$ is again less than $\delta < \varepsilon$ and such that $y(t_1)$ lies as nearly as possible in the same direction from $x_0(t_1)$ as $y_0(t_1)$, the point which it replaces. This procedure continues until the fiducial trajectory has been followed through the experimental time series and the largest exponent is estimated as

$$\lambda_1 = \sum_{t=0}^{r-1} \frac{1}{\Delta t_i} \ln \frac{L'_i}{L_i} ,$$

where r is the number of replacements made.

A variable evolution time program, based on Wolf's algorithm, was used to estimate the largest exponent of the squid data. The diameter of the attractor in phase space was estimated as $\sqrt{d}$ ($x_{max} - x_{min}$), where d is the embedding dimension and x_{max} and x_{min} are the maximum and minimum excursions of any component of the vector time series. The length ε at which a replacement tangent vector is sought was chosen as 10% of the diameter and wherever possible replacement vectors were chosen with a norm less than δ = 2% of the attractor diameter and with an orientation error of less than 20°. Whenever the tangent vector required replacing the fiducial and secondary trajectories were followed backwards for a short time so that replacement was attempts before the tangent vector grew too large.

If a time series consisting of sharp spikes, such as the action potentials considered here, is embedded in Euclidean space using the method of delays the reconstructed attractor will inevitably contain sharp bends, because a projection of the attractor on to any coordinate must reproduce the time series along with its spikes (see figures 1 and 2a). However, the spikes are often an artifact of the measurement process, resulting from the projection of a smooth multi-dimensional attractor down on to one variable, e.g. the membrane potential. Such sharp bends in the reconstruction present difficulties for exponent estimating programs, because trajectories appear to diverge as they round the corner. A reconstruction via the singular value decomposition often alleviates the problem by providing some lowpass filtering and finding a coordinate system in which the attractor is smoother (see Figure 2b). Projections on to the coordinate vectors do not all contain spikes, though, of course, it is possible to find a particular projection to recover the original time series. It was found that our program was far more efficient on attractors reconstructed via a SVD, because spurious replacements at sharp corners were not attempted.

Another major analytical tool we have used is the mutual information. The mutual information, $I(x(t);y(t))$, measures the average information theoretic independence of $x(t)$ and $y(t)$. It is given by

$$I(x;y) = \int p(x,y) \ln \frac{p(x,y)}{p(x)p(y)} \, dxdy,$$

where $p(x,y)$ is the joint probability density generated by $(x(t),y(t))$ and $p(x)$ and $p(y)$ are the marginal densities. Following Shaw [9] and Fraser [5], we calculate the marginal redundancy

$$R'_d(T) = I(X_1^T(t+d); X_d^T(t))$$

Here $X_d^T(t) = \{x(t),x(t+T),x(t+2T),...,x(t+(d-1)T)\}$ denotes the embedding of a scalar time series, $x(t)$, in d-dimensional space, with time delay T and $X_1^T = x(t+dT)$. The marginal redundancy thus measures how much (the number of bits) that can be predicted about the $(d+1)^{st}$ variable from the other d variables in a d-dimensional embedding. It can thus be used to indicate when the embedding dimension is high enough: when $R'_d(T) \approx R'_{d+1}(T)$ one of the variables is dependent upon the other d and nothing is gained by embedding in more than d dimensions. Calculations of marginal redundancies for the data sets used here suggest that 5 or 6 dimensions is adequate.

Also, Fraser has shown that the metric entropy h_μ is given by

$$h_\mu = -\lim_{d \to \infty} \frac{R'_d(T_1) - R'_d(T_2)}{T_1 - T_2} ,$$

A plot of marginal redundancy versus time delay may therefore be used to estimate h_μ, which should equal the largest Lyapunov exponent if there is only one positive exponent.

An algorithm based on Fraser's was used to compute redundancies. The algorithm tends to underestimate redundancies by a multiplicative factor, especially in higher dimensions where data is spread more sparsely. Metric entropies are thus underestimated, but the program may be used to distinguish positive, zero and negative entropies.

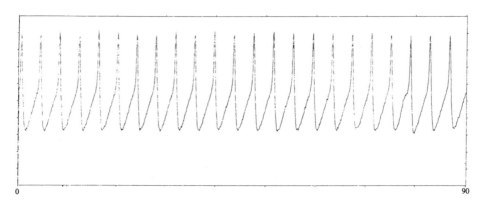

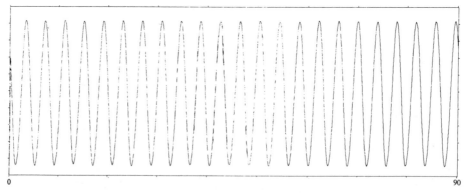

Fig. 1. Stimulation and entrained response (f_N = 213 Hz, f_s = 258 Hz, I = 2.0 μA). In all the figures of time series, the units of time are ms. The ordinates have arbitrary units, but all the stimulations are to the same scale, as are all the responses. The action potentials here have a peak-to-peak amplitude of about 100 mV.

III   ENTRAINMENT

When the stimulating frequency, f_S, is sufficiently close to the axon's natural frequency, f_N, the response may become entrained;   the axon producing a single action potential for each cycle of the stimulating current.   Figure 1 shows the stimulation and response, the power spectrum has peaks at the stimulating frequency and its harmonics only.   Aihara et al. [2] have found that the response remains entrained over a range of currents and frequencies (an Arnol'd tongue in parameter space) and locking with $f_S/f_N = m/n$ (m, n integers) also occurs.   A method of delays reconstruction of phase space is shown in Figure 2a.   The delay was chosen so that the projections on to the two coordinates are most independent,

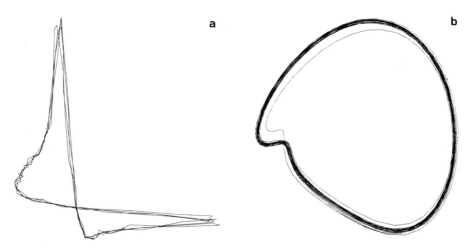

Fig. 2.   (a) Phase portrait of entrained response reconstructed using the method of delays.   The delay time is 1.58 ms, yielding a minimum mutual information between the two coordinates.   (b) Projection of the phase portrait on to the two most significant singular vectors from a singular value decomposition.   The "thickness" of the trajectory is due to drift in the stimulating frequency.

that is, the mutual information between them is at a minimum (I = 1.6 bits), as recommended by Fraser and Swinney [4].   Note that a projection on to either coordinate recovers the original time series.   In contrast, a projection on to the two most significant directions found from a SVD are shown in Figure 2b.   The large loop corresponds to the action potential, while the kink on the left-hand side results from the recovery phase.   Although the attractor is clearly topologically equivalent to a circle, this figure shows its projection on to only two dimensions;   it has significant components extending into the third and fourth dimensions.

The largest Lyapunov exponent should be negative;   our program measures it as slightly negative or positive and very close to zero.   The program is designed to measure positive exponents and these results are indistinguishable from those obtained from a known limit cycle, so we are confident that the largest exponent is negative.

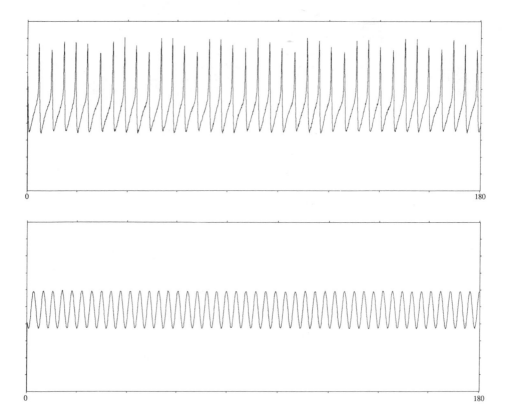

Fig. 3. Stimulation and quasiperiodic response (f_N = 208 Hz, f_S = 261 Hz, I = 0.5 μA).

IV QUASIPERIODICITY

If the ratio of the stimulating and natural frequencies is an irrational number the response is quasiperiodic. Quasiperiodicity, evident as a modulation of the action potential amplitude is most apparent with small forcing. Figure 3 shows the quasiperiodic response for f_S/f_N = 1.255.... and current 0.5 μA, compared with f_S/f_N = 1.211 and I = 2 μA in the entrained case. The quasiperiodicity is evident as a modulation of the action potential amplitude. Reconstructions of phase portraits show the attractor to be a two-torus embedded in four or five dimensions; Figure 4 shows the projections on to the first three most significant singular vectors from a SVD.

The power spectrum (Figure 5) provides strong evidence of quasi-periodicity. All significant peaks are simple combinations, $mf_S \pm nf_N$ of the stimulating frequency and the incommensurate natural frequency f_N. Ideally the trajectory winding around the torus never closes on itself, though with finite precision and in the presence of noise it is hard to distinguish an irrational winding number from the ratio of two large integers.

Further evidence of quasiperiodicity is furnished by redundancy calculations. Marginal redundancies $R'_1(T)$ to $R'_7(T)$ were calculated using a time series of 16384 samples for time delays up to 17.6 ms (about $4\frac{1}{2}$ orbits). With "only" 16384 points the program underestimates redundancy, particularly in high dimensions, but by a constant factor for each dimension. Thus the marginal redundancy too is underestimated, but the

138

Fig. 4. Projections of the quasiperiodic attractor, a two-torus, on
to the three most significant singular vectors from a singular
value decomposition.

gradient with respect to time delay may be used to test for quasiperiodicity. The average gradient for dimensions 3 to 7 was zero, indicating
zero metric entropy and quasiperiodicity. We found $R'_d(T)$ to be a more
sensitive test of quasiperiodicity than the Lyapunov exponent algorithm,
which tends to measure the divergence of trajectories due to noise, producing a spurious positive exponent.

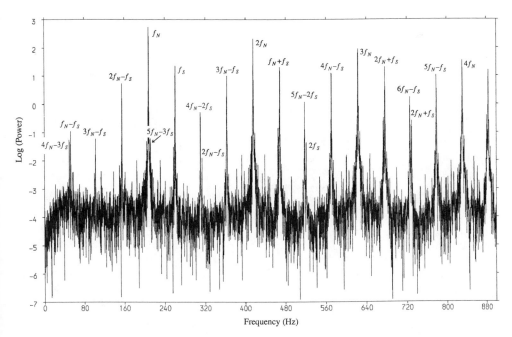

Fig. 5. Frequency spectrum of quasiperiodic response. All significant
peaks are given by $mf_N \pm mf_s$.

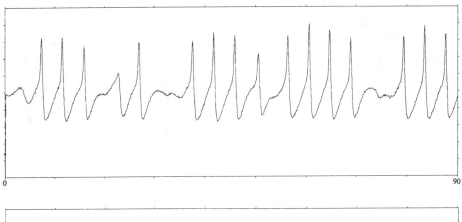

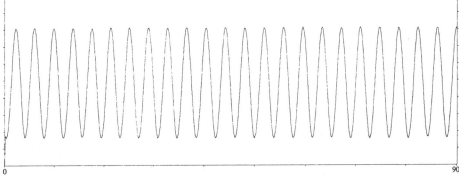

Fig. 6. Stimulation and chaotic response (f_N = 184 Hz, f_S = 264 Hz, I = 0.5 µA).

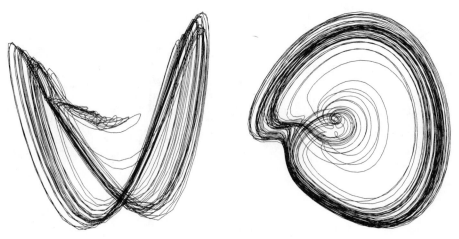

Fig. 7. Projections of the chaotic attractor on to the three most significant singular vectors from a singular value decomposition.

V CHAOS

Forcing the axon at high currents, far from its natural frequency evokes a chaotic response. The time series in Figure 6 shows that the axon usually produces an action potential in response to each stimulating cycle, but occasionally omits one, giving a small oscillation or relaxing towards equilibrium. The amplitude of the action potentials is no longer constant and the power spectrum is broadband with significant energy at all frequencies to beyond 2 kHz.

Figure 7 shows the projections of the reconstructed attractor on to the first three singular vectors from a SVD. The action potentials again form the large loops, while the small loops correspond to the missed cycles. Similar small loops are observed in the phase portraits of numerical integrations of the forced Hodgkin-Huxley equations [7].

Lyapunov exponents were estimated using a time series of 8192 vectors (from a SVD), representing about 380 cycles of the forcing frequency. Convergence close to the final value was obtained after about 100 cycles. The following table shows the exponent estimates for embedding in 3 to 8 dimensions.

Dimension	Exponent bits/orbit
3	1.01
4	0.60
5	0.51
6	0.56
7	0.32
8	0.52

Three dimensions is too small for an embedding, resulting in intersections of the attractor with itself. This causes the program to choose spurious replacements and overestimate the Lyapunov exponent. When the embedding dimension reaches 7 or 8 data is spread sparsely through space and the algorithm begins to underestimate the exponent. Discounting the value from 3 dimensions, we estimate the largest Lyapunov exponent as 97 bits/s or 0.52 bits/orbit (±12%). Estimates of the metric entropy via marginal redundancy yield $h_\mu \approx 0.2$ bits/orbit. We place more reliance on the Lyapunov exponent estimate, because our program underestimates redundancies.

Information is thus produced at about half the rate of the Rössler system. From the time series it is obvious that prediction far into the future is difficult; even gross features, such as whether an action potential will be produced in response to a particular cycle, cannot be predicted accurately. The phase portraits show that the separation of nearby orbits occurs as one makes a cycle of the large loop while the other circles the smaller. Here $f_s/f_N = 1.435$ and $I = 1.5$ µA; analysis of a dataset with $f_s/f_N = 1.344$ and $I = 2.0$ µA has a higher exponent of 0.93 bits/orbit.

VI CONCLUSION

Analysis of behaviour exhibited by a forced axon reveals entrainment, quasiperiodicity and chaos, analogous to a forced relaxation oscillator, such as the van der Pol oscillator. Similar trajectories are exhibited by the forced Hodgkin-Huxley equations and we hope to establish a more quantitative correspondence between the two.

Finally, we note that oscillations were induced by bathing the axon in calcium-deficient seawater. Oscillations and the dynamics analysed here have yet to be found in axons under normal physiological conditions.

ACKNOWLEDGEMENTS

I would like to express my thanks to Dr. John Illingworth for performing the analogue to digital conversion and to Dr. Arun Holden for many helpful discussions and advice.

REFERENCES

[1] K. Aihara, T. Numajiri, G. Matsumoto and M. Kotani, Physics Letters (1986).
[2] K. Aihara, G. Matsumoto and M. Ichikawa, Physics Letters A (1986).
[3] D.S. Broomhead and G.P. King, Physica 20D: 217 (1986).
[4] A.M. Fraser and H.L. Swinney, Physical Review 33A(2) (1986).
[5] A.M. Fraser, In "Dimensions and Entropies in Chaotic Systems", ed. G. Mayer-Kress, 82-91, Springer-Verlag (1986).
[6] A.L. Hodgkin and A.F. Huxley, J. Physiol. 117: 500-544 (1952).
[7] A.V. Holden and M.A. Muhamad, J. Electrophysiol. Tech. 11: 135-147 (1984).
[8] N.H. Packard, J.P. Crutchfield, J.D. Farmer and R.S. Shaw, Physical Review Letters 45(9): 712-716 (1980).
[9] R. Shaw, "The Dripping Faucet as a Model Chaotic System", Aerial Press Inc., Santa Cruz (1984).
[10] F. Takens, In "Proc. Dynamical Systems and Turbulence", eds. D.A. Rand and L.-S. Young, 366-381, Springer-Verlag (1980).
[11] H. Whitney, Ann. Math. 37: 645-680 (1936).
[12] A. Wolf, J.B. Swift, H.L. Swinney and J.A. Vastano, Physica D 16: 285-317 (1985).

CHAOS, PHASE LOCKING AND BIFURCATION IN NORMAL SQUID AXONS

Gen Matsumoto[1], Nobuyuki Takahashi[2] and Yoshiro Hanyu[3]

[1]Electrotechnical Laboratory
Tsukuba Science City
Ibaraki 305
Japan

[2]Tokyo Institute of Technology
Yokohama City
Kanagawa 227
Japan

[3]Faculty of Science and Technology
Department of Physics
Yokohama City
Kanagawa 223, Japan

ABSTRACT

Membrane potential responses of squid giant axons to periodic trains of current pulses were experimentally studied in detail. The giant axon of squid (Doryteuthis bleekeri) was exposed under normal physiological conditions; the intact axon was immersed in natural sea water at $14 \pm 0.01°C$, and stimulated with periodic trains of current pulses with the pulse intensity I and the period T. The firing modes were determined as a function of I/I_{th} and T where I_{th} stood for the threshold current intensity: periodic and chaotic responses were obtained.

I INTRODUCTION

The neural information propagated along an axon is coded as the firing density. This form of coding is quite adequate for transmitting analogue information without decay. In order to gain a better understanding of the coding characteristics, it is desirable to examine the pattern of local and propagated potential responses of the normal axon membrane to periodic current stimulation.

A response characteristic of a mathematical neuron model to periodic current pulse stimulation was theoretically studied by Nagumo and his colleagues [16,23] in order to explain the "unusual" and "unexpected" phenomenon found by Harmon [9] with his transistor neuron model. They found that the mathematical neuron responded by a periodic sequence of action potentials when a periodic train of current pulses with a fixed frequency was applied [9,16,23]. At the same time, they also found that the firing rate n/m, defined as a periodic sequence of n action potentials produced by m successive current pulses, varied in a staircase fashion as the amplitude of the input pulses changed continuously. Nagumo and Sato [16] concluded that the relationship between the firing rate and the amplitude of the input pulses took the form of an extended Cantor function. Recently, Yamaguchi and Hata [22] showed the existence of a Cantor attractor in the mathematical neuron model if and only if the average firing rate is irrational.

We were interested in determining the response characteristic of a real neuron for the following two reasons: one is that the response characteristic to periodic trains of current pulses is indispensable for gaining a better understanding of information processing by action potential production and propagation. Another is that the characteristics specific for the axon under normal physiological conditions should be well understood from the view of non-linear non-equilibrium thermodynamics as has been done for the spontaneous firing state [2,5,7,10,11,12,14,15].

We have described the first experimental demonstration of the occurrence of chaotic responses of the normal axon under periodic stimulation with brief current pulses [13]. The chaotic responses were found between different kinds of periodic (phase-locked) sequences of potential responses. Bifurcation characteristics to the chaos were also examined experimentally [8,18]. Here we review the chaos, phase-locking and its bifurcations, which were experimentally found in normal axons under periodic current stimulation.

II MATERIALS AND METHODS

Giant axons of the squid Doryteuthis bleekeri were used. Large axons, 400 - 700 μm in diameter and 60 - 80 mm in length, were carefully excised under a dissecting microscope and were kept in natural seawater (NSW). The temperature of NSW was well regulated by controlling the temperature of the chamber. The chamber was constructed in such a manner that it was possible to regulate the temperature in the chamber with accuracy of ±0.01°C in the range between 4 and 25°C. Regular trains of current pulses were delivered to the axon through an internal platinized-platinum wire-electrode, which was connected to a pulse generator through a register of 470 KΩ. The amplitude of the current pulse, I, and the interval between individual pulses, T, were taken as bifurcation parameters while the pulse width of 300 μsec was fixed throughout the experiments. By using glass-pipette electrodes of the Ag-AgCl type (filled with 0.6 M KCl solution), the membrane potential was monitored at two different points along the axon. With the tip of the glass-pipette electrode placed at a point 30 - 40 mm away from the end of the current electrode, propagated action potentials of the axon were recorded. With the tip located in the middle of the current electrode, non-propagated potential variations evoked by the current pulses were monitored.

III RESULTS

Potential Response Characteristics as a Function of both Amplitude of Current Pulse, I, and the Interval between Individual Pulses, T

Fig. 1 shows an example of three kinds of potential responses when the time interval between adjacent current pulses, T, was 6.4 msec (A), 5.9 msec (B) and 2.9 msec (C), respectively, while the amplitude, I, was fixed to be 1.54 times the current threshold. Records 1 in the figure are propagated potential responses of the axon measured at 32 mm away from the centre of the current electrode. Records 2 show the time course of non-propagated potential responses recorded at the site of stimulation. It can be seen in Fig. 1A that the potential responses shown in Records 1

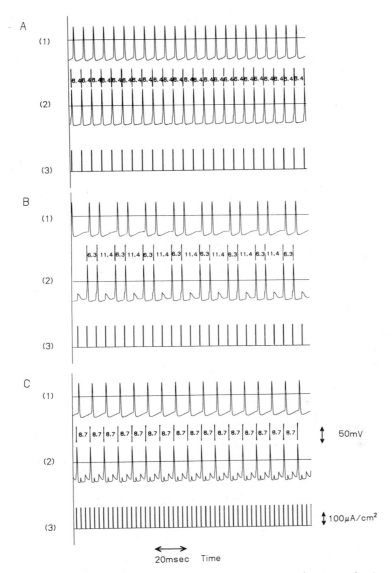

Fig. 1. Periodic potential responses of the periodically stimulated
giant axon of squid. Non-propagated potential responses
obtained at the site where the current pulses were given are
shown in Records (2) and their propagated responses 32 mm
apart from the stimulation site are illustrated in Records (1).
Periodic trains of current pulses are shown in Records (3),
where the pulse intensity normalised by the current threshold
and the pulse width were fixed to be 1.54 and 0.2 msec,
respectively. The interval between adjacent current pulses
was only a variable parameter and 6.4 msec (A), 5.9 msec (B)
and 2.9 msec (C), respectively. Numbers in Records 2 of A,
B and C denote time intervals between firing pulses in units
of msec. Temperature 14.0°C.

and 2 were repeating at a regular interval of 6.4 msec which exactly corresponds to the interval between individual current pulses. In the example of record shown in Fig. 1C, we can also see that the potential responses shown in Records 1 and 2 were repeating at a regular interval of 8.7 msec, corresponding to three times the interval between individual current pulses. In other words, the ratio of the number of applied current pulses to the number of action potentials evoked is exactly 3:1 in this case. In the former case of Fig. 1A, the ratio is 1:1. These patterns of potential responses may be conveniently termed "1:3 (in this case) and 1:1 (in the former case) phase-locked oscillation". A n:m phase-locked oscillation (n and m are both positive integers) is a periodic oscillation with the fundamental period being m times the stimulation pulse period and with n action potentials generated during the fundamental period. In the example of record shown in Fig. 1B, we see that the potential responses are 2:3 phase-locked. Note that the sum of two successive intervals between nearby action potentials, 6.3 and 11.4 msec, is constant and is equal to 17.7 msec (3 times the interval between the stimulating brief pulses).

Repetitive trials of the experiments as shown in Fig. 1 allowed us to determine both the firing patterns of the potential responses and the firing rate of the responses as a function of the normalised amplitude I/I_{th} (the current threshold, I_{th}) and the interval T. Fig. 2 shows typical experimental results of the firing rate as a function of T, which were obtained under the condition that the current amplitudes were 1.54 (A) and 1.29 (B) times the current threshold, respectively. The firing rate n/m, defined as n action potentials produced by m current pulse stimulation when m is reasonably large, changes discontinuously in a staircase fashion in Fig. 2(A) (in the case of $I/I_{th} = 1.5$). The firing patterns are also represented by the notations, 1, $1^\ell 0^m$, $(10)^n 100$ and $(100)^p 1000$, in the figure, where suffixes, ℓ, m, n and p are all positive integers. We principally denote a periodic sequence by one of these notations; for example, $1^\ell 0^m$ stands for a periodic sequence in which 0 (failure of action potential production by current pulse stimulation) appears consecutively m times after 1 (presence of action potential evoked by current pulse stimulation) has appeared consecutively ℓ times. Figure 2A of the firing rate vs. T, together with the firing patterns, shows that the axon responds regularly to the periodic train of current stimulation, and the patterned sequences change discontinuously in a staircase fashion as a function of the continuous variable, T. Within the experimental accuracy, we could not see any internal structures as are found in a mathematical neuron model [22]. However, the firing modes became unstable around their transition region with lower firing rates. The characteristic of the instabilities will be described below in *Chaos* and *Bifurcation*.

In the case of $I/I_{th} = 1.29$ (Fig. 2B), we found the overall characteristics of both the firing rate and patterns vs. T were basically similar to those in the case of $I/I_{th} = 1.54$ (Fig. 2A), except for firing rates < 1/2.

Typical experimental records of propagated (Records 1) and non-propagated (Records 2) potential responses are illustrated in Fig. 3. Fig. 3A$_1$ shows the 100 firing pattern of 1:3 phase-locked oscillation, the same as seen in Fig. 1C. In Fig. 3A$_2$ we can see that the potential responses with the firing rate of 1/3 are 2:6 phase-locked. Note that the sum of two successive intervals between nearby action potentials is constant, 18.6 msec (6 times the interval between the current pulses). Note also that the latency of the action potentials evoked during the fundamental period

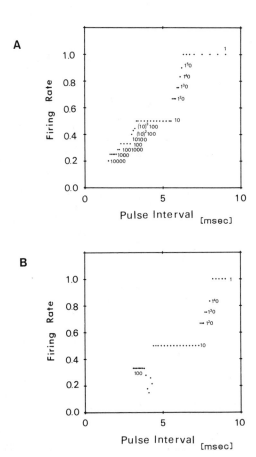

Fig. 2.   Dependence of the firing rate upon the interval between adjacent current pulses when the current amplitude was 1.54 (A) and 1.29 (B) times the threshold, respectively.   Temperature 14°C. The firing rate, n/m, is defined as n action potentials produced by m current pulse stimulation for large value of m. The firing pattern is also represented by the notations, 1, $1^{\ell}0^{m}$, $(10)^{n}100$ and $(100)^{p}1000$, in the figure, where suffixes ℓ, m, n and p, are all positive integers.   We denote a periodic sequence of firing pulses by one of these notations; for example, $1^{\ell}0^{m}$ basically denote a periodic sequence in which 0 (failure of action potential production by current pulse stimulation) appears consecutively m times after 1 (presence of action potential) has appeared consecutively ℓ times.

18.6 msec is fluctuating.   Obviously, the 2:6 phase-locked responses are produced as a consequence of instability existing in the 1:3 phase-locking process, resulting in the period-doubling of the 1:3 phase-locked responses.   Immediately after the train of the records of Fig. 3A$_2$ were photographed, a stable 1:3 phase-locking appeared, suggesting that the 2:6 phase-locked oscillation was unstable.   Fig. 3B shows that, when the current amplitude was increased to 1.48 times the threshold, the regular response pattern was suddenly replaced with irregular bursts of responses.   Detailed inspection of the records 1 and 2 in Fig. 3B

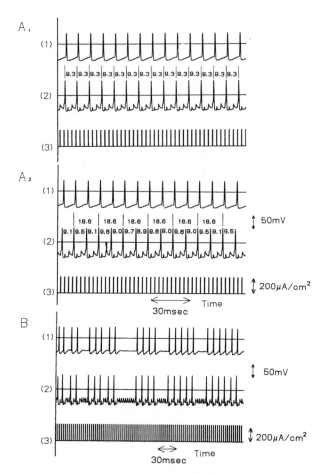

Fig. 3. Periodic and aperiodic potential responses of the periodically
stimulated giant axon of squid. Respective records 1, 2 and 3
represent propagated (30.5 mm away from the stimulation site)
and non-propagated responses, and periodic trains of current
pulses where the interval between individual pulses and pulse
width were fixed to be 3.1 and 0.3 msec, respectively. The
current intensity is the only variable parameter and was
1.32 (A_1), 1.41 (A_2) and 1.48 (B) times the current threshold,
respectively. Numbers in Records 2 of A_1 and A_2 denote for
time intervals between firing pulses in units of msec.
Temperature 14°C.

reveals that irregular bursts appeared in a random fashion between 2:6
phase-locked, regular (laminar) phases, resulting in intermittencies [16].
This caused the firing rate to reduce to less than 1/3 (see also *Chaos
and Bifurcation*).

The overall states of the firing patterns and the firing rate are
shown in Fig. 4, as a function of two independent and continuous bifurcation parameters, I/I_{th} and T. It should be noted that the domains where
the firing rate changes continuously are located in the region with
weaker strengths (smaller amplitude I and shorter interval T).

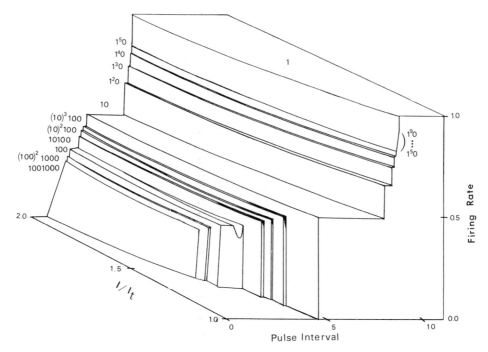

Fig. 4. Schematic diagram showing dependence of firing rate upon both
the current intensity normalised by the current threshold and
the interval between adjacent current pulses. Temperature 14°C.
Firing patterns are also shown in the figure.

Chaos and the Bifurcation Route

The Bifurcation Route to Chaos from the 1:3 Phase-locking. The
characteristics of potential responses were examined in detail by
analysing their power spectra and phase portraits. The phase portraits
were constructed by plotting X = V(t) against y = V(t+τ), where V(t)
represents the time-course of the potential response and τ = T/20 =
0.155 msec. Typical results of analyses of the 1:3 phase-locked, the
period-doubling and the chaotic non-propagated responses are reproduced
in Records A, B and C of Fig. 5, respectively. The respective records A, B
and C exactly correspond to those A_1, A_2 and B in Fig. 3, respectively. In
the case of the 1:3 phase-locking (A), the power density of the spectra (*left*
figure) was represented by several discrete lines corresponding to the
fundamental response frequency and its higher harmonics. The fundamental
frequency, f_1, was exactly 1/3 of the frequency of repetition of the
current pulse, f_0, in this case. The phase portrait (*right* figure) was
represented by a single closed loop. These are consistent with the
potential responses being 1:3 phase-locked. In the case of period-
doubling (B), however, the power density of spectrum (*left* figure) contained discrete lines representing subharmonics, $f_1/2$, as well as the
fundamental frequency, f_1. In this case, phase portrait (*right* figure)
was represented by a double loop. These are consistent with the
situation that the potential responses were the period-doubled ones
originated from the 1:3 phase-locking. In contrast, the power spectrum
in the case of irregular potential responses (*left* record in Fig. 5C)

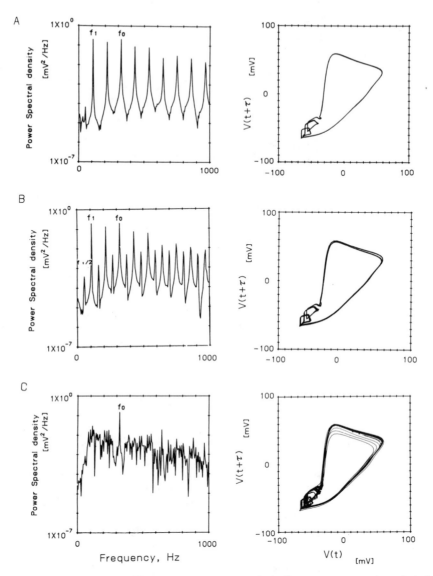

Fig. 5. Power spectra (*left* three records) and phase portraits (*right* three records) for non-propagated potential responses, V(t), at 14 °C. The phase portraits were constructed by plotting x = V(t) against y = V(t+τ) where τ = T/20· = 0.155 msec. (A) In the case of the 1:3 phase-locked potential response, corresponding to Record A_1 in Fig. 3. The fundamental frequency f_1 (107.5 Hz) in the power spectrum is exactly one-third of the stimulation frequency f_0, 322.6 Hz. (B) In the case of the 2:6 phase-locked potential response, corresponding to Record A_2 in Fig. 3. (C) In the case of the chaotic responses, corresponding to Record B in Fig. 3.

contains a broad band, in addition to discrete lines representing f_0 and its higher harmonics. The phase portrait (*right* figure) in this case showed complicated orbits. Thus, taken together with the experimental results described in III, para. 1, the bifurcation to the chaotic potential responses from the 1:3 phase-locked responses can be conveniently termed "subcritical period-doubling bifurcation" [4,19,20].

The route of the subcritical period-doubling bifurcation was examined by analysing the time intervals between adjacent action potentials. This analysis is based upon the hypothesis that the information on the axon is coded as the interval but not as potential variation with time. Fig. 6 shows the variation of the intervals between adjacent firing pulses with the sequence number of firing pulse both for the route of the subcritical period-doubling bifurcation (Records a, b, c, d and e) and for other bifurcation routes (Records f, g, h, i, j and k). In the case of the subcritical period-doubling bifurcation, it can be seen that both frequencies and duration of the bursts increase as I/I_{th} increased from 1.28 to 1.33 times the current threshold while T was set to be 3.5 msec (Records c, d and e). Records a and b illustrate a periodic sequence of the 1:3 phase-locking and the 1:3 phase-locking after a succession of the 2:6 phase-locking, respectively. It is seen in the figure that bursts in Records c, d and e occurred as a result of the growth of the period-doubling, while the 1:3 locking state was stabilised in Records a and b, after the period-doubling ceased. As the state came to the critical point which was present between b and c, the relaxation time for the period-doubling to stabilise to the 1:3 locking state became longer. The correlation exponents, 1.9, 2.4 and 3.2, corresponding to Records c, d and e were obtained by the method introduced by Glassberger and Procaccia [6] to compute the correlation sum according to their notation:

$$C(r,m) = \lim_{N \to \infty} \frac{1}{N^2} \sum_{i \ne j} H(r - |\vec{X}_i(m) - \vec{X}_j(m)|)$$

where we put $\vec{X}_i(m) = (T(i), T(i+1), \ldots, T(i+m-1))$ and the series, $(T(i), T(i+1), \ldots, T(i+m-2), T(i+m-1)$, are the m successive intervals between nearby firing pulses starting from the i-th data of the interval (see Fig. 7 also). The correlation exponents thus obtained are unexpectedly high as compared with the exponent for chaotic potential responses, 2.2 [13]. The latter value was estimated by putting $\vec{X}_i(m) = (V(t_i), V(t_i+\tau), \ldots, V(t_i+(m-1)\tau))$. (Described in detail elsewhere [8,18].)

The Bifurcation Routes to Chaos from the 10100 State. Other bifurcation routes from the 10100 state to the chaos present between the 10100 and 100 states and the 10100 and 1010100 states are also studied by obtaining the relation of firing intervals normalised by T vs. firing pulse sequence, as shown in Records f, g, h, i, j and k of Fig. 6. In Record i, the intervals between adjacent firing pulses change alternately in a regular fashion, suggesting that the firing pattern is exactly 10100. In contrast, in the cases of Records g and h, the patterns change in an alternative way with the firing sequence number, but obviously it has another regularity with a longer period; the intervals are modulated with the number. Record g may suggest this state g oscillates quasi-periodically with the firing rate of 2/5. This should be confirmed by analysing the records in more detail. It should be noted in Record f that the 10100 state once appeared transiently at the initial stage of the periodic stimulation and its stability was lost with the pulse sequence number. When two successive intervals became equal, the 10100 state disappeared after a burst occurred, and afterwards, irregular bursts continued among the 2:6 phase-locked laminar states. Note that the 10100 state in Record g still has a tendency to lose its stability with the

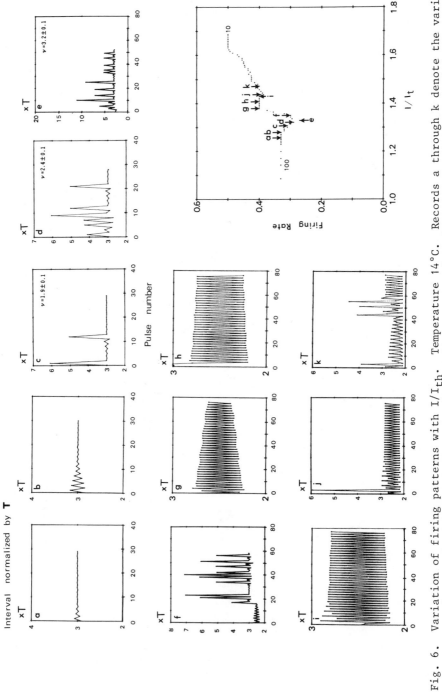

Fig. 6. Variation of firing patterns with I/I_{th}. Temperature 14°C. Records a through k denote the variation of the intervals with the sequence number of firing pulse (see inset of dependence of firing rate upon I/I_{th} where the pulse interval was 3.5 msec and the pulse width was 0.3 msec).

pulse sequence number. In the case of Records j and k, two kinds of the periodic states appeared before and after the intermittencies in the pulse sequence. In Record j, the unstable 10100 state first appeared and then a burst occurred. After the burst, the 10100 state was stabilised through the unstable period-doubling of the 10100 state. In Record k, initially the 1010100 state appeared but became unstable with the pulse sequence number. Then, irregular bursts occurred among the periodic-doubling of the 10100 state. After the bursts, the state was stabilised to evoke the 10100 pattern. These suggest that two periodic states co-exist in these bifurcation regions.

Chaos Propagated along the Axon

Comparative studies on the correlation components for the non-propagated and propagated responses of chaos revealed that the chaotic structure of the pulse interval was stably preserved during propagation along the axon. A typical example of the correlation exponents is illustrated in Fig. 7, where they are 3.2 and 3.4 for non-propagated and propagated responses, respectively. In the computation of the correlation sum, the vector in phase space of dimension m, $X_i(m)$, was put to be $(T(i), T(i+1),\ldots\ldots, T(i+m-1))$ for non-propagated responses, or $(t(i), t(i+1),\ldots\ldots, t(i+m-1))$ for propagated responses, respectively (Fig. 7). The stability of the chaotic response on the propagation could be directly seen in Fig. 8A, where the percentage of the change in the time intervals, $\{t(n)-T(n)\}/T(n)$, was plotted against the firing pulse number. We can easily see that the percentage is nearly zero for all the pulse sequence number. The stability of the chaotic response was compared with that of the response of the unlocked 10100 mode (Fig. 8B). It is seen in the figure that the percentage variation, $\{t(n)-T(n)\}/T(n)$, is more varied with the pulse sequence number in the case of the unlocked 10100 mode (Record A) than that for the chaotic response (Record B).

Power Spectra of the Pulse Density for Chaotic Responses

The power spectrum was calculated for the firing pulses in the chaotic responses, $n(t)$, where $n(t) = 1$ and 0 when a pulse is present and absent at time t, respectively. It was obtained by calculating the Fourier transformation of the auto-correlation function, $\langle n(t)n(t+\tau)\rangle_t$, where $\langle\rangle_t$ denotes the time-average. The spectrum thus calculated is shown in Fig. 9 and can be approximately represented by a function of $A f^{-\delta}$ where $\delta = 0.5$. This could be compared with the theoretical expectation that the power spectra for the intermittent chaos should be in the form of $f^{-\delta'}$ where δ' is equal to $0 - 1$ (3).

IV DISCUSSION

We have experimentally studied potential responses of normal squid giant axons to periodic current stimulations. Normal axons are at rest in the absence of external stimulations. This experimental situation is interesting both from the non-linear, non-equilibrium dynamics and also from physiology.

Periodically stimulated axons responded periodically when the stimulation strengths were large enough (see Figs. 2 and 4). For strengths beyond the critical values, the firing rate changed in a stair-case fashion as a function of the continuous strengths and periodic sequences of the potential responses were phase-locked.

On the other hand, for strengths below the critical values, the

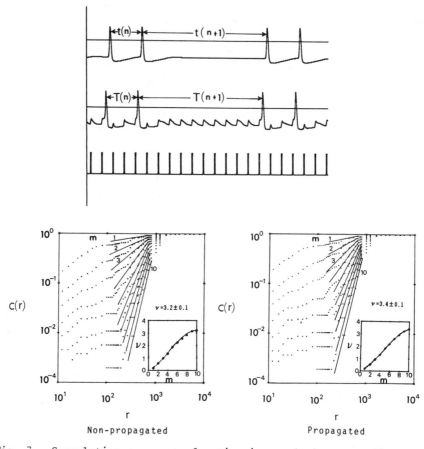

Fig. 7. Correlation exponents for time intervals between adjacent
firing pulses are calculated according to Glassberger and
Procaccia [6] for non-propagated (*left* below) and propagated
(*right* below) chaotic responses of Record e in Fig. 6.

firing rate was observed to change continuously between two phase-
locked states, as a function of the strengths. Generally, the firing
rate once decreased to make a minimum and then increased to another
phase-locked state, as the strenth increased. It was found that the
reduction of the firing rate was due to the production of bursts in an
irregular fashion among regular (laminar) phases of potential responses.

The bursts emerged, followed by instabilities of a phase-locked
state. In the case of the chaos present between the 100 and 10100 locked
state, the route to the chaos from the 100 state was the 100 state → the
period doubling of the 100 state → intermittencies → the chaos; the
bifurcation was termed as a subcritical period-doubling bifurcation.
The intermittency may be categorised as type III of Pomeau-Manneville
[17]. The route to the chaos from the 10100 state was basically the same
as the one from the 100 state. However, in these regions of the stimulation strengths (or the bifurcation parameters), two phase-locked states,
100 and 10100, were clearly overlapped. Emergence of the bursts, followed
by exchanging one locked state to another through the instabilities of
the former state, was observed between the 10100 and 1010100 states.

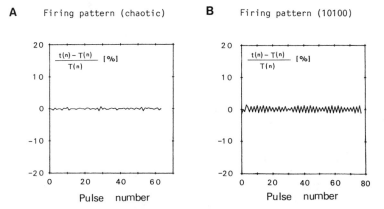

A Firing pattern (chaotic)   **B** Firing pattern (10100)

Fig. 8. Variation fraction (%) of time intervals between firing pulses,
t(n), for propagated responses as compared with those, T(n),
for non-propagated responses, against the firing pulse sequence.
(A) In the case of chaotic responses, corresponding to Record c
in Fig. 6.   (B)   In the case of unlocked 10100 firing-pattern
mode, corresponding to Record g in Fig. 6.

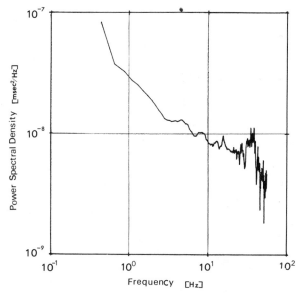

Fig. 9. Power density of spectrum obtained by calculating the Fourier
transformation of the auto-correlation function $\langle n(t)n(t+\tau)\rangle_t$,
where n(t) denotes 1 and 0 when a firing pulse for chaotic
potential responses is found at time t.   Temperature 14°C.
Interval between adjacent current pulses was 3.5 msec and the
current intensity was 1.29 times the threshold.   Sample number
of firing pulses used for this analysis was 1,600.   Sample time
25 μsec.

155

Once chaos occurred, it stably propagated along the axon (Fig. 7). The fact that chaos is stably transformed into chaos on the axon suggests that the chaos may have physiological importance.

Acknowledgements

The authors would like to express their cordial thanks to Dr. K. Aihara for his stimulating discussions.

REFERENCES

[1]   K. Aihara et al., in this volume.
[2]   K. Aihara and G. Matsumoto, in "Chaos", A.V. Holden, ed., Manchester University Press, 1986 and Princeton University Press, 1986, p.257.
[3]   Y. Aizawa, C. Murakami and T. Kohyama, Prog. Theor. Phys., Suppl. 79, 96 (1984).
[4]   P. Bérge, Y. Pomeau and C. Vidal, "Order within Chaos - Towards a Deterministic Approach to Turbulence", translated from the French edition by L. Tukerman, John Wiley & Sons, New York (1986).
[5]   L. Glass and R. Perez, Phys. Rev. Letters 48:1772 (1982).
[6]   P. Grassberger and I. Procaccia, Physica D 9:189 (1983).
[7]   M.R. Guevara, L. Glass, M.C. Mackey and A. Shrier, IEEE Trans. System, Man and Cybernetics SMC-13:790 (1983).
[8]   Y. Hanyu, N. Takahashi, R. Kubo and G. Matsumoto, to be submitted to Phys. Rev.
[9]   L.D. Harmon, Kybernetik 1:89 (1961).
[10] H. Hayashi, S. Ishizuka and K. Hirakawa, J. Phys. Soc. Japan 54:2337 (1985).
[11] H. Hayashi, S. Ishizuka, M. Ohta and K. Hirakawa, Phys. Lett. 88A:435 (1982).
[12] A.V. Holden and M.A. Muhamad, in "Cybernetics and Systems Research 2", R. Trappl, ed., Elsevier, North-Holland, Amsterdam, p. 245 (1984).
[13] G. Matsumoto, K. Aihara, Y. Hanyu, N. Takahashi, S. Yoshizawa and J. Nagumo, Phys. Lett. A (in press).
[14] G. Matsumoto, K. Aihara, M. Ichikawa and A. Tasaki, J. Theor. Neurobiol. 3:1 (1984).
[15] G. Matsumoto, K. Kim, T. Uehara and J. Shimada, J. Phys. Soc. Japan 49:906 (1980).
[16] J. Nagumo and S. Sato, Kybernetik 10:155 (1972).
[17] Y. Pomeau and P. Manneville, Commun. Math. Phys. 74:189 (1980).
[18] N. Takahashi, Y. Hanyu, T. Musha and G. Matsumoto, to be submitted to Phys. Rev.
[19] J.M.T. Thompson, and H.B. Stewart, in "Nonlinear Dynamics and Chaos", John Wiley & Sons, Chichester (1986).
[20] I. Tsuda, Prog. Theor. Phys. 66:1985 (1981).
[21] M. Yamaguchi and M. Hata, in "Competition and Cooperation in Neural Nets", S. Amari and M.A. Arbib, eds., Lecture Notes in Biomathematics 45:171, Springer (1982).
[22] S. Yoshizawa, H. Osada and J. Nagumo, Biol. Cybern. 45:23 (1982).

CHAOS IN MOLLUSCAN NEURON

Hatsuo Hayashi and Satoru Ishizuka[*]

Department of Electronics   *Department of Physiology
Faculty of Engineering   Faculty of Dentistry
Kyushu University   Kyushu University
6-10-1 Hakozaki   3-1-1 Maidasi
Higashi-ku   Higashi-ku
Fukuoka 812, Japan   Fukuoka 812, Japan

ABSTRACT

Irregular responses of Onchidium pacemaker neuron to a sinusoidal current stimulation are classified into three kinds of chaotic oscillations by means of one-dimensional stroboscopic and return maps: chaos, intermittency, and random alternative chaos. Harmonic responses bifurcate to the chaos through the intermittency or the random alternative chaos. All of the chaotic responses are caused by random jumps between two kinds of unstable harmonic responses. Two types of instability exist. Each combination of the two types of instability corresponds to each chaotic response. On the other hand, it is ascertained that spontaneous irregular activities of the neuron are really chaotic. One-dimensional maps of the activities reveal a single-valued function.

I INTRODUCTION

Numerous investigations on chaotic phenomena have been performed in fields of physics and mathematics. Today, many researchers concentrate their attention on significance of chaos in nature, and especially in biological systems. Nervous systems are important for transmitting and processing information in biological systems, and chaotic phenomena can be clearly observed in their electrical behaviour.

The neuron, a unit of nervous systems, is a nonlinear system. It not only acts as a logic element but also responds in various fashions to external signals. For example, the potential of the neuron can synchronize with a periodic current with proper stimulus parameters [4,8]. When the parameters change, the neuron occasionally reveals irregular responses [4-6]. On the other hand, it is well known that many neurons in the brain or ganglia exhibit spontaneous irregular discharges.

In this report, first, the chaotic nature of the voltage responses of Onchidium pacemaker neuron to a sinusoidal current stimulation is elucidated. The irregular responses are classified into three kinds of chaotic responses: chaos, intermittency and random alternative chaos. Harmonic responses bifurcate to the chaos through the intermittency or the random

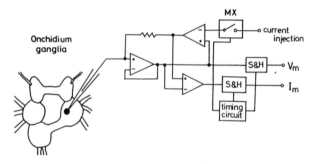

Fig. 1. Experimental method. A glass microelectrode is switched at
2 kHz by an analog switch, MX, for observation of membrane
potential and current injection. S & H is a sample and hold
amplifier.

alternative chaos. All of the chaotic responses are caused by random
jumps between two kinds of unstable harmonic responses. Second, spontaneous irregular activities of the neuron are also investigated. The spontaneous activities are really chaotic.

II METHODS

The nervous system of the esophageal ganglion of Onchidium, marine
pulmonate mollusc, is constructed with comparatively small number of
neurons in contrast to mammalian brain. Moreover, the ganglion has big
neurons whose diameters are about 200 μm. The pacemaker neuron which was
used for experiments is one of the big neurons. Therefore, the voltage
activities were able to be easily observed by an intracellular recording
method with a glass microeleetrode as shown in Fig. 1. The electrode was
switched at 2 kHz for membrane potential observation and current injection. The method is available not only for detection of spikes but also
for observation of membrane potential variation in contrast to an extracellular recording method. Moreover, the current through the electrode
can be used as a stimulation or a control parameter.

The external medium which was used for observing voltage responses
of the neuron to a periodic stimulation was artificial sea water (458 mM
NaCl, 9.6 mM KCl, 10.4 mM $CaCl_2$, 48.5 mM $MgCl_2$, and 10 mM Tris-HCl, pH
7.5). Synaptic inputs to the neuron from the other neurons occur infre-
quently. However, they were blocked by 5 mM Co^{++} ions enough to observe
spontaneous activities of the neuron more clearly. Co^{++} ions in external
medium suppress transmitter release from synapses by blocking inward Ca
current of presynaptic membrane.

A sinusoidal current with dc bias, $I_o + Isin(2\pi f_i t)$, was used as a
periodic stimulation. The frequency of the spontaneous firings, f_o,
depends on the dc level. Therefore, the frequency ratio, f_i/f_o, was used
as a stimulus parameter. On the other hand, the amplitude of the sinusoidal current is not an appropriate measure of the intensity of the stimulation. Because the membrane resistance varies in different neurons.
The parameter, SR/AP, was adopted as another stimulus parameter; SR/AP
is the ratio of the amplitude of the subthreshold response to that of the
action potential.

For investigating spontaneous activities of the neuron, dc current
through the membrane of the cell body was used as a control parameter.
Because it is supposed in general that average membrane potential is con--

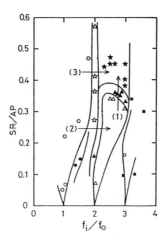

Fig. 2.   Phase diagram of the responses of the pacemaker neuron to a
sinusoidal current stimulation. $\bigcirc$, $\triangle$ and $\square$ are 1/1-, 1/2-
and 1/3-harmonics, respectively. $\star$ is chaos. $\bullet$, $\blacktriangle$ and $\blacksquare$
are intermittencies (laminar phases are quasi-1/1-, quasi-1/2-
and quasi-1/3-harmonics, respectively). $\star$ is random alternation. See text about the arrowed lines (1), (2) and (3).

trolled by summation of synaptic potentials or slow synaptic potential
change.

III   RESULTS

III-I   Chaotic Responses of the Neuron to a Sinusoidal Current Stimulation

The phase diagram of the responses of the pacemaker neuron is shown
in Fig. 2.   The 1/n-harmonic responses which occur around $f_i/f_o = n$ (n:
integer) mean that the neuron fires every   n   period of the stimulation.
The three kinds of chaotic responses occur in proper regions of the stimulus parameters.   In this paper, "chaos" refers to a specific chaos that
shows a cusp-shaped one-dimensional stroboscopic map.   The "intermittency"
is ascribed to gradual phase lag of the action potential with respect to
the stimulation and phase lockings which occur on occasion.   The "random
alternative chaos" consists of unstable 1/n and 1/n+1-harmonic responses
which randomly alternate each other.   When the amplitude of the stimulation increases along the arrowed line (1) the 1/2-harmonics bifurcate to
the chaos through the intermittency.   The random alternative chaos appears
on the route (3) from the 1/1-harmonics to the chaos with increase of the
frequency of the stimulation.   Complicated phase-locked responses occur
along the line (2).   These regular responses compose Farey series;   fifth
Farey series were able to be observed.

The attractors of the chaos, the intermittency and the random alternative chaos in three-dimensional space, (I, V, dV/dt),   are shown in Fig.
3.   I is stimulus current and V is membrane potential.   The trajectory of
each chaotic response forms a complicated two-dimensional surface with
time although each attractor has a different shape.

A stimulus current changes along the I-axis as schematically shown in
Fig. 4.   The trajectory therefore intersects a plane A which is perpendicular to the I-axis every periods of a sinusoidal current.   A stroboscopic
cross-section is obtained from the intersections.

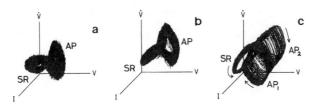

Fig. 3.   Attractors in three-dimensional space, (I, V, dV/dt). (a) The
           chaos observed at f_i/f_o = 2.5 and SR/AP = 0.47.   (b) The in-
           termittency observed f_i/f_o = 2.75 and SR/AP = 0.36.   (c)   The
           random alternative chaos observed f_i/f_o = 2.0 and SR/AP = 3.6.
           The portions of the attractor, SR and AP, correspond to sub-
           threshold responses and action potentials respectively.   In
           regard to the random alternative chaos, the portions of the
           1/1- and 1/2-harmonics are indicated by AP_1 and AP_2 respect-
           ively.

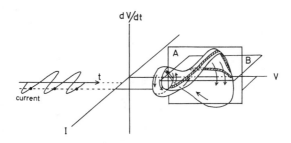

Fig. 4.   Schematic attractor of the chaos in three-dimensional space,
           (I, V, dV/dt).   Trajectory is along the arrowed lines.   One-
           dimensional stroboscopic and return maps are obtained from the
           intersections between the trajectory and planes, A and B,
           respectively.

     Fig. 5 shows the phase-dependence of the stroboscopic cross-section
in regard to the chaos. The cross-section moves clockwise along the contour of the attractor with time. It is expanded, folded and receded with
increase of the phase. This shows a typical mechanism for mapping a
cross-section on to itself. The trajectory of the intermittency or the
random alternative chaos tends to fill up a region of the three-dimensional space in similar manner of the chaos. However, an empty region is
left inside the trajectories which correspond to the action potentials as
shown in Fig. 3(b) and (c). The stroboscopic cross-sections of the
intermittency and the random alternative chaos do not show an ordinary
baker's transformation.

     The one-dimensional stoboscopic map of (V_n) of the chaos is a single-
valued function with a spiky hump as shown in Fig. 6(a). The map clearly
reveals the deterministic nature of the chaos. The stroboscopic map of
the intermittency in Fig. 6(b) is discrete and consists of two branches.
The vertical branch inclines slightly toward the right-hand side and the
horizontal branch is slightly convex. The intermittent responses are also
subject to a deterministic law. Fig. 6(c) shows the stroboscopic map of
the random alternative chaos which reveals the shape of letter L. The map

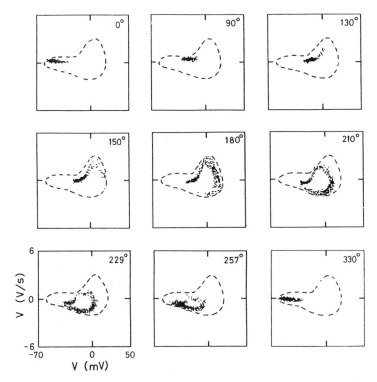

Fig. 5.   Phase dependence of the stroboscopic cross-section of the
chaos.   The dotted lines are the contour of the attractor
projected on the two-dimensional space, (V, dV/dt).

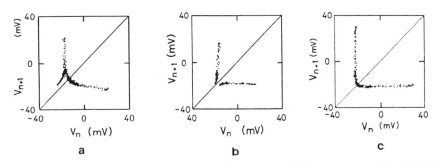

Fig. 6.   One-dimensional stroboscopic maps.   (a) The chaos (at 150°).
(b) The intermittency (at 173°).   (c) The random alternative
chaos (at 136°).

is not useful to ascertain that the random alternative phenomena are
deterministic.   Because a part of the map is exactly vertical.

As shown in Fig. 4, the intersections between the trajectory and a
plane B which is perpendicular to the dV/dt-axis are useful to investigate fluctuations in phase of the action potential.   Because the return
map of $\{I_n\}$ on the cross-section shows fluctuations in the phase with no

161

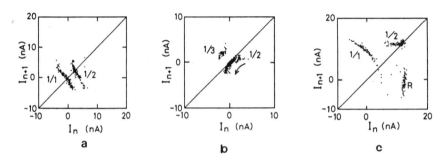

Fig. 7. One-dimensional return maps. (a) The chaos (dV/dt = 0.5 v/s). (b) The intermittency (dV/dt = 1 v/s). (c) The random alternative chaos (dV/dt = 4 v/s).

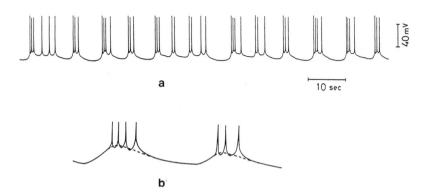

Fig. 8. Spontaneous chaotic activity of the neuron. (a) The bursting activity observed in the neuron at 0.3 nA of dc current. Synaptic inputs are blocked by 5 mM Co^{++} ions. (b) Schematic bursting activity. The dotted line shows slow oscillations of the membrane potential which is caused by discharges.

connection to the fluctuations in amplitude of the action potential. Actually, the currents, I_n, are concerned with the phases, $_n$, of membrane potential at a value of the derivative against the stimulation: $I_n = I_o + I\sin(2 m + _n)$, $0 \leq _n < 2$, m, n:integer.

The return map of the chaos is discrete and consists of two branches which correspond to quasi-1/1- and quasi-1/2-harmonics as shown in Fig. 7(a). The fixed points, that is, the intersections between the branches and the 45° line are unstable because the slopes of the branches at the fixed points are less than -1. Therefore, the orbit on the map spirals out of the fixed points. The map of the intermittency in Fig. 7(b) is also discrete. However, the map has two branches with another type of instability which shows quasi-1/2- and quasi-1/3-harmonics. The branches have no fixed point. The orbit on the map gradually approaches and then leaves from the 45° line and comes back through another branch. The return map of the random alternative chaos in Fig. 7 (c) consists of three branches; two of them correspond to quasi-1/1- and quasi-1/2-harmonics and another one, R, is a branch for jumping from the quasi-1/2- to the quasi-1/1-harmonics. The 1/1- and the 1/2-branches show the same instabilities as the chaos and the intermittency respectively. One-dimensional return maps, in contrast to stroboscopic maps, clearly show

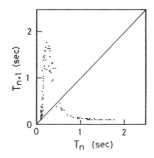

Fig. 9.   One-dimensional map of the interspike intervals of the
          spontaneous chaotic activity.

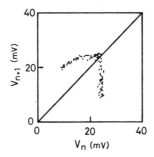

Fig. 10. One-dimensional return map of the spontaneous chaotic
          activity. The map was obtained from a Poincaré cross-
          section of the attractor which was embedded in three-
          dimensional space, $(V(t), V(t+\tau), V(t+2\tau))$. τ = 5 msec.

the deterministic significance of the random alternative chaos.

III.II Spontaneous Chaotic Activities of the Neuron

      Synaptic inputs were blocked by Co^{++} ions in order to observe spon-
taneous voltage activities of the neuron. The ions generally block not
only Ca^{++} channels of presynaptic membrane but also Ca^{++} channels of soma
and axon. However the spontaneous activities of the neuron are not
suppressed with Co^{++} ions [10]. Chaotic bursting activities occur in
some regions of dc current around zero.

      The spontaneous chaotic activity of the neuron is shown in Fig. 8(a).
The interspike interval fluctuates; the interval gradually becomes long
and then comes back to a shorter one after an interburst interval. The
one-dimensional map of the intervals in Fig. 9 reveals an apparently
single-values function. This indicates that the fluctuations in interval
are chaotic.

      The action potential of the chaotic bursting activity also fluct-
uates besides the interspike interval. The attractor embedded in three-
dimensional space, $(V(t), V(t+\tau), V(t+2\tau))$, was obtained. τ is an arbitrary time. The attractor reveals a ribbonlike structure which twists in
a complex way. The Poincaré cross-section of the attractor show that the
deterministic property is based on the fold of the cross-section. Fig.
10 shows the return map obtained from a cross-section. The shape of the

map is different from that of the intervals but it is also a single-valued function. The above maps indicate that the spontaneous irregular activities of the actual neuron are really chaotic.

IV DISCUSSION

The 1/2- and 1/1-harmonic responses bifurcate to the chaos through the intermittency and the random alternative chaos respectively when a stimulus parameter changes. The pacemaker neuron does not show period-doubling bifurcation on the routes. This is attributed to the sharp peak of the stroboscopic map of the chaos. Because of the steep slopes around the peak indicate that new fixed points borne by a tangent bifurcation are unstable from the onset.

Frequency of a sinusoidal stimulus current which cause the chaotic responses is higher than natural frequency of firings. The neuron is frequently stimulated during relative refractory period. A larger stimulation causes action potentials with various amplitudes. Therefore, the stroboscopic cross-section stretches in the direction of the V-axis. On the other hand, the timing of firing depends on the threshold of excitation which varies during relative refractory period. However, if it is decided what trajectory the state point passes along, the motion of the state point along the trajectory is unique. Therefore, according to the trajectories which correspond to the action potentials with same amplitude, the state points sampled stroboscopically gather around a point in the phase space. The cross-section does not spread in the tangential direction of the trajectories. This indicates a hyperbolicity of the attractor.

In comparison with the chaos and the intermittency, frequency of the stimulation which causes the random alternative chaos is lower. Whenever the neuron is stimulated the excitability fairly recovers. Therefore, the fluctuations in amplitude of the action potential are not so large. However, the fluctuations in phase remain and rather increase as shown in Fig. 3 (c); the trajectories which correspond to the action potentials fluctuate in the direction of the I-axis. A stroboscopic cross-section of the attractor that is cut by a plane, I = constant, unfortunately contains the intersections between some of the trajectories and the plane. Stroboscopic mappings are not enough to ascertain the chaotic nature of the random alternative chaos. On the other hand, a cross-section of the attractor that is cut by a plane, dV/dt = constant, contains the intersections between the plane and all of the downward trajectories that are concerned with the action potentials. A return map obtained from the cross-section is suitable to investigate the dynamical nature of the random alternative chaos.

Chaotic nature of the chaos and the intermittency is basically shown by means of stroboscopic maps. However, return maps give more information of their chaotic nature. The return maps indicate that all of the chaotic responses observed in the pacemaker neuron are caused by instabilities of two kinds of harmonic responses. Two types of instability exist: (I) an orbit that spirals out to an orbit from a fixed point, and (II) an orbit that approaches 45° line and then moves away from it. The combinations of the instabilities, (I, I), (I, II) and (II, II), correspond to the chaos, the random alternative chaos and the intermittency respectively.

The neuron causes chaotic responses to a periodic stimulation. However, chaotic activities of the neuron are not limited to the voltage responses. As mentioned in the previous section, it should be stressed that the neuron causes spontaneous chaotic activities.

In general, neurons do not maintain regular beating activities and consequently cause regular or irregular bursting activities. This results from slow oscillations of membrane potential which are caused by discharges of the neuron as shown in Fig. 8 (b). Two causes have been proposed (1) Ca^{++}-activated K^+ channel [3], or (2) voltage-sensitive slow Na^+ and slow K^+ channels [7]. Unfortunately, the cause of the bursting activities of Onchidium neurons has not been elucidated experimentally.

In the case (1), Ca^{++} ions go into the cell body during electrical excitation and are accumulated inside the cell because of slow efflux of the ions. Ca^{++}-activated K^+ channels are therefore activated and then the membrane potential moves to a hyperpolarized state every firing. Consequently, interspike interval gradually become long. At last, a long interval without spikes occurs. The membrane potential recovers as Ca^{++} ions come out of the cell and spikes appear again. In the case (2), similar processes occur. However, the slow Na^+ and slow K^+ channels are directly activated by the firings of the neuron and slow recovery of the membrane potential is due to a time constant of the slow K^+ channel.

An excitable membrane model which is expressed in terms of a Hodgkin-Huxley type formalism has been numerically solved to investigate chaotic bursting activities [2]. One of the variables of the model is intracellular Ca^{++} ion concentration which is connected with voltage-insensitive K^+ channels. The one-dimensional map of the interspike intervals generated by the model quite resembles the map in Fig. 9.

As mentioned above, a single neuron causes spontaneous chaotic activities. The neuron also shows chaotic responses to a periodic stimulation. Moreover, it has been indicated that the electroencephalogram of the human brain observed at a sleep stage reveals an attractor with a low dimension [1]. Spontaneous discharges of neurons in squirrel monkey cortex under anaesthesia have also been investigated; in some neurons, the interspike intervals reveal an attractor with a low dimension [9]. All of the facts indicate that chaotic activities in nervous systems are not specific phenomena that is observed under an unusual condition. Unfortunately, we have no knowledge about the relationship between chaotic phenomena and information processing in the brain at present. However, chaotic phenomena should be considered when the information processing is investigated. Chaos seems to be usually associated with activities of nervous systems.

REFERENCES

[1] A. Babloyantz, J.M. Salazar and C. Nicolis, Evidence of chaotic dynamics of brain activity during the sleep cycle, Phys. Lett. 111A: 152-156 (1985).
[2] T.R. Chay, Chaos in a three-variable model of an excitable cell, Physica. 16D:233-242 (1985).
[3] A.L.F. Gorman and M.V. Thomas, Changes in the intracellular concentration of free calcium ions in a pacemaker neurone, measured with the metallochromic indicator dye arsenazo III. J. Physiol. 275: 357-376 (1978).
[4] H. Hayashi, S. Ishizuka, M. Ohta and K. Hirakawa, Chaotic behaviour in the Onchidium giant neuron under sinusoidal stimulation, Phys. Lett. 88A:435-438 (1982).
[5] H. Hayashi, S. Ishizuka and K. Hirakawa, Chaotic response of the pacemaker neuron. J. Phys. Soc. Jpn. 54:2337-2346 (1985).

[6]   H. Hayashi, S. Ishizuka and K. Hirakawa, Instability of harmonic
      responses of Onchidium pacemaker neuron, J. Phys. Soc. Jpn.   55:
      3272-3278 (1986).
[7]   E.P. Kandel, "Cellular Basis of Behaviour", W.H. Freeman and
      Company, San Francisco, 261-268 (1976).
[8]   D.H. Perkel, J.H. Schulman, T.H. Bullock, G.P. Moore and J.P.
      Segundo, Pacemaker neurons: effects of regularly spaced synap-
      tic input, Science 145:61-63 (1964).
[9]   P.E. Rapp, I.D. Zimmerman, A.M. Albano, G.C. Deguzman and N.N.
      Greenbaum, Dynamics of spontaneous neural activity in the simian
      motor cortex: the dimension of chaotic neurons, Phys. Lett.
      110A:335-338 (1985).
[10] In preparation.

PANCREATIC B-CELL ELECTRICAL ACTIVITY: CHAOTIC AND IRREGULAR BURSTING PATTERN

P. Lebrun

Laboratory of Pharmacology
Brussels University School of Medicine
115 Boulevard de Waterloo
B-1000 Brussels
Belgium

ABSTRACT

Glucose-induced β-cell electrical activity was recorded in islets of Langerhans isolated from Swiss Webster albino mice originating from different suppliers. Twenty-three out of 25 islets obtained from mice bred at the Charles River Breeding Station (CR mice) exhibited irregular or chaotic burst patterns of electrical activity, while 36 out of 40 islets isolated from mice bred locally at the National Institutes of Health displayed the typical bursting activity. The CR mice tended to recover a regular pattern after 1 month on the National Institutes of Health mouse diet. The irregular or chaotic bursting electrical activity is proposed to result from changes in β-cell membrane composition or cellular metabolism, possibly induced by differences in diet.

I INTRODUCTION

Mouse pancreatic β-cells stimulated by glucose display membrane potential fluctuations that have been proposed to represent an early event in the stimulus secretion coupling of glucose-induced insulin release [1]. The general features of the glucose-induced electrical activity consist of a regular succession of depolarised phases with superimposed bursts of spikes (active phase) and silent repolarisation periods. Although the absolute duration of a single burst may vary between 5 and 30 s in cells from different islets, the rhythmical burst pattern recorded from one cell usually remains constant over several hours of experimentation [1]. Also, within an islet, most of the β-cells exhibit an identical burst pattern in the presence of glucose; this characteristic has been proposed to result from electrical coupling between the different cell domains [7].

Recently, an unusual burst pattern, which consists of a slow (∿ 4 min) modulation in the intensity of the β-cell electrical activity, has been reported to occur in <20% of the islets examined [4,5]. Although several hypotheses have been proposed, at present there is no clear explanation for this irregularity. In the present study, we found that the incidence of the atypical glucose-induced β-cell electrical activity was dependent on the origin of the mice.

II MATERIALS AND METHODS

Female Swiss Webster mice, 3-4 months old, originally from the same strain, were obtained either from Charles River Breeding Station (Kingston, NJ) (CR mice) or from the National Institutes of Health (NIH) general animal house, Bethesda, MD (NIH mice). NIH mice were raised on a standard pellet diet (NIH N° 63-8760) containing 4.5% lipids, 23.5% proteins, 4.5% fibres and 54% carbohydrates. The CR mice were raised on another pellet diet (Charles River Breeding Station) containing 6.4% lipids, 21.4% proteins, 3.2% fibres and 49.5% carbohydrates. Both mouse diets were void of any antibiotic or hormonal activity. Animals from both origins and from several shipments were allowed to settle in for at least 1 week in the laboratory before use and received the standard NIH diet. Before decapitation, each animal was weighed and the blood glucose measured using a reagent strip in combination with a glucometer (Miles Ames Div., Miles Laboratories, Elkhart, IN).

A single islet of Langerhans was dissected from the tail portion of the pancreas of each mouse. Islets of the same size were isolated from 40 NIH mice and 25 CR mice. The electrical activity was recorded as previously described [7]. The experiments were conducted from April through September.

III RESULTS

In the presence of 11.1 mM glucose, a typical, regular burst pattern of β-cell electrical activity was observed in 36 out of 40 islets isolated from NIH mice (Fig. 1A). In the remaining islets, the β-cell electrical activity exhibited a slow modulation in the intensity of the burst pattern, which was similar to that described by others [4,5]. In contrast, 23 out of 25 islets isolated from CR mice showed irregular bursting of electrical activity throughout several hours of experimentation (Fig. 1, B and C). However, islets taken from CR mice maintained from 1 month on NIH mouse diet tended to exhibit a more regular burst pattern (Fig. 1D).

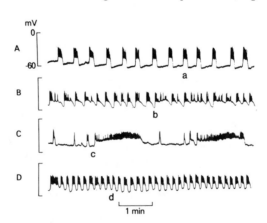

Fig. 1. Steady-state bursting activity from four different mouse islets of Langerhans after 30-min exposure to 11.1 mM glucose. (A) NIH mouse islet; (B,C) CR mouse islets; (D) CR mouse islet after 1 month on the NIH mouse diet. a - d indicate portions of the trace illustrated in Fig. 2.

Two types of irregular pattern of β-cell electrical activity were observed in CR mice; one was chaotic (Fig. 1B), and the other showed regularly repeating groups of bursts of variable duration (Fig. 1C). As shown in Fig. 2, the modification in burst duration did not affect the rate of depolarisation at the onset of each burst, nor did it affect the spike amplitude. However, the repolarisation ending each burst was much slower in the CR mice, even in those maintained for 1 month in our laboratory. Also, the spike frequency distribution along the active phase of the burst was altered with respect to the control experiments. In NIH mice (Fig. 2a) spike frequency was always high at the beginning and decreased gradully during the active phase, as is usually described in pancreatic β-cell electrophysiology. In contrast, in the CR mice, spike frequency was uneven throughout the bursts (Fig. 2b) or increased along the active phase (Fig. 2c). Even in the islets that showed a tendency to recover a regular burst pattern (Fig. 2d), the overall spike frequency was usually lower than in typical islets (Fig. 2a).

In islets isolated from either CR mice or NIH mice, the removal of glucose induced the expected hyperpolarisation with cessation of the electrical activity, whereas increasing the glucose concentration to 22.2 mM induced continuous spiking activity (data not shown).

Blood glucose levels in mice from both origins were not significantly different; their averages were 6.87 ± 0.76 mM (mean ± SEM; n=40) and 7.17 ± 1.08 mM (mean ± SEM; n=25) in NIH and CR mice, respectively. However, body weight differed significantly, averaging 22.75 ± 1.25 g (mean ± SEM; n=40) and 28.45 ± 1.05 g (mean ± SEM; n=25) for NIH and CR mice, respectively (P < 0.005).

IV DISCUSSION

The present data show that 23 out of 25 (∿ 90%) islets of Langerhans isolated from CR mice exhibit an irregular or chaotic bursting electrical activity in the presence of 11.1 mM glucose. This high incidence of irregular or chaotic activity sharply contrasts with our observation of slow modulations in burst intensity in only 4 out of 40 (10%) islets obtained from NIH mice. Previous reports have described a similar pattern in <20% of the islets isolated from albino mice [4,5].

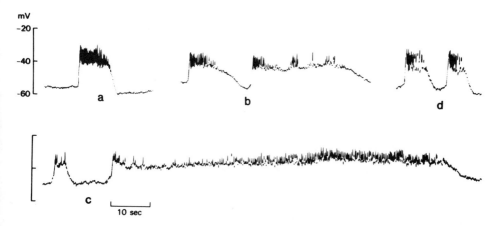

Fig. 2. Details from Fig. 1, shown with an expanded time base.
         a, b, c and d correspond, respectively, to the experiments
         illustrated in Fig. 1, parts A, B, C and D.

The modifications in the repolarisation rate of the burst, in the duration of the active phase and in the spike frequency distribution suggest that islets isolated from the CR mice present alterations in membrane ionic permeabilities. Interestingly, when using a mathematical model that simulates the typical β-cell electrical activity and that contains no stochastic components, chaotic bursting electrical activity, similar to that depicted in Fig. 1B, was reproduced. This chaotic electrical activity was simulated by reducing the rate constants of the ionic permeabilities, and by varying the uptake rate of cytosolic calcium [3]. This latter parameter specifically affects the potassium permeability [1]. Thus, as the glucose-induced periodic oscillations in β-cell membrane potential have been shown to be correlated with cyclic variations in potassium permeability, the chaotic electrical activity observed in islets isolated from CR mice could reflect alterations in the potassium conductance [1].

On the other hand, the occurrence of an irregular bursting activity could also result from alterations in cell-to-cell coupling. This might in turn exaggerate differences in the intrinsic frequency between cell domains and give rise to repeating groups of bursts of different durations (Fig. 1C). Incidentally, since the insulin-releasing process has been shown to be correlated with the glucose-induced bursting activity, it is probable that the islets isolated from the CR mice exhibit irregular secretory oscillations [2].

At present there is no clear explanation for the origin of this irregular or chaotic electrical activity exhibited by the islets isolated from CR mice. The mice were obtained from several shipments and the experiments were conducted during a 6-month period, thus excluding a spurious or seasonal variation. Also, both groups of mice originated from the same strain, had similar blood glucose levels, and were used under the same experimental conditions (see Materials and Methods). However, the CR mice presented a body weight exceeding that of the NIH mice of the same age by ∿25%. In healthy rodents, the body weight has been shown to be mainly correlated with the metabolic rate or with the lipid composition of the pellet diet [8,9]. The fatty acid and sterol content of diet can modify the cell membrane composition and has been proposed to affect the cell membrane fluidity and ionic permeability [6]. As the CR mice were raised on a diet containing more lipids than the standard NIH diet, the erratic β-cell electrical activity could result from modifications in cell membrane composition induced by the diet. This hypothesis is supported by the finding that the islets isolated from the CR mice, which were fed with the NIH diet for 1 month, tended to resume a regular bursting pattern. On the other hand, changes in the cytosolic composition due to alteration of cellular metabolism could also generate such irregular or chaotic electrical activity by affecting the channel properties and/or cell-to-cell coupling.

In conclusion, the incidence of erratic bursting pattern of electrical activity recorded from islets of Langerhans was found to be dependent on the origin of the mice. Alteration in membrane ionic permeabilities or cell-to-cell coupling could result from changes in membrane composition or cellular metabolism. It is tempting to speculate that these changes could result from differences in diet; however, the involvement of some other environmental factors cannot be excluded.

REFERENCES

[1]   I. Atwater, C.M. Dawson, A. Scott, G. Eddlestone and E. Rojas, Horm. Metab. Res. 10 (Suppl.): 100-107 (1980).

[2] I. Atwater, E. Rojas and A. Scott, J. Physiol. (London) 291: 57P (1979).
[3] T.R. Chay and J. Rinzel, Biophys. J. 47: 357–366 (1985).
[4] D.L. Cook, Metabolism 32: 681–685 (1983).
[5] J.C. Henquin, H.P. Meissner and W. Schmeer, Pflügers Arch. 393: 322–327 (1982).
[6] F.A. Kummerow, Ann. NY Acad. Sci. 414: 29–43 (1983).
[7] P. Meda, I. Atwater, A. Goncalves, A. Bangham, L. Orci and E. Rojas, Q.J. Exp. Physiol. 69: 719–735 (1984).
[8] S.C. Peckham, C. Entenman and H.D. Carroll, J. Nutr. 77: 187–197 (1962).
[9] J.B. Storer, Exp. Gerontol. 2: 173–182 (1967).

MULTIPLE OSCILLATORY STATES AND CHAOS IN THE ENDOGENOUS ACTIVITY OF

EXCITABLE CELLS: PANCREATIC β-CELL AS AN EXAMPLE

Teresa Ree Chay* and Hong Seok Kang

Department of Biological Sciences
University of Pittsburgh
Pittsburgh, Pa 15260

ABSTRACT

The behaviour of pancreatic β-cells may be described by a system of three nonlinear differential equations that shows multiple oscillatory and chaotic solutions. Bifurcation diagrams of this system are constructed using AUTO.

I INTRODUCTION

A wide variety of oscillatory patterns have been observed in membrane potential recordings from β-cells of isolated pancreatic islets [1-6]. This burst pattern is believed to be generated by interplay between ionic channels in the β-cell membrane and intracellular calcium ions [7,8]. An interesting feature of the β-cell burst is that the plateau potential appears only with action potentials, and the active phase duration depends on the concentration of glucose. Intracellular acidification can also stimulate β-cell electrical activity in a glucose-like manner [9,10]. Glucose, in addition to inhibiting the ATP-blockable K^+ channels [11] activates the efflux of intracellular Ca^{2+} ions [7].

Based on the recently determined ionic channel properties [12] a simple mathematical model for the burst activity of pancreatic β-cell is presented in this paper. The model is applicable to a "limit-cycle" oscillatory regime, where ATP-blockable K^+ channels are no longer operative. In this regime, glucose is treated as an activator for the rate of efflux of intracellular Ca^{2+} ions; thus, its effect is equated to k_{Ca}, the efflux rate constant. In addition, intracellular H^+ ion, which is a by-product of the glycolytic metabolic process, is treated as a competitive inhibitor for Ca^{2+} ion. Since H^+ is a competitive inhibitor (under our assumption), its effect is equated to the strength of Ca_i dissociation constant K_h. In the model, a Ca^{2+} binding site is assumed to exist in the inner membrane of voltage-gated Ca^{2+} channel [13].

With a bifurcation analysis, we show how multiple oscillatory states and chaos may arise from such a simple system as ours. Since the model

---

*Dr. Chay was unable to attend the workshop because of ill health.

presented here is deterministic, the irregularity in response to the variation of these parameters cannot be explained in terms of any stochastic feature of the cellular processes. Rather, it reflects a not uncommon phenomenon in non-linear dynamical systems referred to as deterministic chaos. We demonstrate the deterministic nature of aperiodicity by constructing one-dimensional maps. Thus, our model is one of a few biophysically realistic models of excitable membranes which exhibit endogenous aperiodic chaos [15-20].

II MODEL

An excitable membrane is considered to contain voltage-sensitive channels which allow Ca^{2+} ions to enter the cell and voltage-gated K^+-channels which allow K^+ ions to leave. The dynamics of the membrane potential V of such a system may be written as [21]

$$-4\pi r^2 \, C_m \, dV/dt = I_{Ca} + I_K + I_L \tag{1}$$

where C_m is the membrane capacitance, r is the radius of a β-cell, and I_{Ca}, I_K and I_L are the Ca^{2+} current, K^+ current, and leak current, respectively. The current I_x of x ions is the product of a conductance g_x and a driving force, the difference between the membrane potential and the reversal potential for the conductance, V_x, i.e.

$$I_x = g_x \, (V_x - V) \tag{2}$$

The detailed knowledge on g_x is essential for our mathematical modelling. Fortunately, the properties of g_{Ca} and g_K channels have recently been elucidated by Rorsman and Trube [12]. These workers have found that the activation time constant of the Ca^{2+} channel was very fast and the degree of steady-state activation (i.e. m_∞) was a sigmoidal function of membrane potential V. They have expressed the steady-state activation term by a Boltzmann equation

$$m_\infty = 1/\{1+\exp[(V_m - V)/slope(m)]\} \tag{3}$$

where V_m is a half-maximal activation potential, and slope (m) is the "steepness" factor. The same experiment revealed that the Ca^{2+} current does inactivate very slowly and weakly during a long period of depolarising pulses. This information and the evidence that intracellular Ca^{2+} ion modulates the inactivation [13] lead us to propose the following equation for the inactivation term P_h:

$$P_h = 1/\{1+\exp(-A(V-V_C)\} \tag{4a}$$

$$V_C = -(n_H/A) \, \ln([Ca^{2+}]_i/K_h) \tag{4b}$$

where A is a constant, n_H the Hill coefficient, and K_h the dissociation constant of Ca_i. The exponential voltage factor $\exp(-AV)$ comes in when a charged Ca^{2+} ion binds to a charged membrane site. In the model, the effect of intracellular H^+ is equated to K_h, to be consistent with our hypothesis that H^+ ion acts as a competitive inhibitor. Thus, by combining the effect of the activation and inactivation terms in Eqs. (3) and (4), an expression for the Ca^{2+} channel conductance can be expressed as

$$g_{Ca} = \bar{g}_{Ca} \, m_\infty \, P_h \tag{5}$$

where $\bar{g}_{Ca}$ is the maximal conductance of calcium channel.

The experiment of Rorsman and Trube [12] also showed that the activation time constant of the K^+ current was slower than that of Ca^{2+} channel and the voltage dependence of activation could also be described by the Boltzmann equation. To put this in a mathematical form, g_K is represented by the probability of opening n of the K^+ gate as

$$g_K = \bar{g}_K n \tag{6}$$

where $\bar{g}_K$ is the maximal conductance, n is the probability of open state at time t. Here, n is a dynamic variable, and the voltage- and time-dependent changes of n are given by a first-order reaction [21]

$$\text{CLOSED } (1-n) \xrightleftharpoons[\beta_n]{\alpha_n} \text{OPEN } (n) \tag{7}$$

where the experiment of Rorsman and Trube [12] indicated that the rate constants α_n and β_n appeared to fit to the following expression:

$$\alpha_n = \lambda \exp[-V_n-V)/slope(n)] \tag{8a}$$

$$\beta = \lambda \tag{8b}$$

$$n_\infty = 1/\{1+\exp[(V_n-V)/slope(n)]\} \tag{8c}$$

and λ is a constant which is independent of V, and n_∞ is the steady-state fraction of open state.

The intracellular calcium concentration $[Ca^{2+}]_i$ is treated as a dynamic variable in the model. Thus, the rate change of $[Ca^{2+}]_i$ is due to the influx of extracellular calcium ions into the cell through the voltage gated calcium channel and the efflux of free intracellular Ca^{2+} ions from the cell to the extracellular medium by the action of Ca-ATPase pump activity. We thus have

$$f^{-1} d[Ca^{2+}]_i/dt = 3I_{Ca}/4 \ r^3F - k_{Ca}[Ca^{2+}]_i \tag{9}$$

where k_{Ca} the rate constant for the efflux of Ca_i, F the Faraday constant, and f is related to the free calcium and the total calcium concentrations in the cell. The effect of glucose on the bursting is equated to k_{Ca}, to be consistent with the experiment of Rorsman et al. [7].

III   RESULTS

The differential equations representing the three dynamic variables developed in Section II were solved numerically on a CRAY XMP. A Gear algorithm [22] has been used to solve these equations, where we set the absolute and relative error tolerances at 10^{-8}. The parameters used for the computation are listed in Fig. 1 caption.

Figure 1 shows time courses of membrane potential, V, (solid) and intracellular free calcium concentration $[Ca^{2+}]_i$ (dashes). The left panels illustrate the mathematically predicted burst pattern as a function of glucose. Changes in glucose are expressed by varying k_{Ca}. This is based on our hypothesis that glucose activated k_{Ca} (the rate at which Ca^{2+} is removed from the cytosol). The right panels of the same figure illustrate the burst pattern as a function of H^+, which is equated by K_h (the dissociation constant of Ca_i from a Ca^{2+} channel site). This is based on our hypothesis that intracellular H^+ ions act as a competitive inhibitor. It is immediately apparent that both hypotheses predict an

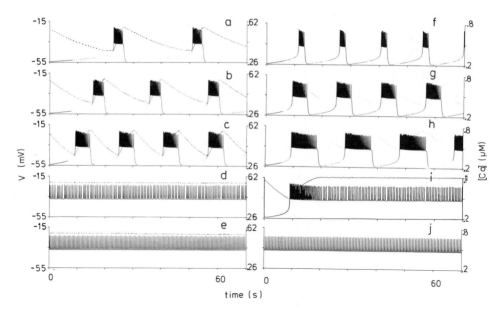

Fig. 1. Burst response in theoretical model of β-cell electrical
activity. Left panels show effect of increasing glucose
modelled as increasing calcium removal rate k_{Ca}, and right
panels show decreased pH_i modelled as increasing K_h, the
dissociation constant of Ca. For left, we used a fixed value
K_h = 90 μM and varied k_{Ca} (a: 0.03; b: 0.04; c: 0.06;
d: 0.0861; e: 0.09 ms^{-1}). For right, we used a fixed value
k_{Ca} = 0.06 ms^{-1} and varied K_h (f: 50; g: 90; h: 120;
i: 127.72; j: 130 μM). Other parametric values are C_m =
1 μF·cm^{-2}, $\bar{g}_{Ca}$ = 250 pS, $\bar{g}_K$ = 1.3 nS, $\bar{g}_L$ = 10 pS, V_{Ca} = 100 mV,
V_K = -80 mV, V_L = -40 mV, V_m = -22 mV, V_n = -9 mV. Slope (m)
= 7.5 mV, slope (n) = 10 mV, λ = 17 s^{-1}, A = 0.1 mV^{-1}, r =
6 μm, f = 0.001 and n_H = 1.

increase in the active phase duration and a decrease in the silent phase
duration as a function of the concentration of these secretagogues. It is
also apparent that the bursting pattern is not significantly different as
predicted by both hypotheses, yet the underlying changes in $[Ca^{2+}]_i$, while
undergoing similar oscillation in time, are quite different in the mean
concentration achieved at different secretagogue concentrations: on left
$[Ca^{2+}]_i$ range unaltered, while on right it shifts upward. It appears that
glucose, at least in the bursting regime, is not effective in raising the
level of $[Ca^{2+}]_i$ (see left panel). Thus, the increase in $[Ca^{2+}]_i$, which
was observed in glucose-induced experiments [7], can only be explained by
incorporating the inhibitory effect of glucose on ATP-blockable K^+
channels into the model [18].

The predited $[Ca^{2+}]_i$ as a function of increasing strength of these
parameters is further studied by constructing a bifurcation diagram, and
this is presented in Fig. 2. A bifurcation diagram is a plot which, for
a given value of the bifurcation parameter (e.g. k_{Ca} and K_h), reveals the
branches of stable periodic states (SPS), unstable periodic states (UPS),
stable steady states (SSS), unstable steady states (USS), Hopf bifurcation
points (HB). In addition, it gives a compact way of showing various modes
of oscillations and different levels of $[Ca^{2+}]_i$, as a function of the
bifurcation parameter. To construct the bifurcation diagram, we used

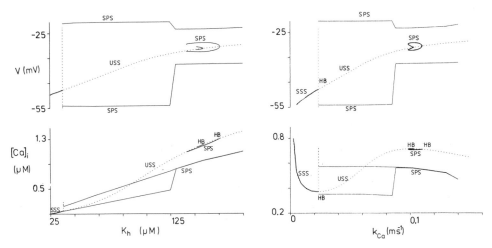

Fig. 2.   Bifurcation diagram revealing many different types of modes as
a function of K_h (left) and k_{Ca} (right). Here, HB stands for
the Hopf bifurcation point, SSS for the stable steady state,
USS for the unstable steady state, SPS for the stable periodic
state, and UPS for the unstable periodic state. Three Hopf
bifurcation points on left are 0.023523, 0.102174 and 0.110107;
those on right are 35.5879, 152.4357, 166.3208.

AUTO, a program for automatic bifurcation analysis developed by Doedel
[23] and the Gear ODE dynamic solutions [22]. As apparent from the bottom
frame, two hypotheses yield different magnitudes and amplitudes of $[Ca^{2+}]_i$
oscillations in the bursting regime: constant values for increasing k_{Ca}
and increasing values for increasing K_h.

Note that in the spiking regime (see Fig. 2), we find stable periodic
states. The top frame of Fig. 3 shows this regime in an expanded scale.
A closer inspection reveals that the two bifurcation diagrams are quite
different: on the left, the periodic branch that originates from the
right HB is connected to an unstable periodic state which in turn terminates at the middle HB. On the right, the two stable periodic branches
which originate from the respective HB's terminate at the homoclinic
orbitals with infinite periods. The bottom frame of the same figure shows
the dynamic results of the membrane potential and $[Ca^{2+}]_i$ at the two HB's.
As shown in this figure, the mode of oscillations are about the same for
both hypotheses. When comparing them to the "normal" spikes (see Fig. 1),
however, they are different in the following respects: (i) the amplitude
of the spikes is much smaller, (ii) $[Ca^{2+}]_i$ is much higher, and (iii) the
period is longer. In this multiple oscillatory state, instantaneous
perturbations, such as brief current pulses, can make a stable repetitive
spiking cell into an almost quiescent cell, i.e. a large amplitude
oscillatory mode to a much smaller one. Note that the level of $[Ca^{2+}]_i$
rises abruptly, as the mode of oscillations changes. Functional implications of the existence of the multiple oscillatory states are not yet
clear.

Aperiodicity occurs during the transition from the bursting mode to
the continuous spiking mode (see the top frame of Fig. 4). Similar
aperiodicity has been observed in some of the experiments on the β-cell
recordings [4,12,24,25]. A useful quantitative tool for studying
deterministic behaviour is a discrete time one-variable representation of

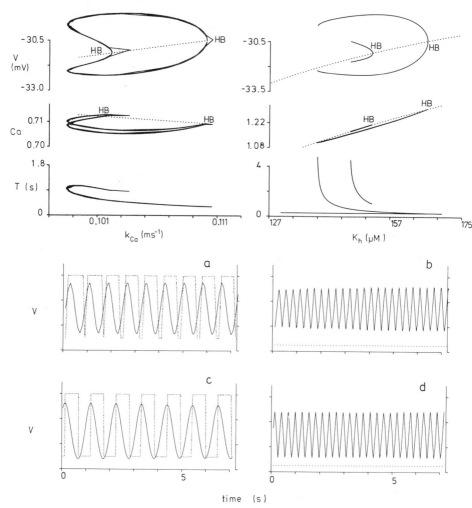

Fig. 3. Bifurcation diagram in an expanded scale around the higher
Hopf bifurcation points for V, $[Ca]_i$ and period T (top) and
the dynamic results on membrane potential spikes and $[Ca^2]_i$
oscillations close to the point of Hopf bifurcations at k_{Ca} =
0.1021739 (a), 0.1101072 (b) and K_h = 152.4347 (c) and
166.3208 (d).

the dynamic quantities shown in the bottom trace of Fig. 4. This map
representation allows a compact and simplified way to display the chaotic
behaviour of the spike-to-spike interval (left) and the minimum-to-
minimum $[Ca^{2+}]_i$ (right). Here, the n-th burst period was recorded on the
x-axis and the n+1-st one on the y-axis. The spike-to-spike intervals
exhibit intermittency in which periods of relative regularity (i.e.
spikes) are abruptly and irregularly interrupted by bursts of quite
different activity. The one-dimensional $[Ca^{2+}]_n$ maps exhibit period-
doubling chaos having a unimodal single-hump.

    The nature of period doubling chaos can be seen clearly by studying
en route to chaos from the stable continuous spiking regime. Figure 5

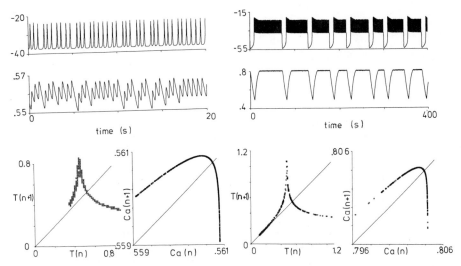

Fig. 4.   The dynamic results showing chaos in spikes (left) and bursts
(right). One-variable map $g_{n+1} = F[g_n]$ constructed from
discrete time dynamics of spike-to-spike interval lengths
(left) and the minimum-to-minimum $[Ca^{2+}]_i$ (right).   The
values of the parameter used for the computation are k_{Ca} =
0.0861 (left) and K_h = 127.72 (right).

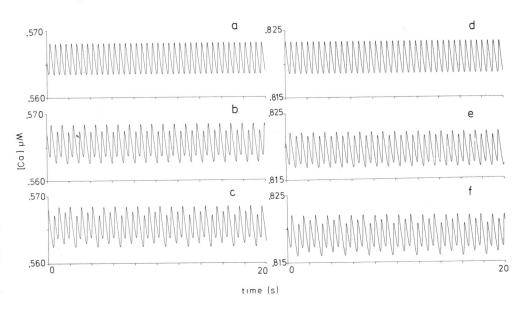

Fig. 5.   Period-doubling in $[Ca^{2+}]_i$ as a function of k_{Ca} (left) and of
K_h (right). From the top trace to bottom, the values used
for the computation are: k_{Ca} = 0.088, 0.0866, 0.0865 (left)
and K_h = 130, 129, 128.2 µM (right).

reveals the fine structure of the successive doublets on the calcium records, as k_{Ca} (left) and K_h (right) are decreased from the stable continuous spiking regime to the bursting regime. Note that as the bifurcation parameter is decreased, we observe a cascade of period doublings; the approximate periods are T_0, $2T_0$, $4T_0$, where T_0 is the interspike interval of the basic spiking pattern in the uppermost trace. At bifurcation, the period exactly doubles. The bifurcation phenomenon is caused by a loss of stability in a periodic solution. This period-doubling cascade is evidently the route that leads to the chaotic bursting pattern of Fig. 4.

IV DISCUSSION

We have presented a theoretical model for the burst activity of insulin-secreting pancreatic β-cells. The model shows that combined spike and burst pattern can be created by using only single species of inward and outward currents. The inward current is inactivated by $[Ca^{2+}]_i$ and activated by $[H^+]_i$. Our simulated results look very similar to those observed in glucose-induced electrical bursts of the β-cells [1-6] when k_{Ca} is varied. Also, the results are consistent with pH_i-induced electrical activity of the β-cell [10,11] when K_h is varied. Thus, it appears from our simulations that our dynamical model provides a fairly accurate quantitative description of steady state burst in the pancreatic β-cell in response to glucose (see left panels in Fig. 1) and to pH_i (right panels). The model, furthermore, predicts different levels of intracellular Ca^{2+} concentration, depending on the strength of the electrical parameters, k_{Ca} and K_h. The controlling influence of these parameters on the magnitude of $[Ca^{2+}]_i$ makes it particularly interesting, since levels of $[Ca^{2+}]_i$ are implicated in glucose-induced insulin release [8]. The strength of pH_i is also implicated in insulin release rate [9,14].

In the transition regime between the continuous bursts and continuous spiking mode, our model exhibits aperiodicity, i.e. irregular spiking (left of Fig. 4) or irregular bursting (right). The chaotic electrical burst was observed in the whole-cell recordings of single β-cells by the patch-clamp technique [12]. In some cases, such as cardiac arrythmias or neuronal epileptic activity, physiological consequences of chaos are obvious, and aperiodic behaviour from deterministic biological systems may be direct functional consequences. In other cases such as chaos in the β-cell [3,12,24,25] functional meaning is less clear. For biophysical interpretation of experimental data like this, however, one need not invoke hypotheses about environmental noise to account for observed aperiodic behaviour. Nevertheless, it is tempting to speculate that the chaotic bursts simulated in our model are inherent in the electrical activity of single cells and the stable oscillatory cycle observed in intact β-cell owes it origin to electrical coupling or chemical coupling among the synchronised β-cells. The coexistence of multiple oscillatory states with different levels of $[Ca^{2+}]_i$ implies that when some β-cells are in spiking modes, $[Ca^{2+}]_i$ can be raised abruptly by brief perturbations. The current belief that intracellular Ca^{2+} ion is involved in insulin release may raise an interesting speculation about the existence of these multiple oscillatory states.

ACKNOWLEDGEMENT

This work was supported by NIH RO1 HL33905-01 and NSF PCM82.

REFERENCES

[1]   P.M. Dean and E.J. Matthews, J. Physiol. 210: 255-264 (1970).
[2]   H.P. Meissner, J. Physiol. Paris 72: 757-767 (1976).
[3]   P.M. Beigelman, B. Ribalet and I. Atwater, J. Physiol. Paris 73: 201-217 (1977).
[4]   B. Ribalet and P.M. Beigelman, Am. J. Physiol. 237: C137-C146 (1979).
[5]   I. Atwater, C.M. Dawson, A. Scott, G. Eddlestone and E. Rojas, in "Biochemistry Biophysics of the Pancreatic β-cell", ed. Georg Thieme, Verlag, New York, 100-107 (1980).
[6]   D.L. Cook, Fed. Proc. 43: 2368-2372 (1984).
[7]   P. Rorsman, H. Abrahamsson, E. Gylfe and B. Hellman, Fed. Eur. Biochem. Soc. Lett. 170: 196-200 (1984).
[8]   C.B. Wollheim and T. Pozzan, J. Biol. Chem. 259: 2262-2267 (1986).
[9]   C.S. Pace, J.T. Tarvin and J.S. Simith, Am. J. Physiol. 244: E3-E18 (1983).
[10] G.T. Eddlestone and P.M. Beigelman, Am. J. Physiol. 244: C188-C197 (1983).
[11] D.L. Cook and N. Hales, Nature 311: 271-273 (1984).
[12] P. Rorsman and G. Trube, J. Physiol. 374: 531-550 (1986).
[13] R. Eckert and J.E. Chad, Prog. Biophys. Molec. Biol. 44: 215-267 (1984).
[14] P. Lebrun, I. Atwater, L.M. Rosario, A. Herchuelz and W.J. Malaisse, Metabolism 34: 1122-1127 (1985).
[15] T.R. Chay, Biol. Cybern. 50: 301-311 (1984).
[16] T.R. Chay, Physica D 16: 233-242 (1985).
[17] T.R. Chay and J. Rinzel, Biophys. J. 47: 357-366 (1985).
[18] T.R. Chay, in "Biomathetics, Non Linear Oscillations in Biology and Chemistry", Lecture Notes in Biomathematics, Springer Verlag, NY (1986).
[19] T.R. Chay and Y.S. Lee, Biophys. J. 45: 841-849 (1984).
[20] T.R. Chay and Y.S. Lee, Biophys. J. 47: 641-651 (1985).
[21] A. Hodgkin and A.F. Huxley, J. Physiol. (London) 117: 500-544 (1952).
[22] A.C. Hindmarsh, "Ordinary Differential Equations Systems Solver", Lawrence Livermore Laboratory, Livermore, CA, Report (1974).
[23] E.J. Doedel, "AUTO86 User Manual, Software for continuation and bifurcation problems in ordinary diferential equations", second printing, California Institute of Technology, February (1986).
[24] J.C. Henquin, H.P. Meissner and W. Schmeer, Pflügers Arch. 393: 322-327 (1982).
[25] P. Lebrun and I. Atwater, Biophys. J. 48: 529-531 (1985).

BIFURCATIONS IN THE ROSE-HINDMARCH MODEL AND THE CHAY MODEL

Christian Kaas-Petersen

Centre for Nonlinear Studies and Department of
Applied Mathematical Studies
University of Leeds
Leeds LS2 9JT, England

ABSTRACT

We have examined a model of a neuron: the Rose-Hindmarch model and
a model of an excitable membrane: the Chay model. Chaos was known to
exist in both models. As chaos is (often) reached after a sequence of
bifurcations of periodic solutions, we have traced curves of bifurcation
points of periodic solutions, and thereby determined the regions within
which chaos can be found.

I INTRODUCTION

Rose and Hindmarch [5] observed chaos in a system of 3 ordinary
differential equations (ODEs) modelling neuronal behaviour for certain
values of the two parameters I and r. Chay [2] observed chaos in a
system of 3 ODEs modelling an excitable membrane for certain values of
the parameters I and $g_{K,C}^*$. We shall in both cases try to give an
account of how this chaos is established. We do it using computational
methods to map out curves of bifurcation points in these two-parameter
families of ODEs. In this way we can determine in which regions of the
parameter space we can expect chaos. It turns out that it is possible to
treat the Rose-Hindmarch model geometrically [1] giving some qualitative
insight. This model might then turn out to be the model which can give
use an in-depth understanding of how chaos is established.

II THE ROSE-HINDMARCH MODEL

We consider the system of 3 autonomous non-linear ODEs [5]

$\dot{x} = y-ax^3+bx^2+I-z$

$\dot{y} = c-dx^2-y$

$\dot{z} = r(s(x-x_1)-z)$ (1)

Time t is the independent variable, and the dependent variables are x,
the membrane potential, y, the recovery variable, and z, an adaptation
current. We have examined some of the bifurcations in dependence of the

two parameters I and r for b = 3.0 and b = 2.95. The constants were a = 1.0, c = 1.0, d = 5.0, s = 4.0, x_1 = -1.6.

III THE CHAY MODEL

We consider the system of 3 autonomous non-linear ODEs [2]

$$\dot{V} = g_I^* m_\infty^3 h_\infty (V_I - V) + g_{K,V}^* n^4 (V_K - V)$$
$$+ g_{K,C}^* C/(1+C) \cdot (V_K - V) + g_L^* (V_L - V) + I$$
$$\dot{C} = \rho \{ m_\infty^3 h_\infty (V_C - V) - k_C C \}$$
$$\dot{n} = (n_\infty - n)/\tau_n \qquad\qquad (2)$$

Time t measured in msec is the independent variable. The dependent variables are V, the membrane potential in mV, C, the dimensionless calcium concentration, and n, a probability of activation.

If we let y stand for h, m or n, then the explicit expressions for h_∞, m_∞ and n_∞ can be written

$$y_\infty = \alpha_y/(\alpha_y + \beta_y)$$

with

$$\alpha_h = 0.07 \exp(-0.05V - 2.5)$$

$$\beta_h = 1/(1+\exp(-0.1V-2)$$

$$\alpha_m = 0.1(25+V)/(1-\exp(0.1V-2.5))$$

$$\beta_m = 4.0\exp(-(V+50)/18)$$

$$\alpha_n = 0.01(20+V)/(1-\exp(-0.1V-2.0))$$

$$\beta_n = 0.125\exp(-(V+30)/80)$$

Also

$$\tau_n = 1/(230(\alpha_n + \beta_n)).$$

With these definitions we have $\dot{n} = 230((1-n)\alpha_n - n\beta_n)$. The parameters are I, the applied current in mV, and $g_{K,C}^*$, the maximal conductance of the Ca^+-sensitive K^+ channel divided by the membrane capacitance. The values of the constants are

$$g_I^* = 1800 \text{ s}^{-1} \quad g_{K,V}^* = 1700 \text{ s}^{-1} \quad g_L^* = 7 \text{ s}^{-1}$$

$$V_I = 100 \text{ mV} \quad V_K = -75 \text{ mV} \quad V_L = -40 \text{ mV} \quad V_C = 100 \text{ mV}$$

$$k_C = 3.3/18 \text{ mV} \quad \rho = 0.27 \text{ mV}^{-1}\text{s}^{-1}.$$

IV METHODS

First we shall consider methods for examining stationary solutions, next for periodic solutions.

We have developed a technique to trace curves of saddle node and Hopf bifurcation points of stationary solutions in two parameter families

184

of autonomous ODEs, formally written

$$\dot{x} = f(x;c), \quad x \varepsilon R^n, \quad c \varepsilon R^2, \quad f:R^n x R^2 \to R^n. \tag{3}$$

It is a step by step method to proceed along the curve. At each step a bisection is performed to locate the zero point of $Re\lambda$. Here λ is the eigenvalue near to criticality of $Df(x_s;c)$ (D is the derivative of f with respect to x) and x_s is the stationary solution satisfying $f(x_s;c) = 0$. These stationary solutions are found using Newton's method. Newton's method is an iterative process, where approximations to the root of $f(x;c) = 0$ is sought in a direction orthogonal to the 2 dimensional surface of stationary solutions implicitly given by $f(x;c) = 0$. Because we always move orthogonal to the surface, saddle node bifurcations and Hopf bifurcations can be treated with equal ease. In the case of Hopf bifurcations we have used the program BIFOR2 [4] to examine the stability of the bifurcating periodic solution. A thorough description of the method is under preparation [6].

We have developed a similar technique to trace curves of bifurcation points for periodic solutions of (3). In this case a point on the curve is given as a zero point of the function $|\lambda| - 1$ where λ is the eigenvalue of the derivative $DP(x_p;c)$ of the Poincaré map P at the periodic solution x_p. x_p is found as a zero point of the residual map $Q = P-I$ using Newton's method. As we are dealing with autonomous ODEs, both P and Q are defined on a hypersurface H in the state space R^n. Q is then a function of the n-1 variables in the hypersurface and the two parameters. The solutions of the ODEs were obtained using the code LSODA [8] with absolute error tolerance of 10^{-9}. A thorough description of the method is under preparation [7].

The bursting solutions we obtained by brute force simulation of the ODEs for 0<t<1000 starting at an appropriate initial condition and then plotting the solution for 1000<t<2000 whereupon the number of bursts were counted.

V   RESULTS FOR THE ROSE-HINDMARCH MODEL

First we consider b = 3.0 for which we have obtained the curves of bifurcation points shown in Fig. 1. Let us fix our attention to r = 0.015 and describe the sequence of bifurcations that takes place, when I is varied. For I above approximately 25 only stationary solutions exist. When I is decreased to 25 a Hopf bifurcation takes place, and a stable periodic solution bifurcates. If we follow the unstable stationary solution it becomes stable in a Hopf bifurcation at I = 1.51 - the bifurcating periodic solution is unstable.

The periodic solution bifurcating at I = 25 has become a one burst solution for I = 3.7, see Fig. 2. This periodic solution becomes unstable in a flip (period doubling) bifurcation at I = 3.57. If we follow the unstable one burst solution it becomes stable in a flip bifurcation at I = 2.32. The solution for I = 1.9 is shown in Fig. 2. This stable solution becomes unstable in a flip bifurcation at I = 1.42. We have not followed the unstable solution below I = 1.35. If we start close to the unstable one burst solution the trajectory is attracted to the stable stationary solution.

In the bifurcation at I = 3.57 a stable two burst solution is generated. It is shown in Fig. 3 for I = 3.4. This solution becomes unstable in a flip bifurcation at I = 3.36. If we follow the unstable two burst solution it becomes stable in a flip bifurcation at I = 2.78.

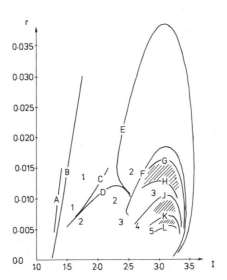

Fig. 1. Curves of bifurcation points in the Rose-Hindmarch system for
b = 3.0. "2", say, refers to a region of stable two burst
solutions. Chaos has been found in the hatched regions. The
curves of bifurcation points are labelled: A: 1 burst flip,
B: Hopf bifurcation, C: 2 burst flip, D: 1 burst saddle
node, E: 1 burst flip, F: 2 burst flip, G: 4 burst flip,
H: 3 burst saddle node, J: 3 burst flip, K: 4 burst flip,
L: 5 burst flip.

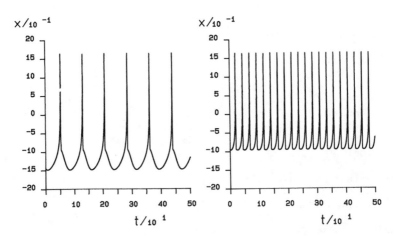

Fig. 2. The x-component versus time of the stable periodic solution
with 1 burst. To the left I = 1.9 and to the right I = 3.7;
r = 0.015 for both.

186

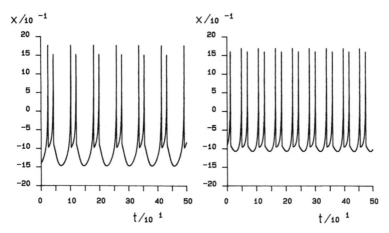

Fig. 3. The x-component versus time of the stable periodic solution
with 2 bursts to the left I = 2.5 and to the right I = 3.4;
r = 0.015 for both.

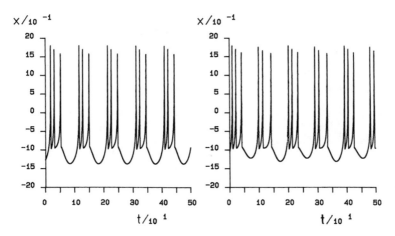

Fig. 4. The x-component versus time of stable periodic solutions.
To the left a 3 burst solution for I = 2.9, r = 0.01; to the
right a 6(= 3+3) burst solution for I = 3.1, r = 0.01

The stable solution at I = 2.5 is shown in Fig. 3. This solution
becomes unstable in a saddle node bifurcation at I = 2.17. We have not
followed the unstable two burst solution further. Although the one
burst and the two burst solutions were created in succession when decreasing I from above, they have different fates for low values of I. In
the region 2.17<I<2.32 they co-exist as stable solutions.

In the bifurcation I = 3.36 a stable four burst solution is generated, which becomes unstable in a flip bifurcation at I = 2.85, beyond
which point we have not followed it further.

We believe to have at full Feigenbaum sequence leading to chaos,
which has been observed for I = 2.9, 3.0, 3.1 and 3.2.

187

If we take the stable one burst solution at r = 0.010, I = 1.6 it becomes unstable in a saddle node bifurcation at I = 1.95. How this one burst saddle is linked up (if linked up at all) with the one burst solution existing for I>3.55 has not been resolved yet.

For r = 0.010, I = 2.9 we have stable three burst solution, see Fig. 4, which became unstable in a flip bifurcation at I = 2.99 and stabilizes again in a flip bifurcation at I = 3.21, later followed by a saddle node bifurcation at I = 3.29. The stable six burst solution for I = 3.1 is shown in Fig. 4. If we follow this solution for constant I decreasing r it meets a flip bifurcation explaining the chaos which we have observed. For r = 0.007, I = 3.0 a four burst solution exists. If r is decreased a period doubling takes place leading to chaos; if r is increased a saddle node bifurcation takes place at r = 0.0079.

For r = 0.005, I = 3.0 a five burst solution exists, which meets a flip bifurcation at I = 3.02.

The chaotic solutions at I = 3.0, r = 0.008 and I = 3.0, r = 0.006 are shown in Fig. 5.

The appropriate initial conditions for obtaining burst solutions were x = 0.0, y = 5.0, z = I-0.1.

When b = 2.95 parts of the one, two and three burst flip bifurcation curves are shown in Fig. 6.

VI   RESULTS FOR THE CHAY MODEL

Chay studied the equations (2) for I = 0. For $g_{K,C}^{*}$ = 30 a stable stationary solution exists. This solution becomes unstable at $g_{K,C}^{*}$ = 27.2 in a Hopf bifurcation. The bifurcations periodic solution is unstable. If we follow the stationary solution, it becomes stable in a Hopf bifurcation at $g_{K,C}^{*}$ = -7.8, where the bifurcating periodic solution is stable, see Fig. 7.

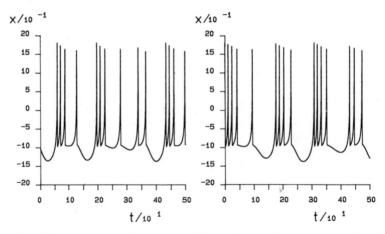

Fig. 5.   The x-components versus time of stable chaotic solutions. To the left r = 0.008 and to the right r = 0.006; I = 3.0 for both.

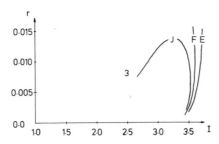

Fig. 6. Some curves of bifurcation points in the Rose–Hindmarch system for b = 2.95.

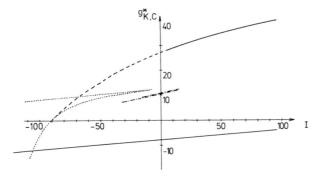

Fig. 7. Curve of bifurcation points for the Chay system. Solid line means Hopf bifurcation with stable periodic solution bifurcating, dashed line is Hopf bifurcation with unstable periodic solution bifurcating, dotted line is saddle node bifurcations and dash–dotted lines are period doublings.

For I = -20 the unstable stationary solution goes through two saddle node bifurcations near $g^*_{K,C}$ = 11.5. For I = +20 the bifurcating periodic solution at $g^*_{K,C} \cong$ 30.5 is stable.

The curve of saddle node and Hopf bifurcations join near I = 90, $g^*_{K,C}$ = 0. It is known [3] that a curve of homoclinic orbits starts from that point. We have not traced that curve.

The stable periodic solution generated at I = 0, $g^*_{K,C}$ = -7.8 becomes unstable in a flip bifurcation for $g^*_{K,C}$ = 10.53. The stable periodic solution with the double period becomes unstable at $g^*_{K,C}$ = 10.73.

We do not know at present, whether the periodic solution generated at $g^*_{K,C}$ = -7.8 is related to the periodic solution generated at $g^*_{K,C}$ = 27.2.

VII DISCUSSION

Almost all the curves of bifurcation points are open ended, as are the paths of solutions. The curves stops because we have not pursued them further. For small values of r in the Rose–Hindmarch system computational difficulties arise because of differences in order of magnitude of the (x,y) and z components of equation (2).

There are several regions of chaos in the Rose-Hindmarch system. The region of chaos is bounded partly by curves of flip bifurcations and partly by a curve of saddle node bifurcations. The chaos reported earlier [5] was found at I = 3.25, r = 0.005, and seems to be a result of a large number of competing unstable solutions more than a single Feigenbaum sequence as I is decreased. There are also several regions of different co-existing stable bursting solutions. The chaos found by Chay seems to follow directly from the period doubling cascade taking place when $g_{K,C}^*$ is increasing. We have not yet mapped out the bursting patterns of the Chay system. We see that a study of the equations for unphysiological values of $g_{K,C}^*$ is helpful to get an overall picture.

We hope by this kind of analysis to establish data, which might be helpful towards understanding the behaviour of the solutions.

ACKNOWLEDGEMENTS

I would like to thank Dr. Arun Holden, CNLS, Leeds for bringing these equations to my attention and for all ensuing discussions.

REFERENCES

[1] T. Bedford, Private communication (1986).
[2] T.R. Chay, Chaos in a three-variable model of an excitable cell, Physica 16D:233 (1985).
[3] J. Guckenheimer and P. Holmes,"Nonlinear oscillations, dynamical systems, and bifurcations of vector fields", Springer-Verlag, New York (1983).
[4] B.D. Hassard, N.D. Kazarinoff and Y.-H. Wan, "Theory and applications of Hopf bifurcation", London Mathematical Society Lecture Notes Series No. 41, Cambridge University Press, Cambridge (1981).
[5] J.L. Hindmarch and R.M. Rose, "A model of neuronal bursting using three coupled first order differential equations", Proc. R. Soc. Lond. B221:87 (1984).
[6] C. Kaas-Petersen, Technique to trace curves of bifurcation points of stationary solutions (in preparation).
[7] C. Kaas-Petersen, Technique to trace curves of bifurcation points of periodic solutions (in preparation).
[8] L. Petzold, Automatic selection of methods for solving stiff and nonstiff systems of ordinary differential equations, SIAM J. Sci. Stat. Comput. Vol. 4:136 (1983).

DENDRITIC BRANCHING PATTERNS

A.K. Schierwagen

Carl-Ludwig-Institut fur Physiologie
7010 Leipzig
Liebigstrasse 27, DDR

Collicular neurones impress strongly by the specific branching mode
of their dendrites. In dependence on localization within the superior
colliculus (SC) generalized (isodendritic) and specialized (allodendritic)
branching patterns can be found in the deep and superficial layers, respectively. These structural differences are correlated with distinct
functional roles in information processing: deep layer neurones (TRSNs)
get multimodal afferences, whereas superficial layer neurones (SLNs)
receive exclusively unimodal (visual) inputs [11,12]. As in biology in
general, the structure-function relationship is the central question in
neurobiology. The paradigm of self-organization currently being discovered in physics and biology should allow to overcome the undialectic
separation of structure (or form) and function which arose historically
in biology due to the absence of precise concepts [7]. The unity of
structure and function is a fundamental of any theory of self-organization: the developing structure corresponds to its function, and vice
versa.

The study of non-equilibrium growth models based on simple diffusion
processes [9] may contribute to a better understanding of self-organization in biological systems, especially the nervous system. Since there
is some evidence that neuronal growth is controlled by the diffusion of
biochemical substances as nutrients or nerve growth factors [10], the
diffusion-limited aggregation (DLA) model of Witten and Sander [14] may
provide new insights into the process of self-organization at the single-
neurone level.

Many variants of DLA models have been studied by computer simulations in order to describe such diverse phenomena as dendritic solidific-
ation, electodisposition and viscous fingering [13]. Common to all of
these phenomena is their fractal nature, i.e. they exhibit invariance
under changes in spatial length scale [8]. The following findings are
of particular importance in the present context:

1. There is numerical evidence that DLA shows remarkable anistropy
    effects which can be quantified by fractal exponents related to
    growth directions [1].

2. A direct dependence of dynamic transport properties on the geometry
    of a tree-like fractal structure can be stated [5].

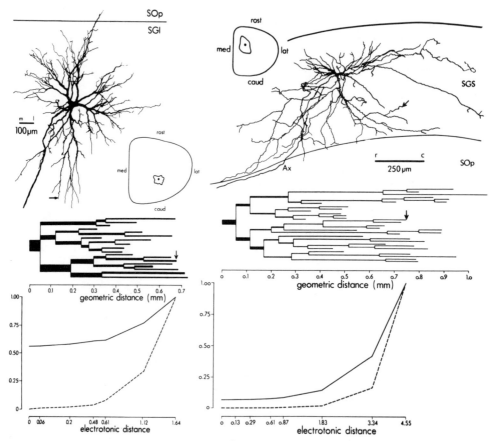

Fig. 1. Dendritic arborization and transfer properties of a TRSN (left)
as compared with a SLN (right). The distinct structural patterns (above) are mirrored by the corresponding transfer ratios
which has been calculated for a tonic synapse located on a
dendritic termination (arrow). Whereas voltage transfer
(dashed line) is negligible in both neurones, current transfer
(continuous line) is considerably higher, but significantly
different in TRSN (56%) and SLN (7%).

The first result could serve as basis for a new, clear interpretation of the term 'positional information' introduced by Wolpert [15] in
order to explain cell differentiation and morphogenesis: once the migra-
tion of a postmitotic neurone has finished at its definitive location in
the SC, neuronal cytodifferentiation will take place according to the
special isotropic or anisotropic diffusion conditions imposed by that cell
position. In fact, the large-scale shape of SLNs resembles that of DLA
clusters grown with anisotropic sticking rules, where that of TRSNs
corresponds to DLA under isotropy conditions, as one could expect it if
neuronal growth were governed by DLA-like rules.

Point 2 above is important because it shows a way how to relate the
geometry of neurone to its integrative, signal-processing capabilities
which are believed to underlie the characteristic behaviour of the nervous
system [6]. Calculations performed with a cable segmental model show that
SC-neurones of different types exhibit distinct synaptic transfer properties (Fig. 1).

Any application of DLA model results to neurone growth must, however, take into account that neurones are not self-similar. They might be nevertheless, non-uniform fractals, i.e. composed of parts of varying fractal dimensions. This view is supported by findings both from in-vitro and in-vivo experiments which suggest that both dendritic growth occurs in two stages. In the first phase, intrinsic factors (of unknown nature as yet) dominate, where in the second phase environmental cues bring about the definitive branching pattern by remodelling [2,3]. Determination of 'effective' fractal dimensions of neuronal dendrites refers, therefore, to the first growth phase. Preliminary results demonstrate an effective dimension of $D = 2.1$ and $D = 2.4 - 2.7$ for TRSNs and SLNs which should be compared with $D = 2.5$ for 3-dimensional DLA [9].

In conclusion, DLA may provide an appropriate description of the first phase of neuronal dendritic growth, showing at the same time 'genetic' factors to be dispensible in explaining neuronal form development. Instead, simple diffusion processes with special boundary restrictions imposed by the embedding brain structure can lead to a variety of physical forms, in accordance with the requirements of an adequate morphogenetic theory [4].

REFERENCES

[1]   R.C. Ball, R.M. Brady, G. Rossi and B.R. Thompson, Anisotropy and cluster growth by diffusion-limited aggregation, Phys. Rev. Lett. 55:1406-1409 (1985).

[2]   M. Berry, P. McConnel and J. Sievers, Dendritic growth and the control of neuronal form, in 'Current Topics in Developmental Biology', Vol. 15, ed. K. Hunt, pp. 67-101, Academic Press, New York (1980).

[3]   W.M. Cowan, Aspects of neural development, in International Review of Physiology, Neurophysiology III, Vol. 17, ed. R. Porter, pp. 149-191, University Park Press, Baltimore (1978).

[4]   B.C. Goodwin, What are the causes of morphogenesis? Bioessays 3: 32-36 (1986).

[5]   S. Havlin, Z.V. Djordjevic, I. Majid, H.E. Stanley and G.H. Weiss, Relation between dynamic transport properties and static topological structure for the lattice-animal model of branched polymers, Phys. Rev. Lett. 53:178-181 (1984).

[6]   J.J.B. Jack, D. Noble and R.W. Tsien, 'Electric Current Flow in Excitable Cells, Clarendon Press, Oxford (1975).

[7]   R. Levins and R. Lewontin, 'The Dialectical Biologist', Harvard University Press, Cambridge, Mass. and London, England (1985).

[8]   B.B. Mandelbrot, 'The Fractal Geometry of Nature', Freeman & Co., San Francisco (1983).

[9]   P. Meakin, A new model for biological pattern formation, J. theor. Biol. 118:101-113 (1986).

[10]   A. Prochiantz, Neuronal growth and shape, Dev. Neurosci. 7:189-198 (1985).

[11]   A. Schierwagen, Segmental cable modelling of electronic transfer properties of deep superior colliculus neurons in the cat. J. Hirnforsch. 27:679-690 (1986).

[12]   A. Schierwagen and R. Grantyn, Quantitative morphological analysis of deep superior colliculus neurons stained intracellularly with HRP in the cat. J. Hirnforsch. 27:611-623 (1986).

[13]   T.A. Witten and M.E. Cates, Tenuous structures from disorderly growth processes, Science 232:1607-1612 (1986).

[14]   T.A. Witten and L.M. Sander, Diffusion-limited aggregation, a kinetic critical phenomenon, Phys. Rev. Lett. 47:1400-1403 (1981).

[15]   L. Wolpert, Positional information and the spatial pattern of cellular differentiation, J. theor. Biol. 25:1-47 (1969).

CHAOS AND NEURAL NETWORKS

E. Labos

Semmelweis Medical School
1st Dept. of Anatomy
Tuzolto-u 58, Budapest, Hungary

ABSTRACT

Nervous systems are strongly connected networks of building modules the neurons. The irregular or periodic neural autoactivity might take its origin either from units or networks.

The formal concept of network presented here is a separable system of a finite number of units or component variables (cells). These 'state-variables' interact, similarly to the nerve cells. This means that their future is determined by the past history of a set of other variables. In models the different sets of formal variables may correspond either to real units (cells) or real networks of neurons.

In specific cases regular (stable) units may become irregular or unstable when coupled into nets. In other examples unstable or irregular unit activities may turn into stable or periodic functioning. The prediction of the fate of the interconnected units in a network in some cases is possible. Nevertheless, no general theory of the possible consequences of interconnections is now available.

INTRODUCTION

Nervous systems are strongly connected networks of cellular units, the neurons. The electrical signs of the propagated neuronal activity are called spike potentials. A special and relatively frequent class of neuronal firing is regarded 'spontaneous'. Such an intrinsic activity might take its origin either from networks or units or even parts of neurons [1]. These factors are called pacemakers of the autoactivity.

Spike sequences normally show simple or complicated 'regularities' or 'patterns'. Although the term 'periodic' is widely used in neurophysiology, mathematically exact periodicity in fact is not observable. Nevertheless, various kinds and degrees of 'regularities' are distinguishable in practice.

The main objective of this paper is to investigate the role of unit- or network-factors in the generation of both regular and more or less irregular spike-patterns. The lack or presence of 'chaotic' mechanisms at normal or pathological conditions in the neural dynamics is still an undecided problem [6].

Experiments with model neural nets [7,17] supports the intuitive view, that more lengthy described patterns require for generation either larger or more densely and specifically interconnected networks [10]. On the other hand chaotic interval maps [3,24] as spike generator units may have universal capability of generating interspike interval patterns [9, 12], which apparently contradicts to the former view.

II A GENERAL CONCEPT OF NETWORK

A general and formal concept of network corresponds to a system, separable into a finite number of units or component variables. These state-variables interact, similarly to nerve cells. Using the terminology of abstract machine theory, the networks are machines built of parts [18] or are essentially multivariable multivalue (explicit) functions. Their behaviour is obtained either by iteration of this function or by solution of a finite system of ordinary differential equations (ODEs). This means that the future of each component is unambiguously determined by the past history of a set of other variables.

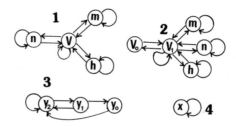

Fig. 1.   Examples of 'dependence-graphs' or 'network-structure':
1 - Hodgkin-Huxley (HH) nerve equation, current clamp; 2 - HH-module in travelling wave case; 3 - non-linear iteration of order 3 defined by equation (4) and 4 - PLMI unit of equation (3).

This concept is general enough to enclose even certain ODEs like some versions of the Hodgkin-Huxley [5] (HH) nerve-equations. Thus the gap between the concepts of units and their network is made narrower.

The wiring or directed 'dependence' graph of a net is defined by the vertex-set of variables. Edges are drawn to variable x from a set of variables if the next value of x is explicitly and directly depends on the values of the states in the given set.

When solving ODEs the iteration determined by the Euler-method also defines a 'wiring graph'. Such graphs of the HH equations are depicted in Fig. 1. Although these 'structural graphs' contribute they do not provide powerful help to the forecasting of behaviour of (non-linear) dynamics. This is partly due to the fact that each 'unit' may have individual equilibrium and stability pattern.

III VARIOUS MODELS OF NEURAL UNITS

The following functions are used in this paper as building modules of networks:

(i) McCulloch-Pitts (MCP) neurons, i.e. threshold gates - see [19,11]: a neuron is a truth-function computable as

$$f(b) = u(bh-T) \tag{1}$$

where $b \in B^n$ Boolean cube, $h \in R^n$, bh is an inner product, $T \in R^1$ is the threshold number and $u(r) = 1$ iff $r>0$ or it is zero otherwise. Their network, a $B^n \to B^n$ map consists of n independently specified neurons. Its interactions generate the state transitions as follows

$$b_{n+1} = u(b_n N - \theta) \tag{2}$$

where $N = (a_{ij})_{n \times n}$ square matrix of interconnections, $\theta \in R^n$ is the threshold vector and the u is applied by coordinates.

(ii) Piecewise linear maps of interval (PLMI): the neurons are defined on R^1. The next state function is computed by the iteration as follows [12,9,20,21]

$$x_{n+1} \begin{cases} = m_0 x_n b_0 & \text{if } x_n \ (-\infty,0) \ \text{(line } L_0) \\ = m_1 x_n + b_1 & \text{if } x_n \ (0,a) \ \text{(line } L_1) \\ = m_2 x_n + b_2 & \text{if } x_n \ (a,+\infty) \ \text{(line } L_2) \end{cases} = \mathcal{F}(x_n) \tag{3}$$

where $0<m_0<1$ and b_0 is small positive value. These conditions guarantee the stability below the 'resting zero value' and jump from zero to a positive value. The L_2 line is defined as follows: $m_2 = 1/(a-1)$ and $b_2 = -m_2$. The L_1 line is defined either as 'pacemaker' or 'silent' corresponding to $m_1>1$ or $m_1 \leqslant 1$, while $b_1 = 0$.

(iii) A non-linear 3-order iteration for generation of spike sequences [9,13]

$$y_{n+1} = y_n + a(y_n-p)(y_n-r)(y_n-s) + cy_{n-2}^3 + by_{n-1}^4 + k/(y_n-s) \tag{4}$$

where a typical set of parameters resulting in 'rhythmic discharges' is as follows: $a = -1.2$, $p = 1$, $r = 0$, $s = -0.2$, $c = -0.675$, $b = 0.225$, $k = 0.0016$. The system might have 3, 4 or 5 real equilibrium points.

Such units are called polynomial spike generators (PSG). Depending on the b, c and k parameters it shows a broad spectrum of more or less regular spiking, non-damped or damped oscillations and complicated irregular strange, exotic or chaotic motions. The units were coupled into nets either by an additive term or the values of b, c or k were controlled directly and selectively.

IV THE COMPLEXITY OF NEURONAL BEHAVIOUR: PSEUDORANDOMNESS AND PSEUDO-COMPLEXITY?

Is it a tenable view that the 'simple' or 'complex' spike sequences are outputs of 'simpler' or 'complicated' machines respectively? The Kolmogorov-Chaitin (algorithmic) complexity of an Y (binary) sequence with respect to an X input tape is the shortest one of the programs capable of generating Y [8,2]. In this theory Y is regarded random if its complexity is near to its size, that is if Y is not 'compressible' by a shorter program.

The following questions emerge: (1) in Section V binary sequences as 'activities of component neurons' are to be mentioned and generated with the smallest threshold gate nets [10,9,14]. This sequences are the 'most complicated' ones available by MCP-networks of a given size. Although the size of their generating net i.e. the number of its components is a 'measure' of their complexity, the component-flows may have different recursive order, i.e. the length of past history which makes the next digit computable. This is an additional parameter to characterize their complexity. (2) In case of PLMI exact periodic solutions may be universally generated in a very unstable ephemeric manner because of the extreme sensitivity to initial values [9,12]. The initial value computable for a given 'simple' or 'complex' periodic spike pattern is essentially a translated version of the interspike interval pattern to be obtained. Thus their generation needs an additional computation i.e. the encoding of the pattern into an initial value. Are these very unstable output patterns more complicated in some sense than the 'irregular' but very short program generated output number? A question is how to include the initial conditions or stability into the complexity measures?

The interval maps defined by short algorithms to generate spikes and thus represent a low Kolmogoroff-complexity, could display as complex dynamics as we desire at least in terms of finite interspike interval lengths [9,12]. What kind of complexity measures are involved in this dilemma of 'simple versus complex' raised by May [12]?

V MAXIMAL AND OPTIMAL CYCLES IN Mc CULLOCH-PITTS NETWORKS. TRANSIENTS OR LONG CYCLES AS SPURIOUS APERIODICITIES.

The state space of these systems is the n-dimensional Boolean cube. State-transitions are defined by the network-relationship (2).

Why have such finite systems relevance to an issue of 'networks and chaos'? It is because in the practice of 'empirical computer dynamics' the very long-lasting transients and the very large cycle lengths cannot effectively be distinguished from the kinds of aperiodic behaviour.

Table 1. A set of threshold gate network specifications: matrices and threshold vectors for dimensions n = 2,..,9. Using the iteration specified in (2), the lengths of cycles generated are 2^n.

```
MATRIX      MATRIX       MATRIX        MATRIX        MATRIX
0-1          1 1 1        1 1-2 1       3-1-1 1 1     -1 1 1-1-1 4
1 0         -1 1 1       -1-1-1 2       1-3 1-1-1    -1 1 1 4-1 1
THRESHOLD   -1 1-1       -2-1-1-1      -1-1-1 1 3    -1 1 1 1 4 1
0-1          THRESHOLD    .1-2 1 1      1 1 3-1 1    -1-4 1 1 1 1
            -1 1 0        THRESHOLD     1 1-1-3 1     4-1-1 1 1 1
                         -1-2-2 1       THRESHOLD     1-1-4 1 1 1
                                        2-2 0-2 2     THRESHOLD
MATRIX                                                0-2 0 3 2 4
-1-1-1-1 1 1 7-1 1       MATRIX
-1-1 1 7 1 1 1-1 1       1-1-1-1 1 6-1-1              MATRIX
-1-1 1 1 1-7 1-1 1       1-1-1-1 1 1 6-1              5-1 1 1 1-1-1
-1-1 1 1-7-1 1-1 1       1-1 6-1 1 1 1-1             -1 1-1 5-1 1-1
 7-1 1 1-1-1 1-1 1       1-1 1-1-6 1 1-1              1 1-1-1-1 1-5
 1-1 1 1-1-1 1 7 1       1-1 1-1-1 1 1 6              1 1-5-1-1 1 1
 1 7 1 1-1-1 1 1 1      -6-1 1-1-1 1 1 1              1 1 1-1-1 5 1
 1 1 1 1-1-1 1 1-7      -1-1 1 6-1 1 1 1              1 5 1-1-1-1 1
 1 1-7 1-1-1 1 1-1      -1 6 1 1-1 1 1 1              1-1 1-1-5-1 1
THRESHOLD               THRESHOLD                     THRESHOLD
 3 1-1 6-5-6 7 2-1      -2-1 4 0-4 6 5 2              4 3-2 0-5 2-2
```

198

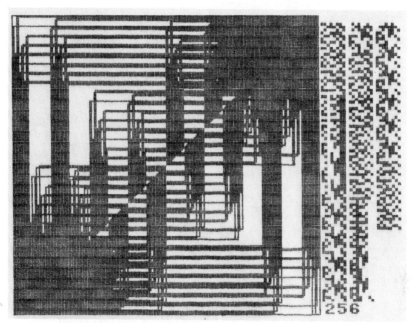

Fig. 2. State transition diagram of an 8 dimensional threshold-gate
network which produces single and L = 256 (i.e. maximum)
length of cycle. Decimal codes of consecutive binary vectors
are plotted and the (x,x), (x,y) or (y,y) points are connected
by lines. Outside the box the punch-tape of states in the
cycle is shown. The network is 'self-dual'. Specifications
of design: columnar permutation = (42856137), partial
negation code = 249 (instructions in ref. [15]).

This section will demonstrate or mention maximum-cycle ($L = 2^n$) and
'optimal' ($L > 2^{n-1}$) length generator McCulloch-Pitts networks of which the
design problem was posed in 1975 by the author. The first results of
dimension 3 and 4 were published in 1980, cases of n = 5 and 6 in 1984,
while n = 7,8,9 in 1985 and 1986. The optimal case is solved for $n \leqslant 72$.
These results were achieved on the basis of a necessary and sufficient
condition for threshold gate nets to be 'free of transient states' and
also with extensive computer-aided searching for special networks. The
low speed of getting these results illustrates the 'practical tract-
ability' of similar synthesis problem(cf. [4]). Table 1 shows solutions
up to 9 dimension. A case of dimension 8 is also depicted in Fig. 2.
Higher dimensional optimum nets were obtained by introducing a new method,
the selective modification of the transitions of the so-called 'marginal
states' [13,14,15].

In the Appendix it also will be proven that cycle lengths being
exponential function of the number of neurons may be designed for every
n by the iteration of two other methods (1 - doubling or prolonging the
cycle-length by 1 or 2 additional state-recognizer or command neuron;
2 - independent clocks of relative prime lengths). For each n a cycle-
length $L = c\sqrt{(2^n)}$ is available.

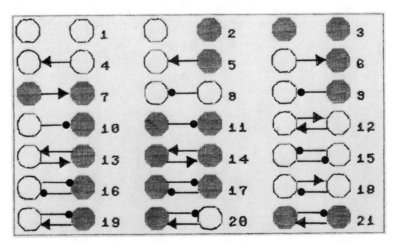

Fig. 3.   The 21 different 2v2e coloured directed graphs at n = 2 as
          defined in Section VI. They represent 'detailed structure'
          of nets: 2 sorts of interconnections and 2 sorts of units.
          At n = 3 1032 such cases exist.

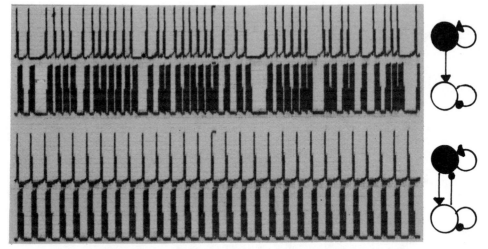

Fig. 4.   A pacemaker-PLMI (trace 1) - with parameters 0.1, 0.02; 1.20,
          0; -1.111, 1.1111; 0, 0.1 - excites (by a 0.08 coefficient)
          a similar but silent PLMI (trace 2) without feedback. The two
          last traces show the same but with -0.08 negative feedback.
          Other data: m_1 = 0.9 a = 0.02 for silent unit. Initial
          values are: 0.05, 0. Observe the 'regularization' of
          originally chaotic pacemaker.

VI SMALL NETS BUILD UP FROM INTERVAL MAPS AS UNITS

The units are the PLMIs, each defined by 3 lines in 3 intervals, i.e. by 9 parameters. Networks of 2,3 and 4 units are investigated. At first the 'structure' or 'wiring' graphs are specified assigning two 'colours' both to vertices and edges. The colours may correspond to pacemaker versus silent units or excitatory and inhibitory lines respectively. This species of graphical objects are called 2v2e labelled or coloured di(rected) graphs. At n = 1,2,3,4 it can be enumerated 1,2,4 or 11 graphs, 1,3,10 or 66 2-multigraphs, 1,3,16 or 218 digraphs, or finally 2, 21,1032 or about 360000 2v2e-digraphs or 'structures'. The 21 structures of 2 neurons are depicted in Fig. 3. The immense diversity of essentially different structures involves also that the number of 'interesting' cases cannot be significantly reduced. In addition these 'structures' provide only ambiguous information about the activity patterns.

Some interesting nets were investigated in detail:

(1) A chaotic pacemaker activates a silent unit whose activity is fed back by inhibition to the pacemaker. At suitable choice of parameters, the chaotic driver becomes 'regular'. A case is depicted in Fig. 4. At weaker interaction the synchronization appears without 'regularization', at stronger coupling both of the activities become regular, i.e. 'almost periodic'. When positive feedbacks applied together with pacemakers, then irregularities may appear (Fig. 5).

(2) Silent units were coupled by excitatory lines into a ring and one of them was activated. At small coupling coefficients the activity dies out ('dissipates') but it can be maintained by adequate parameter assignment

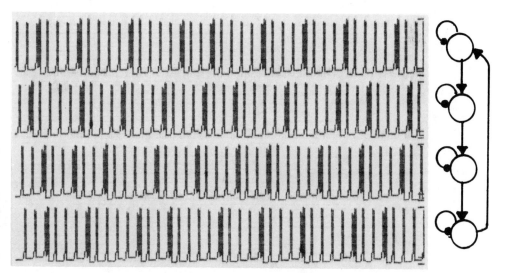

Fig. 5. Coupling four identical non-pacemaker PLMI units into a single ring by excitatory term with factor 0.12 added to the unit's equations. Parameters of the three lines of units (see relations (3)): 0.1, 0.01; 0.99, 0; -1.111, 1.111; thresholds are 0 and a = 0.1. Only spikes were left propagating and during spikes a refractoriness was introduced. The same holds for interactions of Fig. 4. The case demonstrates non-vanishing 'patterned' activity.

(Fig. 5). The lifetime increases with the size of network. The case corresponds to the classical reverberation.

(3) Pacemakers of different or identical degree of 'resting instability' were coupled to mutually inhibit each other. In such nets an alternation of activities occur, a kind of 'competition'. It may occur also that the unit started with handicap could never discharge ('handicap effect' analogous to the hairy behaviour of locally coupled growth equations..).

(4) Nogradi [22] observed the sensitivity to time constants and delays: e.g. parameter choice even for simple On- or Off- circuits is unexpectedly delicate.

(5) Beyond wiring cases four types of interactions were studied: (a) a positive propagation threshold was or was not introduced and (b) the receiver unit was or was not settled unresponsive (refractory) during the spike. The main conclusion with respect of the 4 combinations of these 2 factors is that in some cases the patterns are very sensitive; in other cases are not, or are hardly sensitive to the introduction of refractoriness or progagation threshold. The conditions of this sensitivity are not yet completely clarified. The coupling term (a matrix entry) is multiplied by two 'threshold term' expressing the effectiveness of emitted signal only above a value and also the effectiveness only at non-spike values of the signal-acceptor neuron.

(6) Evoked burst activity can be obtained with low threshold of burster (Fig. 4).

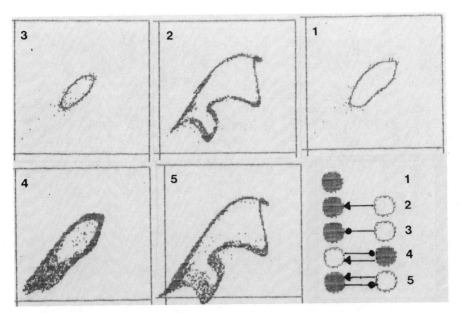

Fig. 6. New behaviours appear in networks of two identical PSG units defined in equation (4). Parameters: p = 1, r = 0, s = −0.2, a = −1.2, b = 0.225, c = −0.55, k = 0.019. Comparison of uninfluenced (1), excited (2 and 5) or inhibited (3 and 4) units. The coupled terms were multiplied by 0.1 or −0.1 and added to the generator term. Next values are plotted against their predecessors. Trajectories of dark units are displayed.

(7) The order of linear recursions in the relation (3) was increased
also to order two. This permitted to introduce various degrees of
accommodation of the units. The new factor seems to be significant in
the fine 'shaping' of patterns.

VII   EXAMPLES OF INTERACTING PSG UNITS: NEW BEHAVIOURS IN NETWORK

Two identical PSG units were coupled by adding terms of the other's
effect multiplied with inhibitory or excitatory (- or +) coefficients.
Five stroboscopic plots were compared: those of the uninfluenced (1),
the excited (2), the inhibited units (3) and finally the behaviours of
those units which were coupled into a net where the excitatory unit is
inhibited by a feedback term (4 and 5). In Fig. 6 the remarkable differences of the 5 cases are demonstrated. The plots of controlled cases are
'similar' but in fact are new compared to the free-running behaviours.
When two non-identical units are coupled then not 5 but 10 cases have to
be compared in order to reveal the novel motions. Experiments with such
cases support the previous findings.

If the control of a unit in a network is not a constant shift of a
parameter but it is an influence by values changing 'in time' then new
behaviour may emerge. This is evoked by a 'wandering in the original
parameter-space' or even by the emergence of radically new features not
explainable in terms of the original dynamics that is by those of the
units.

VIII   THE WAYS OF INTERACTION: PARAMETER-CONTROL AND ADDED INTERACTIONS

The difference between the added term and parameter-control is the
best demonstrated by a simple linear process of first order. Let the
uncontrolled process be as follows

$$x_{n+1} = mx_n + f(1-m) \tag{5}$$

where x_i is the state-variable, m is the speed-constant and f is the
fixed point of the system. Two kinds of parameter control are applicable,
either separately or together: (i) in the equilibrium-control, f is
substituted by a changing f_n process and (ii) in the speed-control, the
same is made with m by an m_n-process.

The parameter control makes the forecasting better, while adding
a y_n process multiplied by an 'amplification-factor' or 'synaptic-
coefficient' often leads to surprising results. All control cases are
summarized in a single expression for shortness

$$x_{n+1} = m_n x_n + f_n(1-m_n) + cy_n \tag{6}$$

The m_n, f_n or y_n processes may represent 1,2 or 3 other units of
which the activity might be also influenced by feedback from x.

In non-linear cases the equilibrium or follower control cannot be
well separated from the speed-regulation. This contributes to the
difficulties of forecasting in non-linear dynamics.

IX   NEUROBIOLOGICAL RELEVANCE: NORMAL IRREGULARITIES AND ABNORMAL
DISCHARGES

Model experiments do not permit the exclusion of either 'chaotic' or
'quasiperiodic' dynamics even in the case of normal irregularities of

spike sequences. Whether the chaos is an easily recognizable phenomenon [23] or not, remains a problem. The abnormal (mainly 'epileptic') neural discharges occurring in the clinical practice or experimental pharmacology represent fields for application of chaos theory as well.

APPENDIX

Theorem

For every n there exists periodic sequence of states in the iteration of threshold-gate network of form (2) of which the length is an exponential function of the dimension n.

Proof

Table 1 shows that both for small odd and even n numbers of neurons the $L = 2^n$ maximum length can be reached. Let N be such a net of n neuron and $L = 2^n$ cycle length. Choose s and t, two successive states in its cycle. Synthesize a new (n+1-th) neuron A, which would selectively recognize the state s and activates itself with a sufficiently high positive weight. Call the new net M. In this net a transient of length L ended in a cycle of length L will appear: $(t,0),\ldots(r,0),(s,0),(t,1)$, $\ldots(r,1),(s,1),(t,1)$. Introduce another (n+2-nd) neuron – called B – which 'observes' the M extended net and is activated selectively by the state $(r,1)$ which precedes $(s,1)$. The new net of n+2 neurons goes through the following state sequence: $(t,0,0),\ldots,(r,0,0),(s,0,0),(t,1,0),\ldots,$ $(r,1,0),(s,1,1)$. This permits that as soon as B is first activated it can control the observed net, i.e. a 'sufficiently strong' positive and negative innervations have to be designed to reset this $W = N \cup A \cup B$ net into the state $(t,0,0)$ just after the state $(s,1,1)$.

The selective recognition of an arbitrary state s can be designed by +1 or -1 weights at 1 or 0 coordinates| of the state to be recognized respectively. The threshold have to be settled to the number of 1-s minus 1 of the state s. The control is also designed simply using high + or - coefficients to evoke absolute activation or depression of the desired components.

Starting with the (n,L) values we get by the outlined process (n+2, 2L). Iterating this process $(n+2k, 2^k L)$ can be reached. It follows that if u is the number of neurons, then the reachable length

$$L(u) = c.2^{int(u/2)} = d.2^{u/2} \quad \text{where } d = 2^{n/2}, \; c = 2^{int(u/2)+r} \qquad (7)$$

and $u = n+2k$, $r = 0.1$. Table 1 proves that $c = 16$ or $c = 32$ is reachable depending on the start (n,L) which may be either (8,256) or (9,512). The proof evidently holds both for even and odd values of u.

The constructions of this proof are illustrated in Fig. 7. The design which results in doubling of the original length of cycle by two supplementary units has to be repeated with the resulting net again ad infinitum. A similar method permits to increase the exponent of 2 until 0.9u. An L and an L+1 length (primes to each other) was iteratively designed: start with (n,L) and step to (2n+1,L(L+1)). Any length polynomial function of dimension is outdone with these constructions but the maximum is not yet reached for arbitrary large n.

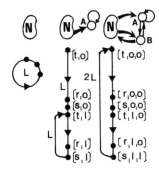

Fig. 7. Synthesis of one network in a sequence of nets whose
behaviour includes exponentially increasing length of
cycle. The synthesis is decomposed into three steps:
take an initial net (N), supplement by an A and a B
recognizer unit. The critical state-transitions are
depicted below the nets. See Appendix.

REFERENCES

[1]   T.H. Bullock, The origins of patterned nervous discharges,
      Behaviour 17:48-59 (1961).
[2]   G.J. Chaitin, Randomness and Mathematical Proof, Sci. Am. May:
      47-52 (1975).
[3]   P. Collet and J.P. Eckmann, Iterated Maps on the Interval as
      Dynamical Systems, Progr. Physics 1, Birkhauser, Boston (1980).
[4]   M.R. Garey and D.S. Johnson, "Computers and Intractability",
      Freeman & Co., San Francisco (1979).
[5]   A.L. Hodgkin and A.F. Huxley, A quantitative description of mem-
      brane current and its application to conduction and excitation
      in nerve, J. Physiol. 117:500-544 (1952).
[6]   A.V. Holden, Why is the nervous system not as chaotic as it should
      be? in "Dynamic phenomena in Neurochemistry and Neurophysics:
      Theoretical Aspects", Proceedings of a Workshop, ed. P. Erdi,
      pp 6-11, Press of KFKI, Budapest 1984.
[7]   S.C. Kleene, Representation of events in nerve nets and finite
      automata, in "Automata Studies", eds. C.E. Shannon and J.
      McCarthy, 34-41 (1956).
[8]   A.N. Kolmogoroff, Three approaches for defining the concept of
      information quantity, Information Transmission 1:3-11 (1965).
[9]   E. Labos, Periodic and non-periodic motions in different classes
      of formal neuronal networks and chaotic spike generators, in
      "Cybernetics and System Research 2", ed. R. Trappl, 237-243,
      Elsevier, Amsterdam (1984).
[10] E. Labos, Optimal Design of Neuronal Networks, in "Neural Commun-
      ication and Control", eds. Gy. Szekely et al., Adv. Physiol.
      Sci. 30:127-153 (1980a).
[11] E. Labos, Effective Extraction of Information Included in Network
      Descriptions and Neural Spike Records, in "Biomathematics in
      1980", eds. L.M. Ricciardi and A.C. Scott, North-Holland,
      Amsterdam (1980b).
[12] E. Labos, A Model of Dynamic Behaviour of Neurons and Networks,
      Lecture Abstracts of Annual Meeting of Hungarian Physiological
      Society, Budapest, I.S4:117 (in Hungarian), (1981).
[13] E. Labos, Spike Generating Dynamical Systems and Networks,
      Proceedings of a Symposium on The Mathematics of Dynamic
      Processes held in Sopron (Hungary), 1985 (in press).

[14] E. Labos and E. Nogradi, Examples of computer-aided exploration and design of dynamical systems in neurosciences: design of optimal nets, Proceedings of a Workshop on Dynamical Systems and Environmental Models held in Wartburg (DDR), 1986 (in press).

[15] E. Labos, Self-dual and other structures of optimal cycles in neural network behaviour, in "Cybernetics and Systems '86" ed. R. Trappl, pp 367-374, Reidel, Dordrech (1986).

[16] R. May, Simple mathematical models with very complicated dynamics, Nature 261:459-467 (1976).

[17] W.S. McCulloch and W.H. Pitts, A Logical Calculus of the Ideas Imminent in Nervous Activity, Bull. Math. Biophys. 5:115-133 (1943).

[18] M.L. Minsky, "Computation: Finite and Infinite Machines", Englewood Cliffs, N.J., Prentice Hall (1967).

[19] S. Muroga, "Threshold Logics and Its Application", Wiley-Interscience, New York (1971).

[20] E. Nogradi and E. Labos, Simulations of Spontaneous Neuronal Activity by Pseudo-Random Functions, Lecture Abstracts of the Ann. Meeting of Hungarian Phys. Soc., Budapest I.P. 74:151 (1981).

[21] E. Nogradi and E. Labos, Pseudo-Random Interval Maps for Simulation of Normal and Exotic Neuronal Activities, in "Cybernetics and Systems '86", Reidel, Dordrecht, pp 427-434 (1986).

[22] E. Nogradi, personal communication.

[23] L.F. Olsen and H. Degn, Chaos in Biological Systems, Quarterly Review of Biophysics 18:165-225 (1985).

[24] E. Ott, Strange attractors and chaotic motion of dynamical systems, Rev. Mod. Phys. 53:655-671 (1981).

NOTE

The synthesis of maximum-length generator threshold gate nets has become much more tractable following an observation. Previously the author did not know which columnar permutations are effective in this sense. The conjecture, which is supported, is as follows: If the 'cycle decomposition' of the columnar permutation does not include too many 'fixed columns' and the lengths of column-cycles are prime to each other, then the chance to reach maximum length seems definitively higher, supposed that the start matrix to be transformed was adequately chosen. This observation is important above dimension 6. Now at $n = 7$ more than 200 and at $n = 8$ 52 such nets are available. Nevertheless this observation does not solve completely the maximum length synthesis which seems different in odd or even dimensions as it can be proven by $n = 4$, 5 and 6 and supported by $n = 7$, 8 and 9. At the moment of writing no network of dim 1 or higher have been found with maximum length of state-cycle.

DATA REQUIREMENTS FOR RELIABLE ESTIMATION OF CORRELATION DIMENSIONS

A.M. Albano[1], A.I. Mees[2], G.C. de Guzman[3] and P.E. Rapp[4]

[1]Department of Physics   [2]Department of Mathematics
Bryn Mawr College   University of Western Australia
Bryn Mawr, PA 19010   Nedlands, 6009 WA

[3]Center for Complex   [4]Department of Physiology and
Systems   Biochemistry
Florida Atlantic   Medical College of Pennsylvania
University   3300 Henry Avenue
Boca Raton, FL 33431   Philadelphia, PA 19129

ABSTRACT

It is not always possible to resolve the dimension of an attractor
from a finite data set. The number of data points required depends on the
structure of the attractor, the distribution of points on the attractor,
and the precision of the data. If the chaotic component of a system's
behaviour is sufficiently small relative to its large scale motion, and
if orbits seldom visit the region of the attractor with small scale
fractal structure, any method will fail to resolve the attractor's
dimension. It is simple to construct abstract mathematical examples that
present this behaviour. However, while these limitations should be
explicit recognised, it should also be noted that a growing body of
empirical experience suggests that experimentally encountered physical
and biological systems do not invariably display these behaviours. It is
possible to estimate reliably the dimension of these attractors with
comparatively small data sets.

I INTRODUCTION

A growing body of theoretical and experimental evidence suggests that
biological systems can enter domains of chaotic behaviour. Proposed
examples include glycolytic metabolism [31,32], cyclic nucleotide signalling in Dictyostelium [33], periodically stimulated cardiac cells [24,17]
and neural systems [2,25,26,37,38]. Just as in the physical sciences,
these results have raised questions concerning the reliable quantitative
characterisation of chaotic behaviour. A number of measures of dynamical
behaviour have been considered; they include the correlation dimension
[21,22], Kolmogorov entropy [23] and Lyapunov exponents [47,43]. Each of
these measures is complementary and introduces its own advantages and
difficulties. In this contribution we restrict attention to the

correlation dimension. A didactic presentation of a procedure based on results of Grassberger and Procaccia is given in the next section. This is followed by an examination of the reliability of these calculations.

Calculation of the correlation dimension: A didactic presentation

Over the last century the intuitive concept of topological dimension has been made more precise. Recently this has resulted in an ever proliferating collection of interrelated definitions of dimension that includes the Kolmogorov dimension, the Hausdorff-Besicovitch dimension, the Renyi information dimension, the Lyapunov dimension and more. The literature is additionally complicated by a nonuniform nomenclature. Admirable reviews have been provided by Farmer [13,14], Farmer, Ott and York [15] and Eckmann and Ruelle [12].

A problem with many of these measures is that they are difficult, or effectively impossible, to implement computationally. Grassberger and Procaccia [22,32] introduced a measure, the correlation dimension, which is closely related to the information dimension and the Hausdorff dimension. The correlation dimension has the important advantage of being more readily computable from time series data. The reader is referred to [22] for the demonstration of the relation between correlation dimension and other definitions. The discussion here is limited to an examination of operational questions of computation reliability.

The computation begins with M measured values of a dependent variable. These values are denoted by v_1, v_2, v_3, ... v_M. These values are embedded to construct points, X_j, in an N-dimensional embedding space.

$$X_1 = (v_1, v_2, \ldots v_N)$$
$$X_2 = (v_2, v_3, \ldots v_{N+1})$$

$$X_i = (v_i, v_{i+1}, \ldots v_{i+N-1})$$

A total of K, K=M-N+1, points are constructed. (Alternative embedding protocols are equally valid.)

For embedding dimension N, the correlation integral $C_N(r)$ is calculated as a function of r, the correlation length.

$$C_N(r) = (1/N_p) \sum_{i=1}^{K} \sum_{j=i+1}^{K} H(r - |X_i - X_j|)$$

H is the Heaviside function.

$$H(x) = 1, \quad x \geq 0$$
$$H(x) = 0, \quad x < 0$$

N_p is the number of distinct pairs of points X_i and K_j;

$$N_p = K(K - 1)/2$$

$C_N(r)$ is monotone increasing from 0 at r=0 to the saturation value 1 which is obtained whenever $r \geq \max|X_i - X_j|$. Grassberger and Procaccia show that for N and M sufficiently large and r sufficiently small, the logarithm of the correlation integral as a function of the logarithm of correlation length will have a linear region, called the scaling region, with slope D_c

$$\ln[C_N(r)] = D_c \ln r + \ldots\ldots$$

D_c is the correlation dimension. They further demonstrate that

$$D_c \leq \text{information dimension} \leq \text{Hausdorff (fractal) dimension}$$

The relationship between the slope of the correlation integral and D_c is obtained only if the embedding dimension is greater than the dimension of the attractor. In practice $C_N(r)$ is calculated for a sequence of embedding dimensions. If the N's are sufficiently large, D_c will not change when N is increased. If N is less than the dimension of the attractor, the slope should be equal to the embedding dimension. (In practice, a slope equal to embedding dimension is obtained only if M, the number of data points, is large.) Similarly, if the procedure is applied to a random, infinite dimensional, signal, then the slope increases with each increase in embedding dimension. Thus, the method never distinguishes between a random signal and a time series with a high, but finite, dimension.

It is possible to reduce the implementation of the Grassberger-Procaccia result to a step-by-step procedure.

1. Obtain M data points by either experiment or computation. In the example treated here, data were generated computationally from the Hénon equations [27]

$$x_{t+1} = 1 - ax_t^2 + y_t$$

$$y_{t+1} = bx_t \quad a=1.4 \quad b=0.3$$

Calculations were performed with the x values.

2. The embedding is constructed.

3. The correlation integral is calculated. The results of these computations are presented in Fig. 1 for 1000 data points and embedding dimensions N=3, 4, 5 and 6.

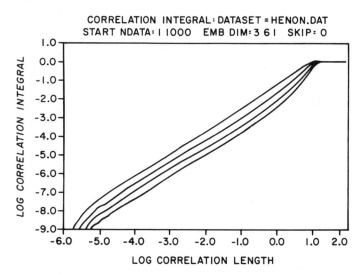

Fig. 1. The logarithms of the correlation integral for the Hénon attractor calculated with 1000 data points for embedding dimensions n=3, 4, 5 and 6.

4. The extent of the scaling region, if it should exist, must be determined. The correlation integral is differentiated, Figure 2, and the upper and lower bound of the scaling region is found for each embedding dimension. Several alternative numerical criteria can be used to define the scaling region. The simplest is a region in which the derivative does not vary by more than some specified amount, for example 10%.

Finally D_C is determined for each embedding dimension. It should be stressed that D_C is not determined from the numerical values of the derivative. The derivative is used to establish the existence and boundaries of the scaling region. D_C is found by returning to the logarithm of $C_N(r)$ and finding the slope of the least squares straight line that spans the scaling region.

It should perhaps be noted explicitly that this procedure has been reduced to a numerical algorithm. No subjective assessments are made.

The reliability of this procedure has been improved by imposing a series of criteria that must be satisfied by the computation.
1. The degree of divergence of the derivative in the scaling region can be controlled. Typically we use a 10% criterion.
2. A minimum length of scaling region is required. In most computations we require at least two orders of magnitude in r.
3. The value of D_C must be stable, to some allowed variation, over several successive embedding dimensions. We require agreement to within 10% over at least four embeddings.
If the computation fails to meet any of these three criteria, a "Dimension Unresolved" result is returned.

In the example of Figs. 1 and 2 the value of D_C is 1.20, 1.22, 1.23 and 1.21 for N=3, 4, 5 and 6 respectively. This gives an average value of 1.21 ± 0.1. The literature value of D_C for this attractor obtained with fifteen thousand points is 1.22 ± 0.01 [22].

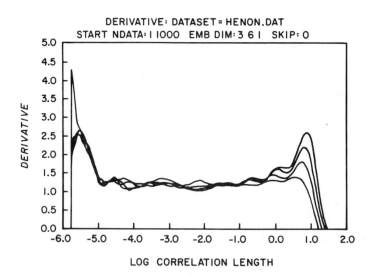

DERIVATIVE: DATASET = HENON.DAT
START NDATA: I 1000   EMB DIM: 3 6 I   SKIP: 0

Fig. 2. The derivative of the logarithm of the correlation integral as a function of the logarithm of the correlation length, $dC_N(r)/d\ln r$, for the Hénon attractor. One thousand data points were used. The embedding dimensions are N=3, 4, 5 and 6.

   A central problem that must be addressed in the design of any experiment is the data requirement. How often should the signal be sampled?
How long must the total sampled time period be? What noise level is
acceptable? If the object of an experiment is to determine an attractor's
correlation dimension, the question becomes: what are the data requirements for correlation dimension measurements? There is, unfortunately,
no general answer to this question. A number of warnings should be made.

   1. The answer to the question depends on which dimension is being
calculated. For example, the data requirement of correlation dimension
and Lyapunov dimension calculations can be very different.

   2. Once the type of dimension is chosen, the amount of data required
will depend, at least to some extent, on the algorithm used to estimate
that dimension.

   3. The number of data points required depends on the distribution
of those points on the attractor. If the orbit rarely visits the region
of the attractor with fractal structure, very long time series will be
required.

   4. The resolving power of the experiment depends on the scale of the
fractal component. If the magnitude of the fractal component is less than
the experimental noise, it will not be resolved by an experiment of arbitrary duration.

   Given the absence of a general answer to the data requirement
question, and with these warnings in mind, the best approach is to consider specific examples. The first is the Hénon attractor considered in
Figs. 1 and 2. These results suggest that D_c can be found to within 1%
of the literature value with only one thousand data points. Abraham and
his colleagues [1] calculated the dimension of the Hénon attractor using
500, 1200, 4000 and 10,000 data points and found that they could compute
D_c to 6% of the literature value with 500 data points. Upon reflection
this is perhaps not to be greatly wondered at; the correlation dimension
is a property of an integral, and integrals are numerically robust
objects. A striking example of this is given in Fig. 3. This shows the
correlation integral of the Hénon attractor calculated for 50, 100, 200
and 500 data points. While the extent of the scaling region decreases
with decreased data, the slope within the scaling region remains reasonably unchanged.

   Now, we do not wish to suggest that the dimension of the Hénon
attractor can be calculated with fifty data points. However, the results
in this diagram, and more pertinently Abraham's systematic study, do
strongly argue that the previously made claim that several million data
points are required to calculate the correlation dimension of the Hénon
attractor is in error by a factor of 10^3 to 10^4. So it goes.

   It could be argued that the results with the Hénon attractor were
obtained because these calculations were performed with computer-generated
data, that is, with data that was accurate to double precision. Would
the method inevitably fail to meet resolution criteria if noisy experimental data were used? Apparently not. A study of this question has been
undertaken by Swinney and his colleagues at the University of Texas
(Swinney, personal communication). These experiments measured Couette-
Taylor flow. With a twenty thousand point data set, the value $D_c = 2.4$
was obtained. The number of data points was successively reduced. It was

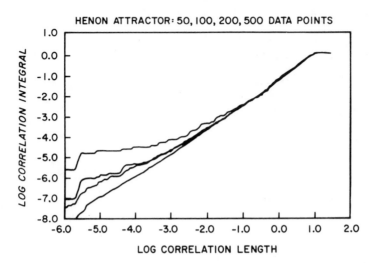

Fig. 3. Logarithms of the correlation integral for the Hénon attractor.
Fifty, one hundred, two hundred and five hundred data points
were used. In each case, the embedding dimension is three.

still possible to resolve the dimension of the attractor with one thousand
points. On the basis of experience with both physical and chemical systems, Swinney has proposed an approximate empirical guideline for the
number of data points required to resolve the correlation dimension of an
attractor. When considering this guideline, all of the qualifications and
warnings made previously apply. As a rough rule of thumb, Swinney has
found that 10^{D_c} points of at least 0.5% accuracy can give a reasonable
estimate of the dimension of the attractor for D_c up to about 5. However,
it should again be stressed that if the folding occurs on a small scale
and orbits rarely visit the region of the attractor with small scale
fractal structure, then millions of points will be inadequate for determining the dimension of a low dimensional attractor.

Other examples of successful calculations of D_c using small data sets
have been published. Puccioni and his colleagues [2] measured the output
intensity of a CO_2 laser. A total of 6000 data points were sampled at a
rate of 16 per modulation period. Initially the dimension was calculated
with all 6000 data points. It was then recalculated using subsets of 500
successive data points. The results of the dimension calculations were
not significantly altered. Albano [3] reported similar studies of
instabilities in a xenon laser. They successfully calculated the fractal
dimension of the associated attractor using 512 data points.

Kadanoff and his group [29] examined data from a forced Rayleigh-
Benard system using mercury as the fluid. They report dimension calculations using 2500 points. Their calculations were more extensive than
those described here as they calculated the continuous spectrum of scaling indices. They report an accuracy in the region of worst convergence
of 10-20%.

As part of their study of solar radio pulsations, Kurth and Herzel
[30] analysed computed data from the Lorenz attractor and from a periodic
signal using 640 points. They found a high degree of agreement between
the literature and their values. For the Lorenz attractor, using their
parameter values, the literature value of the dimension is 2.06; their

estimate is 2.09. Their estimated dimension of the periodic signal was
1.03.

How many data points are needed?: The bad news

The preceding section presents an unwarrantably optimistic picture
of dimension calculations. While it can be comparatively easy to take a
data set and produce a numerical value for D_C, producing an accurate value
can be surprisingly difficult. When the three convergence criteria
listed in the previous section are imposed, many seemingly well behaved
sets will return a "Dimension Unresolved" result. Further difficulties
should also be considered. Happily, at least in some cases, there are
remedies.

The correlation dimension is determined by finding the best fit
straight line between r_L and r_U, the lower and upper bounds for the
scaling region. Specification of the scaling region is done numerically.
No subjective criteria are used. However, several alternative numerical
criteria can be used to calculate r_L and r_U. In the case of an extremely
well behaved data set like that in Fig. 1, different criteria give
essentially the same results. With noisy experimental data, this isn't
necessarily the case. Changes in the numerical criterion that specifies
r_L and r_U can then result in different values. For some correlation
integrals, small changes in r_L and r_U can result in significant changes
in D_C. No universally acceptable scaling region criterion has been
established. Caswell and Yorke [7] suggest including r_L and r_U as part of
the dimension estimate. Thus, rather than report D_C, $D_C(r_L,r_U)$ is
reported. This is admittedly not an ideal solution in the sense of determining D_C as an absolute number, but it does allow systematic comparisons
of independent calculations.

Often we are not interested in the absolute value of D_C. Interest
may centre on how D_C changes in response to differing physiological
conditions. Examples include the changes in electroencephalographic
activity in response to changes in cognitive activity, and changes in
neuronal spike trains in response to changes in extracellular K^+. For
experiments of this type, the ratio of D_C values is of interest. We have
found that though there can be differences in D_C estimates obtained with
different algorithms and different signal sampling protocols, the ratio
of D_C's is robust.

It is possible to combine these two additions to the basic procedure.
Namely, it is possible to report the ratio of D_C values obtained over a
specified range of r_L and r_U.

More bad news: The sampling interval

The estimates of the correlation dimension can be sensitive to the
sampling interval. The sampling interval, T_S, is the time between
successive measurements. Decreasing T_S does not necessarily improve the
accuracy of D_C estimates. In fact, if the sample interval is too small,
spuriously low values of dimension result. Artifactual correlations
result that are the result of high frequency sampling. It is possible to
see that this is so by considering the limiting case as T_S approaches
zero; all data points approach some constant value. D_C approaches the
value zero. Thus, a finite number of data points sampled at a sufficiently high rate can result in an arbitrarily low D_C estimate.

This raises the question, what is the optimal sample interval? An
even more fundamental question would be, is there such a thing as an
optimal T_S? At this point, care must be taken to distinguish between

scalar data values that are directly measured (v_1, v_2, ... v_M, in our notation), and the N-dimension points X_i that exist in the embedding space. To make this distinction and its implications clear, the theoretical foundations underlying dimension calculations should be reviewed.

Measured variable v is a single variable of possibly many variables that make up the original system. The object of the exercise is to determine the dimension of the attractor of that large dimensional system. However, not all variables are measured; in fact, usually only one is measured. As previously outlined, the scalar data set $\{v_i\}$ is used to construct a set S in N-space,

$$S = \{X_i, \mid X_i = (v_i, v_{i+1}, ... v_{i+N-1})\}$$

The structure of the original attractor is then inferred from the structure of set S. This inference is justified by a theorem of Takens' [41] which is derived from the Whitney embedding theorem [45]. In the proof of the theorem, it is assumed that all data points are on the attractor; that is, it is assumed that transient initial behaviour has died away. It is also assumed that an infinite number of measured values that are uncorrupted by noise are also available. Given these assumptions, the theorem shows that an embedding exists between the attractor and set S whenever the embedding dimension, N, is sufficiently large. An embedding is a proper injective immersion. Whenever an embedding exists between two sets, they are structurally very similar. Specifically, the dimension of the original attractor is equal to the dimension of S. There are some technical qualifications to this statement in Takens [41]. In practical applications, the conditions of the theorem are never satisfied because data are corrupted by noise and an infinite number of values are not measured.

Correlations are not between X_i and X_j in N-space. In order to ensure that a correlation is the consequence of the geometry of set S and not an artifactual consequence of high frequency sampling, we would want X_i and X_j to be separated by the greatest possible time interval. If this were the only consideration, it would follow that the bigger T_s, the better.

However, the sample interval is also related to the efficacy of the approximated embedding relationship between set $S=\{X_i\}$ and the original attractor. In the proof of one variant of the theorem, the existence of an embedding function in N-dimensional space is demonstrated analytically for the case of infinite sets of noise-free data by showing that this data can be used to specify the (N-1)-derivatives of v(t), dv/dt, d^2v/dt^2, ... $d^{N-1}v/dt^{N-1}$, from the measured data. In practical applications, this ideal analytic case is approximated by finite sets of noise-corrupted data. The sample interval T_s must be small enough to allow a local approximation of these derivatives. Thus an arbitrarily large value of T_s is not the ideal response. The best compromise between these conflicting demands would be to measure N values of v at some appropriate T_s, construct a point X_i, and then only after some longer waiting period, measure another N values to construct the next point in N-space. This is the resolution proposed independently by Theiler [42].

However, this still leaves the question of what an appropriate value of T_s might be. The arguments presented here suggest that arbitrarily short and arbitrarily long values of T_s are unsatisfactory. How should an intermediate value be selected? Albano et al. [4] based the selection of T_s on the characteristic time, T_c, defined by the first zero of the time series' autocorrelation function. The sample interval was fixed by the relationship $T_c = N \cdot T_s$. Broomhead and King [5,6] proposed using a characteristic time defined by $1/f*$ where $f*$ is the largest frequency contributing significantly to the power spectrum. They refer to this as the frequency of the noise floor. Though this frequency is often readily apparent to a human observer examining a power spectrum, it is not easily implemented numerically. In any case $1/f*$ is related to the zero of the autocorrelation function by the Wiener-Khinchine theorem (the power spectrum is the Fourier transform of the autocorrelation function). Simm et al. [40] suggested that the sample time should be

$$T_c/N < T_s < T_c$$

where N is the embedding dimension and T_c is the characteristic time, which they define as the quarter period for pseudoperiodic signals. Thus, for periodic signals the lower bound of the Simm et al. range is identical to the value specified by Albano.

An alternative procedure for selecting T_s proposed by Shaw and systematically implemented in [16] is constructed on a time scale defined by the first minimum of the mutual information. Fraser and Swinney have demonstrated that at least for some signals, some of the time, this offers significant improvement over criteria based on autocorrelation functions.

It should be noted that the concern over sample interval selection arises in measurements of flows, that is, in the measurement of continuous waveforms like EEG signals. It does not arise when measuring maps, for example the interspike intervals of neuronal spike trains. In the latter case one simply measures the event interval to the highest possible accuracy. In experimental neural systems the average interspike interval is on the order of 1 second. These intervals can be measured to an accuracy of 15-20 microseconds.

Interpolation of points between elements of the original data set [36] results in spurious dimension estimates [39]. Depending on the number of interpolated points and the interpolation protocol, it is possible to tune the estimated value of D_c obtained from a random (that is, infinite dimensional) signal to any desired finite value.

Still more bad news: Nonstationarity of attractor dimension

All of the preceding discussion incorporates the assumption that the dimension of the attractor is constant throughout the recording period and that difficulty in estimating the dimension results solely from experimental noise and finite data sets. There is no a priori reason to suppose that this is a legitimate assumption. Indeed, in highly complex biological systems it is reasonable to anticipate frequent, irregular changes in the dimension. This has particularly important implications in the analysis of EEG data. In these studies, changes in D_c are possibly more interesting than its value at any given time. For example, does the dimension of EEG signals display greater variability in schizophrenics and can this variability be used to monitor the efficacy of treatment with psychotropic drugs?

The frequency analysis of nonstationary signals has a long history. One approach is to compute spectra from data sets recorded over short epochs. The length of the assumed short time stationarity determines the frequency resolution. If the epoch is too long, the stationarity assumption is violated. If the epoch is too short, the frequency resolution deteriorates. This is the classical signal analysis trade-off between temporal resolution and frequency resolution. A possible response

to this is the Wigner distribution [46] in which the instantaneous energy
for a given time and frequency is approximated. Though introduced over
fifty years ago, widespread interest in this approach is comparatively
recent [9,10,11]. For example, Gevins [35] has calculated Wigner distributions of human event related potentials.

The Wigner distribution is an extension to frequency domain analysis.
Can the basic idea be employed in the analysis of attractor dimension?
An obvious way in which the analogy between frequency analysis and
dimension analysis fails is that the frequency constant of a signal forms
a continuous spectrum, and D_c is a single number. However, Hentschel and
Procaccia [28] have shown that the concept of an attractor's dimension
can be extended to create an infinite number of generalised dimensions.
With the extension of dimension to a continuum, the analogy with frequency
is complete. This suggests that the construction of a dimension analysis
analog to the Wigner distribution should be possible in principle. The
computational feasibility with noisy data is less certain.

Some encouraging news: Singular value decompositions

Singular value decomposition is an important noise reduction technique. Related, essentially identical, techniques include principal
component analysis, factor analysis and the Karhunen-Loeve expansion [44].
The use of singular decompositions as a complement to dynamical analysis
was introduced by Broomhead and King [5].

We begin with a fundamental result from matrix algebra [20]. Let A
be an r x d matrix with r>d (if r<d, the operation can be carried out on
the transpose). Matrix A can be decomposed into a matrix product of the
form

$$A_{rxd} = W_{rxd} \cdot S_{dxd} \cdot U^T_{dxd}$$

W and U are orthogonal matrices.

$$W^T_{dxr} \cdot W_{rxd} = 1$$
$$U \cdot U^T = 1$$

S is a diagonal matrix of non-negative elements that are the singular
values of matrix A. By convention the singular values are denumerated in
descending magnitude

$$s_1 \geq s_2 \geq \ldots\ldots \geq s_d \geq 0$$

There are a number of numerical procedures for effecting the decomposition.
We employ a method originally developed by Golub [18,19]; see also
extensions by Chan [8].

In dynamical analysis this general theorem is applied to a very
specific matrix A, namely E^N, the matrix formed by embedding scalar data
(v_i) in an N-dimensional embedding space.

$$A_{rxN} = E^N = \begin{bmatrix} X_1 \\ X_2 \\ . \\ X_r \end{bmatrix} = \begin{bmatrix} v_1 & v_2 & \cdots & v_N \\ v_2 & v_3 & \cdots & v_{N+1} \\ . & . & & . \\ v_r & v_{r+1} & \cdots & v_{r+N-1} \end{bmatrix}$$

Using the theorem there are matrices W, S and U such that

$$E^N = W \cdot S \cdot U^T$$

Matrix U is a rotation in N-space. Therefore, the correlation dimension

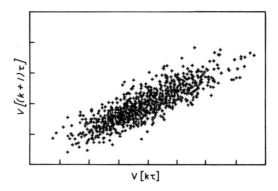

Fig. 4. The delay map of an EEG signal of a normal subject recorded at
site O_z. During the recording the subject's eyes were closed
and the subject was silently counting backwards in steps of
seven. The signal was sampled at 500 Hz. In this figure, the
(i+1)-st measured value is plotted against the i-th value.

obtained using elements of the set $\{X_i\}$ should be the same as the estimate
obtained by using the elements of the rotated set $\{(X_R)_i\}$

$$X_R = \{(X_R)_i = X_i \cdot U\}$$

In an ideal universe where experimental and computational noise are not
present, there is no advantage to using rotated data. To see why this
might be advantageous when computations are performed with noisy data,
consider the decomposition of E^N.

$$E^N = W \cdot S \cdot U^T$$

Thus,

$$E^N \cdot U = W \cdot S$$

S is the diagonal matrix of singular values.

$$E^N \cdot U = \begin{pmatrix} s_1 \overset{\uparrow}{\underset{\downarrow}{W}}_{j1} & s_2 \overset{\uparrow}{\underset{\downarrow}{W}}_{j2} & \cdots\cdots\cdots & s_N \overset{\uparrow}{\underset{\downarrow}{W}}_{jN} \end{pmatrix}$$

The k-th column of W is multiplied by the k-th singular value. If all of
the singular values are of uniform size, this rotation is of negligible
significance. This is what happens if a matrix of random numbers is
decomposed. However, suppose there is a marked nonuniformity in the
singular values, for example, it sometimes happens that there is an abrupt
decrease in the singular values, that is:

$$s_1 \sim s_2 \sim \ldots \sim s_j >> s_{j+1} \geq \ldots \geq s_N \geq 0$$

Most of the information would be contained in the first j components of
$(X_R)_i$. With most of the information concentrated in the first components
of $(X_R)_i$, noise will be most significant in the final components. In the
cases where the smallest s_j's are much smaller than the largest, the
corresponding components of $(X_R)_i$ can be discarded and a significant
measure of noise reduction is obtained.

The effect of singular value decompositions as a noise reduction
technique can be dramatic. Figure 4 is constructed from an EEG record
obtained from a normal subject, recorded at site O_z. During the recording
the eyes were closed and the subject was performing mental arithmetic

217

(silently counting backwards in steps of seven). In Fig. 4, the ordered pairs (v_i, v_{i+1}) are plotted. This delay plot shows that there is a structure to the data; points do not fill the plane uniformly, but the fine structure of the data set is not readily discerned.

The original data set was embedded in a seven dimensional space. The singular values were calculated. The first four singular values were significantly larger than the last three. The ratio s_5/s_4 was on the order of 10. The transformation U was calculated and the rotated set $\{X_R\}$ was constructed. Figure 5 shows the ordered pairs formed by the first two elements of the rotated vectors: $((X_R)_1, (X_R)_2)_j$. A complex and delicate structure is immediately seen.

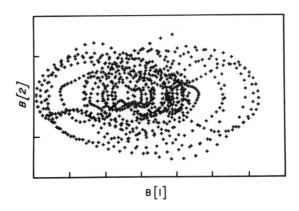

Fig. 5.   The EEG data of Fig. 4 transformed by a singular value decomposition. The original data was embedded in a seven dimensional embedding space and rotated by the transformation defined by matrix U. In the diagram, the second components of the transformed vectors are plotted against the first components.

The effect on the dimension calculations is similarly dramatic. Prior to rotation, the correlation dimension could not be resolved. The correlation integrals were recomputed using the first 4, 5, 6 and 7 components of the rotated vectors. Figure 6 shows the derivative of the logarithms of the correlation integral. A well defined scaling region is now apparent. The agreement between the 4, 5, 6 and 7 component calculations is such that the calculations superimpose in the scaling region.

CONCLUSIONS

Calculating the correlation dimension is a deceptive numerical procedure. Getting a number is easy. Getting a dynamically meaningful number can be difficult, except in those instances where it is impossible. It is possible to introduce reliability criteria which prevent spurious values, at least in some cases. However, while these difficulties should be recognised, a growing number of successful applications of these concepts justifies cautious optimism. As a complement to classical forms of signal analysis, dynamical analysis has much to offer.

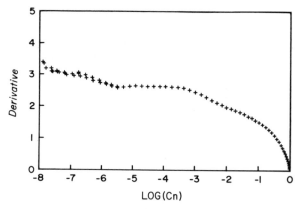

Fig. 6. The derivative of the logarithm of the correlation integral
for the EEG signal of Figs. 4 and 5 following transformation
to the singular value basis. Embedding dimensions 4, 5, 6
and 7 are shown. The values obtained at different embedding
dimensions superimpose in the scaling region.

ACKNOWLEDGEMENTS

The authors would like to thank Professors B. Onaral and C.V.K.P.
Rao of Drexel University for introducing us to the Wigner distribution.
PER would like to acknowledge support from NIH Grant NS19716.

REFERENCES

[1]   N.B. Abraham, A.M. Albano, B. Das, G. de Guzman, S. Yong, R.S.
      Gioggia, G.P. Puccioni and J.R. Tredicce, Phys. Lett. 114A:
      217-221 (1985).
[2]   K. Aihara and G. Matsumoto, in "Chaos", A.V. Holden, ed., 257-269,
      University Press, Manchester & Princeton (1986).
[3]   A.M. Albano, J. Abounadi, T.H. Chyba, C.E. Searle and S. Yong,
      J. Opt. Soc. Amer. 2B: 47-55 (1985).
[4]   A.M. Albano et al., in "Dimension and Entropies in Chaotic Systems.
      Quantification of Complex Behaviour", G. Mayer-Kress, ed., 231-
      240, Springer-Verlag, Berlin (1986).
[5]   D.S. Broomhead and G.P. King, Physica 20D: 217-236 (1986a).
[6]   D.S. Broomhead and G.P. King, in "Nonlinear Phenomena and Chaos",
      S. Sarkan, ed., Adam Hilger, Bristol (1986b).
[7]   W.E. Caswell and J.A. Yorke, in "Dimensions and Entropies in Chaotic
      Systems", G. Mayer-Kress, ed., 123-136, Springer-Verlag, Berlin
      (1986).
[8]   T.F. Chan, ACM Trans. Math. Software 8: 84-88 (1982).
[9]   T.A.C.M. Claasen and W.F.G. Mecklenbrauker, Phillips J. Res. 35:
      217-250 (1980).
[10] T.A.C.M. Claasen and W.F.G. Mecklenbrauker, Phillips J. Res. 35:
      276-300 (1980).
[11] T.A.C.M. Claasen and W.F.G. Mecklenbrauker, Phillips J. Res. 35:
      372-389 (1980).
[12] J.-P. Eckmann and D. Ruelle, Rev. modn. Phys. 54: 617-656 (1985).
[13] J.D. Farmer, in "Evolution of Order and Chaos", H. Haken, ed.,
      228-246, Springer-Verlag, Berlin (1982a).
[14] J.D. Farmer, Z. Naturforsch. 37A: 1304-1325 (1982b).
[15] J.D. Farmer, E. Ott and J.A. Yorke, Physica 7D: 153-180 (1983).
[16] A.M. Fraser and H.L. Swinney, Phys. Rev. 33A: 1134-1140 (1986).

[17] L. Glass, A. Shrier and J. Belair, in "Chaos", A.V. Holden, ed., 237-256, University Press, Manchester & Princeton (1986).

[18] G. Golub and W. Kahan, SIAM J. Numer. Anal. 2: 205-224 ((1965).

[19] G.B. Golub and C. Reinsch, Numer. Math. 14: 403-420 (1970).

[20] G.H. Golub and C.F. Van Loan, "Matrix Computations", Johns Hopkins University Press, Baltimore (1983).

[21] P. Grassberger and I. Procaccia, Physica 9D: 189-208 (1983a).

[22] P. Grassberger and I. Procaccia, Phys. Rev. Lett. 50: 346-349 (1983b).

[23] P. Grassberger and I. Procaccia, Phys. Rev. A 28A: 2591-2593 (1983c).

[24] M.R. Guevara, L. Glass and A. Shrier, Science, Wash. 214: 1350-1353 (1981).

[25] H. Hayashi, S. Ishizuka and K. Hirakawa, Phys. Lett. A 98A: 474-476 (1983).

[26] H. Hayashi, S. Ishizuka, M. Ohta and K. Hirakawa, Phys. Lett. 88A: 435-438 (1983).

[27] M. Hénon, Commun. math. Phys. 50: 69-78 (1976).

[28] H.G.E. Hentschel and I. Procaccia, Physica 8D: 435-444 (1983).

[29] M.H. Jensen, L.P. Kadanoff, A. Libchaber, I. Procaccia and J. Stavans, Phys. Rev. Lett. 55: 2798-2801 (1986).

[30] J. Kurth and H. Herzel, Physica D, submitted (1986).

[31] M. Markus and B. Hess, Proc. natn. Acad. Sci. U.S.A. 81: 4394-4398 (1984).

[32] M. Markus, D. Kuschmitz and B. Hess, FEBS Lett. 172: 235-238 (1984).

[33] J.L. Martiel and A. Goldbeter, Nature, Lond. 313: 590-592 (1985).

[34] A.I. Mees, P.E. Rapp and L.S. Jennings, "Singular value decomposition and embedding dimension", Phys. Rev., in press.

[35] N.H. Morgan and A.S. Gevins, IEEE Trans. Biomed. Eng. BME-33: 66-70 (1986).

[36] C. Nicolis and G. Nicolis, Nature, Lond. 311: 529-532 (1984).

[37] P.E. Rapp, I.D. Zimmerman, A.M. Albano, G.C. de Guzman and N.N. Greenbaum, Phys. Lett. 110A: 335-338 (1985a).

[38] P.E. Rapp, I.D. Zimmerman, A.M. Albano, G.C. de Guzman, N.N. Greenbaum and T.R. Bashore, in "Nonlinear Oscillations in Chemistry and Biology", H.G. Othmer, ed., Springer-Verlag, NY (1985b).

[39] W.M. Schaffer, IMA J. Maths. appl. Med. and Biol. 2: 221-252 (1985).

[40] C.W. Simm, M.L. Sawley, F. Skiff and A. Pochelon, "On the analysis of experimental signals for evidence of determinstic chaos", preprint (1986).

[41] F. Takens, in "Dynamical Systems and Turbulence, Lecture Notes in Mathematics, Volume 898", D.A. Rand and L.S. Young, eds., 365-381, Springer-Verlag, NY (1980).

[42] J. Theiler, Phys. Rev. A 34: 2427 (1986).

[43] J. Vastano and E.J. Kostelich, in "Dimensions and Entropies in Chaotic Systems", G. Mayer-Kress, ed., 100-107, Springer-Verlag, Berlin (1986).

[44] S. Watanabe, Proc. Conference on Information Theory, Prague (1965).

[45] H. Whitney, Ann. Math. 37: 645-680 (1936).

[46] E. Wigner, Phys. Rev. 40: 749-759 (1932).

[47] A. Wolf, J.B. Swift, H.L. Swinney and J.A. Vastano, Physica 16D: 285-317 (1985).

CHAOTIC DYNAMICS IN BIOLOGICAL INFORMATION PROCESSING:

A HEURISTIC OUTLINE

John S. Nicolis

Department of Electrical Engineering
University of Patras
Greece

ABSTRACT

A given (but otherwise random) environmental time series impinging
on the input of a certain biological processor passes through with overwhelming probability undetected. A very small percentage of environmental
stimuli, however, are "captured" by the processor's non-linear dissipative
operator, as initial conditions, that is as solutions of the processor's
dissipative dynamics. The processor in such cases is instrumental in
compressing or abstracting those stimuli, thereby making the external
world collapse from a previous regime of a "pure state" of suspended
animation on to a set of stable eigenfunctions or "categories" - chaotic
strange attractors. The charateristics of this cognitive set depend on
the operator involved and the hierarchical level where the abstraction
takes place. In this paper we model the physics of such a cognitive
process and the role that the thalamocortical pacemaker of the (human)
brain plays in both stimulating the individual attractors and permutating
them on a time division multiplexing basis. A synthesis of the Markovian
processes taking place within each individual attractor-memory and the
Markovian or Semi-Markovian process involving the intermittent jumping
among the different attractors-memories is discussed.

I WHERE IS INFORMATION STORED?

Although the overwhelming majority of environmental stimuli are
incompressible by a given biological processor, a small subset of them
("a Basin") are in principle compressible up to a given number of bits
[26]. For the members of such a subset it is therefore meaningful to ask
the above question. In our opinion the answer has emerged from a very
simple and very original experiment performed some years ago by Glass and
Perez [23].

These authors took a blank sheet of paper and xeroxed it in an
imperfect machine. Eventually some black dots appeared here and there.
They repeated the process on the same sheet (up to ~20 times) until they
obtained a rather homogeneous set of dots - a perfectly random (noisy)
and informationless pattern, it seems. Then they made a transparency of
this sheet and superimposed this transparency on the original thereby
forming a linear (a Cartesian) two-dimensional map.

Playing with the three available parameters of this map (the scale a along the x-axis, the scale b along the y-axis and the angle of rotation Θ) they were able to see patterns which consisted of dynamical flows (eigenfunctions) in the neighbourhood of the singularities as deducted from the eigenvalues of the map, i.e. depending on a, b and Θ, circles, spirals, rays emerging from a centre or hyperbolic trajectories (saddle). Then we deduce: information is neither to be confined to an impinging time series nor to the internal activity of the processor: both are by themselves equally meaningless. It is rather emerging from the iterated map that the processor applies by way of conjugating externally impinging stimuli with internal activity; it consists of the set of (stable) eigenfunctions of such a map - which for a dissipative non-linear operator is simply the set of the coexisting attractors. Biological processors of a given species, possessing identical central nervous systems tend to generate more or less identical maps or flows - thereby creating, as it were by "social consensus", univeral pattern classification schemes.

II COMMUNICATION BETWEEN TWO BIOLOGICAL PROCESSORS

Figure 1 sketches the essentials of such a communication process. Each correspondent possesses two hierarchical levels. In general the process of stimulus reduction is a multihierarchical one (for example from the retina to the occipital area of the brain). Level(s) H_i are the "hardware" ones where the processor's dynamical activity is manifested, say, in a mosaic of coupled cortical neurons as a non-linear continuous dynamics. S_i stand for the software level(s) of discrete symbolic activities.

If t is the time then the coupling between the two systems may be expressed with a two-dimensional hierarchical map, e.g. as a set of coupled non-linear difference-integral equations [27, 19, 9, 10, 13, 14, 16] of the form:

$$S_1(t+1) = f_1(S_1(t)) + <\lambda_{1,2} \, H_1^\rho(t) \, S_2^\xi(t)>$$

$$S_2(t+1) = f_2(S_2(t)) + <\lambda_{2,1} \, H_2^\nu(t) \, S_1^\tau(t)>$$

(1)

where λ_{12}, λ_{21}, ρ, ξ, ν, τ are fixed parameters and the brackets denote

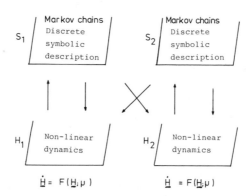

Fig. 1. A general layout of two communicating hierarchical systems. "Recognition of the partner" takes place readily only when the hierarchical two-dimensional map involved (see text) possesses stable eigenfunctions (attractors).

integration in the interval o...t (in case $\rho = \xi = \nu = \tau = 1$ the collective property influencing S_1, S_2 is just the cross-correlation function.
Higher moments however are in general expected to contribute as well).

The communication process then involves a cascade of iterations of
the above map. If there exist stable eigenfunctions (attractors) we may
say that the two partners recognise each other or keep memories of each
other. Moreover in such a case of existence of asymptotic stability the
interaction turns to a stable symbiosis (successful mutual simulation).
Otherwise the system of the two partners may stand for a game of
attrition where each partner tries to "break to code of the opponent"
first, that is form attractors about the other system while the other
partner is still in a transient, inconclusive regime. In case of coexisting repellors the communication may turn "paradoxical". In such a
case the perception of each system about the partner involves a mechanism
of intermitten jumping amongst asymptotically unstable regimes.

III A "BIOLOGICAL" PROCESSOR POSSESSING MULTIPLE COEXISTING ATTRACTORS

The processor amounts to a non-linear dissipative operator of the
form $\dot{x} = f (x ; \mu)$, $\nabla . f < 0$ whose dynamical ongoings are embedded in an
N-dimensional state space.

Suppose there are Λ coexisting attractors (steady state, limit
cycles, tori and strange attractors) with basins divided by fractal
separatrixes - either fully connected or disconnected ("Mandelbrot dust").
Note that a dynamical memory has to be contrasted from a static one. In
the latter all possible environmental stimuli are prestored in the
processor. For large memories such a technique becomes impractical.
Hence instead of memorising all possible "answers" to a set of environmental inputs one is rather storing an algorithm which once triggered
generates on the spot all possible strings which may match the incoming
message provided they share the same attractor, i.e. they belong to the
same basin. From a biological processor we expect the formation of multiple dynamical memories (attractors) about which we have a number of
demands - if they have to emulate at all biological structures.

Such demands are: Input sensitivity, Fault tolerance, Content
(context) addressability, Large storing capacity, Compressibility, as well
as criteria for evaluating the self-consistency of (Linguistic) self-
referential schemes. But before examining all the above pre-requisites
and modelling the memories accordingly, let us note that since all
linguistic schemes are by necessity discrete (in order to allow for error
detection and correction) there is a pressing need of knowing how it is
possible to deduce them in principle from continuous dissipative flows.
There are two general ways.

a) Concerning a single chaotic strange attractor in N-dimensional
space we first perform an (arbitrary) analog-to-digital conversion by way
of Poincare return mapping which ammounts to cutting the trajectory with
a hyperplane, recording the isophonic crossings and deriving a set of ·
discrete coupled nonlinear difference equations in N-1 (or less) dimensions.

Further we choose an (arbitrary) partition of β on the discrete
map above and then derive a Markov chain of single or multiple memory
with number of states the partition number and transitional probabilities
determined by the local slopes of the map [16]. Specifically,
the conditional probability of jumping from the jth partition

element to the ith one in exactly 1 iterations is given by the ratio of the value of the invariant measure of the map at the intersection of the 1th iterate of the map of the element β_i with the element β_j versus the value of the invariant measured at β_j or,

$$P_1(\beta_i/\beta_j) = \frac{\bar{\mu}[f^{-1}(\beta_i) \cap \beta_j]}{\bar{\mu}(\beta_j)} \qquad (2)$$

This formalism holds true not only for the cases where division points of the partition are mapped on to division points: A great deal of partitions in fact are non-Markovian in the sense that division points are not always mapped on division points. This means that instead of having β_i to be mapped either on β_j and/or β_k etc., it is mapped in part of β_j, in part of β_k etc. The analysis becomes very involved [4] and a numerical program carried out by R. Feistel (1983, personal communication) seems to indicate that in the case of such non-Markovian partitions a long range coherence in the digits time series $\beta_i \ \beta_j \ \beta_k \ \beta_\lambda \ \beta_\mu \ \dots\dots$ may result, namely the appearance of a given digit may influence the appearance of a remarkable number of those which follow. At the same time, however, the overlapping of the partition intervals results in increasing the entropy production in cascade iterations.

b) Concerning now the case of multiple coexisting attractors (Fig. 2) we may, following Arecchi et al. [14] write down a system of "kinetic" equations with respect to the probabilities P_i of occupancy of attractor i and accept the possibility of jumping from attractor to attractor as a result of external noise which is instrumental in shifting an initial condition from Basin i (where it normally belongs) to Basin j. In this formalism the noise appears via the individual transient times τ_{0i} for leaving attractor i - until landing on to another - and the transition probabilities P_{ij} of jumping from attractor j to attractor i. The shape of the separatrixes does also influence the above parameters. For Λ coexisting attractors then Arecchi's equations read:

$$\dot{P}_i = -\frac{1}{\tau_{0i}} P_i + \sum_{j=i}^{\Lambda} \frac{1}{\tau_{0j}} P_{ij} \ P_j = \text{(Escape term)} + \text{(Reinjection term)} \qquad (3)$$

Note: it is physically meaningless to search for asymptotic stability in Arecchi's equations since they essentially describe an intermittent mechanism which is structurally unstable. Further we cannot really talk about a priori probabilities P_i. The notion of a priori probability presumes the existence of asymptotic stability (convergence) in a trial process. Their solutions would fully determine a semi-Markov chain having now to do with a linguistic scheme among the coexisting "memories" of the system.

So we have in a "biological" model-processor two distinct types of linguistic schemes:

a) One which refers to one global memory, e.g. the memory of a Basin (attractor). This scheme is instrumental in relegating single messages into a particular attractor. This provokes a slow diffusion (mixing) from one part of the attractor to another. The memory of the specific message inevitably fades (unless the transinformation fluctuates with time).

b) Another scheme refers to establishing a connectivity and communication among otherwise independently existing global memories (coexisting attractors). In cases where the noise catalysing the transitions between attractors can be provided from within, the processor itself, so much the better. Happily, as we are going to see in the next section, neurophysiological evidence from the (human) central nervous system rather supports the view that the above "scanner" of the memories might be identified as the thalamocortical pacemaker whose "one-dimensional signature" is the recorded E.E.G. Let us now return to the group of characteristics a memory should possess in order to qualify as "biological": Input sensitivity, Fault tolerance, and Content addressability can all three be handled by accepting as a memory a chaotic attractor. Recognition of an object from a distorted or partial input – which implies a large Hamming distance of the initial stimulus from the attractor – can be accomplished since the underlying dissipative dynamics will "pull-down" this initial condition on to the attractor in a transient time amounting to progressive abstraction of the signal from the initial dimensionality N to the final, information dimensionality of the attractor D_i, thereby ensuring an average compression factor $C_i = N-D_i$.

The process is also context-addressable since only signals belonging to the basin of the particular attractor will be affected by the processor. Next the attractor must be strange in order to ensure large (dynamical) memorisation capacity. (Steady states and limit cycles possessing dimensionality 0 and 1 respectively are excellent devices as compressors but poor gadgets as memory banks.) In a recent paper (Nicolis and Tsuda, 1985) the issue of best compromise has been raised: we found that there is an optimum non-zero resolution ε^* under which the dynamical capacity for memory storage by a strange attractor (that is its information dimensionality) becomes maximum – without jeopardising compressibility. Of course for ε finite D depends on the partition and no more expresses a topological property as in the case $\varepsilon \to 0$. Still the derived expression of D^* is dynamically meaningful and determines the partition which is most parsimonial for the processor involved and also the optimum code or length of string

$$M^* = \log_2 \frac{1}{\varepsilon^*} \qquad (4)$$

which may describe the attractor-memory.

The fact that in some examples M^* turns out to be near the "magical number seven ± two" – known from experiments in cognitive psychology – may not perhaps be entirely accidental.

IV NEUROPHYSIOLOGICAL EVIDENCE ABOUT THE ROLE OF THE E.E.G. AND MODELLING OF THE ROLE OF THE THALAMOCORTICAL PACEMAKER

Synoptic description of the state of the art in E.E.G. research is given elsewhere (see Nicolis, 1982, 1985, 1986 and references therein, especially the contributions by R. Elul, W. Ross-Adey and M. Verzeano). Here we will comment only on the essentials.

Specific thalamic nuclei are capable of self-sustained oscillatory activity. Via fibres emanating from these nuclei and projecting on various cortical areas as well as fibres leading back from cortical areas to others non-specific thalamic nuclei, a thalamocortical-corticothalamic feedback oscillatory activity – the so-called thalamocortical pacemaker – is established. A one-dimensional macroscopic manifestation of this activity is the E.E.G. It appears that the thalamocortical pacemaker acts as a scanner of the population of cortical neurons on a time-division multiplexing basis: in the absence of external stimulation, this non-linear oscillator takes up, on a forced, intermittent basis, individual subsets of cortical postsynaptic membrane dendritic oscillations and entrains or phase locks them thereby forming semicoherent neuronal groups which stand for the Hardware of the attractor involved.

The sequence of jumping then from attractor to attractor may be
simulated as a semi-Markov process. The pacemaker here plays two roles:
on the one hand it is responsible for stimulating the attractors one-by-
one and on the other hand it is instrumental in making them commute as if
under the spell of external noise. One might ask what happens to these
attractors – memories when the pacemaker leaves them and "grabs" another
subgroup of cortical neurons. Do they dissolve into oblivion? The
answer is that the pacemaker simply "wakes up" these memories since by
making the cortical group involved coherent it helps it elevate itself
above the ambient neuronal noise. The consolidation of memory, however,
may be achieved via synaptic-membrane-genome interactions [27,16] - by
stimulating genes in the neuronal genome, which genes manufacture proteins
that renew ("recoat") the postsynaptic membrane sites of the population
involved - thereby ensuring a long-term "engram" of the particular memory.

When now stimulation comes from the environment (via the peripheral
nervous system) the degree of arousal of the ascending branch of the
reticular formation increases and the specific thalamic nuclei responsible
for the generation of the initial oscillatory activity get polarised by
amounts of time which are roughly proportional to the speed of information
pumped from the peripheral nervous system or proportional to the intensity
of the impinging external stimuli on the peripheral receptors. The result
is that the simple semi-Markov sequence so far describing the intermittent
processing by the thalamocortical pacemaker turns via a phase-transition
as it were to a composite semi-Markov process.

This means that during the time interval τ_{hi} the specific thalamic
nuclei are polarised, the oscillatory activity of the pacemaker stops,
and the system gets stuck at the previous attractor memory in excess of
the usual residence time τ_{Ri} that one should expect from simple intermittency. After selecting the next attractor but before moving to it the
pacemaker holds at the previous one by an amount of time equal to the
interruption interval of the specific thalamo-nuclei activity. So now the
scanner works on a different modus, namely the modus of metastable chaos.
The distribution of holding times is of course unknown. In some applications [19] it is feasible to consider a geometrical distribution.

Two questions arise now. Does the thalamocortical oscillator itself,
in spite of its manifestly irregular (noisy) activity, possess a low-
dimensional attractor, and secondly, how could one determine the dimensionality of the individual attractors-memories of the system? The issue in
the first question has been quite recently addressed [6,7] by collecting
E.E.G. strings during epilepsy, deep sleep, awaken regime, etc., and then
by trying to infer the dimensionality of the underlying dynamics. A
basic difficulty in such an enterprise is the essentially non-stationary
character of the E.E.G., especially during the awaken regime.

Nonetheless, people came up with low dimensionalities: about 2.1
for the epileptic, 4.1 for the deep sleep regime, and with high-
dimensionalities a ± b, b ∿ 0(a), a ∿ 7 to 8 [7] in the case of the
awake regime, which de facto shows the unreliability of the underlying
method when the modus operandi of the pacemaker is the metastable chaotic
(partial covering of many attractors). So the general model emerging
from the combination of the dynamical processes going on, within each
individual strange attractor-memory, and between coexisting attractors-
memories is the following. For a given "Bio-processor" a small subset
of environmental stimuli is partitioned in a number Λ of coexisting
attractors-memories whose formation is mediated by a non-linear dissipative

operator (map). These attractors are separated by fractal basin boundaries. The entropy of such a partition (of the messages to the attractors) and the degree of compressibility afforded by each attractor $C_i = N - D_i$ (where N the dimensionality of the raw environmental messages and D_i is the dimensionality of the individual memory) give the two essential macro-parameters characterising the cognitive channel between environment and processor at a given hierarchical level. (This process may not be accomplished in one single hierarchical step; in a second step the attractors (if numerous at a lower level) will play the role of the members of a hyper-Basin towards a new hierarchy of fewer hyper-attractors and so on.)

The attractors-memories involved establish further communication via the thalamocortical pacemaker - within the processor. This activity has two aspects:

a) The intra-memory activity (which involves rehearsal and consolidation) refers to a Markovian process within each attractor; this amounts to a slow diffusion from one part of the attractor to another, that is a progressive "smearing out" (mixing) of the specific initial stimulus. The memory of the Basin as a whole though remains intact. In the long run, we do not memorise specific events but rather sets of events unless we possess a phase coherent attractor where regeneration of trans-information is possible, thereby ensuring persistence of memory of any individual member of the basin as well.

b) The inter-memory activity refers to establishing an intermittent connectivity between the individual stable memories, and in the absence of external excitation this is a simple semi-Markov process; in the presence of stimulation it turns into a composite semi-Markov process with total holding times distribution depending not only on the intrinsic residence times but also on the statistics of the external stimulus, modulating the pacemaker itself. Let us end these series of speculations with one last topic: namely the spectrum of inter-memory jumping activity. If we consider the slow process of self-diffusion within each individual attractor-memory as resulting to an autocorrelation function exp $(-\frac{t}{\tau_i}$ i (1, ...,Λ) involving a single time scale,τ_{i2}, then the corresponding spectrum is simply a Lorenzian $\tau_i/(1+\omega^2\tau_i)$.

If we possess too many attractors and in some way we weight them in a scale-invariant fashion (e.g. like $\frac{1}{\tau_i}$) we might get as the spectrum of the processing activity

$$S(\omega) \sim \int_{\tau_1}^{\tau_2} \frac{\tau}{1+\omega^2\tau^2} \frac{d\tau}{\tau} = \frac{\tan^{-1}\omega\tau}{\omega} \Big|_{\tau_1}^{\tau_2} \tag{5}$$

and if $\frac{\tau_2}{\tau_1} \gg 1$ then $S(\omega) \sim \frac{1}{\omega}$ over a correspondingly large range of frequencies.

Is there any physical model which could justify the existence of this notorious "1/f-noise" for a biological processor involving the sequential jumping among coexisting memories? Arecchi et al. [1], inspired by the work of Montrol and Shlesinger [5] have suggested an interesting possibility: whenever an event is conditioned by a sequence of previous ones, assuming the probability per unit time of each step independent of the others, the probability of the final event is the product of the probabilities of the individual preliminary events

$$P = \prod_{i=1}^{\Lambda} P_i \text{ and } \log_2 P = \sum_{i=1}^{\Lambda} P_i, \; \Lambda \gg 1$$

So the density of $\log_2 P$ follows the normal distribution provided that the central limit theorem holds.

A variable X obeying a long-normal distribution has a density $\sim 1/X$ over a wide range. Apply this argument to the time constant $\tau \sim 1/P$ of an event conditioned by a chain of previous ones; you get a $1/\tau$ distribution providing the necessary scaling assumed above. To apply this theory to our physical model of jumping from attractor to attractor Arecchi et al. [1] consider a case where there is a "leakage" from one attractor to the next one (perhaps via a crisis) and the last one can be reached only via a unique chain and not via parallel ways; also the basin of stimuli feeds mainly the first attractor and all other attractors have negligible basins of attraction.

Alternatively, it is possible to reach a 1/f spectrum [3] even if the processor possesses only two attractors separated by a fractal boundary. Then the noise responsible for making the attractors commute will force the trajectory of any initial condition to perform a random-random walk. Spectra of E.E.G., however, under all behavioural conditions studied so far do not seem to conform with 1/f noise. This could be the case if one could identify τ_i as $\tau_{Ri} + \tau_{hi}$, e.g. accept that full memory mixing does take place before the jumping to the next attractor and that $\tau_{0i} \gg \tau_i, \tau_{Ri}, \tau_{hi}$. Note that we have four (different) time scale distributions in our problem:

a) The average relaxation times τ_i within each attractor. We may define operationally this "relaxation time" τ_i within an individual attractor as the time required for the transinformation to fall, say, by 1/2.

This transinformation I(t) is not uniquely defined, but depends on the map or flow (f) we choose and the given partition β with M states − subintervals β_i on this map (or quantized flow). The transinformation is given by the expression

$$I(t) = \sum_{i=1}^{M} \sum_{j=1}^{M} \bar{\mu}[f^{-t}(\beta_i) \cap \beta_j] \log_2 \frac{\bar{\mu}[f^{-t}(\beta_i) \cap \beta_j]}{\mu[f^{-t}(\beta_i)]\bar{\mu}(\beta_j)} \text{ bits} \qquad (6)$$

I(t) is then the information contained in an initial condition about the state of the system t time units ahead. The slope of I(t) is a measure of the loss rate of initial information, that is the rate of fading of the specific initial condition.

b) The transient time τ_{0i} from one attractor to another.

c) The residence times τ_{Ri} which depend on the mechanism of the forced intermittency − before the pumping of messages from the outside. In this regime the thalamocortical pacemaker is continuously on.

d) The holding times τ_{hi} − after the biasing of the specific thalamic nuclei and the selective interruption of their pacemaking function takes place. The way of synthesising τ_{Ri} and τ_{hi} to get an overall "residence time" on the ith attractor is an open problem.

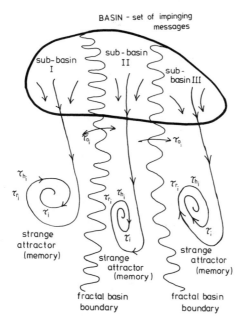

BASIN - set of impinging
messages

sub-basin
I

sub-basin
II

sub-
basin III

τ_{o_i} τ_{o_i}

τ_{h_i} τ_{r_i} τ_{h_i}

τ_{r_i} τ_{r_i} τ_{h_i}

τ_i τ_i

strange
attractor
(memory) τ_i

strange
attractor
(memory)   strange
attractor
(memory)

fractal basin
boundary

fractal basin
boundary

Fig. 2.   Sketch of a cognitive channel working after the dynamics of
"generalised intermittency" (see text); within each attractor
a "micro"-intermittency may go on as well (as for example in
the Lorenz system).

τ_i =   Relaxation time on an attractor.

τ_{r_i} =   Residence time before the interruption of the thalamocortical pacemaker.

τ_{h_i} =   Holding time, after the interruption of the thalamocortical pacemaker.

τ_{0_i} =   Transient time between attractors after leaving
attractor i.

The thalamocortical pacemaker is responsible for the jumpings
among the coexisting attractors-memories.

V   INFORMATION TRANSFER THROUGH GENERALISED MULTIHIERARCHICAL INTERMITTENCY

Random (semi-Markov) alternations from long residence in one attractor to short or long transients involving jumps between coexisting attractors

Look again at Fig. 2. We have the basin of all possible environmental stimuli (say 2^N strings of length N) partitioned amongst Λ
attractors. Most of the above series are incompressible transients and
the rest are falling in some of the coexisting attractors. This problem
of static partitioning becomes very difficult due to the (in general)
fractal character of the basin boundaries. In biological information
processing, however, the problem lies elsewhere. Since there exists an
active "internal agent" (the thalamocortical pacemaker) which provides an
intermittent regime of jumps from attractor to attractor, our interest
focusses on the time persistence of a given attractor-memory or the
estimation of transinformation $I(\xi;t)$ that is the average amount of
information contained in a prediction at time t corresponding to ξ jumps
into the future (don't forget the process is non-stationary). If
$P^{(\xi)}(i;j,t)$ is the (joint) probability that at time t the processor is
at the attractor j and at the attractor i ξ jumps later, the non-

stationary transinformation is given as

$$I(\xi;t) = \sum_{i=1}^{\Lambda} \sum_{j=1}^{\Lambda(\xi)} P(i,j;t) \log_2 \frac{P(i,j;t)^{(\xi)}}{P(i;t)P(j;t)} \text{ bits} \qquad (7)$$

All above entities can be in principle calculated from the solutions of Arecchi's kinetic equations. [Eq. (7) shows that I is "epoch-dependent".] Of course the fundamental practical difficulty in solving these equations lies in the estimation of the transitional probabilities P_{ij} which again can be traced to the difficulty in knowing the shapes of the individual basin boundaries.

Concerning now a single attractor (resulting from a flow or map, f) we introduce a stroboscopic process $f_t \rightarrow f_{t_n}$, $t_n = n(\Delta T)$, n=0,1,2,... and choose a partition β on the attractor of M,β_i elements. Let $\bar{\mu}$ be the invariant measure.

Then the entropy of the partition β with respect to the measure $\bar{\mu}$ will be

$$S_0(\beta) = - \sum_{i=1}^{M} P_i \log_2 P_i \text{ where } P_i = \bar{\mu}(\beta_i) \qquad (8)$$

The conditional entropy of a partition $f^{-t}(\beta)$ with respect to the partition β will be

$$S_c(f^{-t}(\beta)/(\beta)) = - \sum_{i=1}^{M} P_i \sum_{j=1}^{M} P_{j/i}(t) \log_2 P_{j/i}(t) \text{ with} \qquad (9)$$

$$P_{j/i}(t) = \frac{\bar{\mu}[f^{-t}(\beta_j) \cap \beta_i]}{\bar{\mu}(\beta_i)} \qquad (10)$$

This last relation (10) is identical to (2).

The transinformation of f standing for the amount of memory persistence of an initial condition on the attractor involved, is

$$I(t) = S_0 - S_c(t) \qquad (11)$$

where t = 0,1,2,.... stands for the discrete time or the iteration number (if we have one iteration per unit time). Expression (11) is identical to (6).

The above analysis assumes that all transients have already died out, so that one can speak of processes which statistics are given by the invariant measure $\bar{\mu}$. Under this regime, S_0 also is time-independent. $I(t)$ may decay monotonically or not depending on the partition chosen and the degree of coherence of the attractor involved that is on its mixing properties. $I(t)$, then, is a measure of the memory persistence. For $t > t_n$ for which $I(t_n) = 0$, we may say that the memory of the initial condition involved $I(0)$ has been completely erased.

Note finally that the regime within each attractor may also be intermittent (example: the Lorenz attractor for fairly large regions of the control parameters). We have two different categories of intermittency: "Common" intermittency involves a jumping process among

coexisting repellers (semi-attractors). "Forced" (or "generalised") intermittency involves a jumping process among coexisting stable attractors.
Hence "noise" is required here to provide for the jumps. So in general
we have the superposition of two intermittent regimes (intra- and inter-
attractor). Last but not least let us mention that the commuting process
may be semi-Markovian even in the absence of holding times (due to the
polarisation of the specific thalamic nuclei).

The semi-Markovian character of any intermittent process is due to
the fact that after reinjection from one regime on to another, the
residence time in it is not fixed, but it comes out from a mass density
function - characterising the type of intermittency involved.

All strange attractors perform "phase mixing" along the transverse
directions. Along the (longitudinal) direction of the flow itself,
however, mixing and loss of phase coherence varies according to the
specific attractor, the values of its control parameters and the specific
partition. "Regeneration" of transinformation due to phase coherence or
imperfect mixing is an interesting manifestation of "memory persistence"
and results in an oscillatory behaviour of $I(t)$ - in attractors whose
continuous spectra are interspersed with occasional high spikes (like the
Rössler system). For further details see Farmer et al. [29]. This
phenomenon of memory persistence of an individual initial condition on a
given attractor should be further pursued for the modelling or redesigning
of biological information channels based on the dynamics of intermittency.

Let us finally point at three (still unresolved) problems as
illustrated in the sketch of Fig. 2.

a) The problem of (static) partition of 2^N strings-messages of
length N into Λ coexisting attractors with dimensionalities of $D_i \ll N$.
The difficulty lies in the fractal character of the basin boundaries
(connected or, even worse, unconnected). In such cases no attractor can
claim around him a clean "sphere" of undisputed jurisdiction of radius
ε, however small, since the finite probability of intrusion of a "tongue"
of the fractal boundary relegates immediately any initial condition
within ε on to another attractor. However, since this happens for all
attractors the trajectory may "hesitate" for a long time thereby creating
very long, random walk-transients.

b) The problem of estimating the residence time mass function
$r(i,j,\tau_j)$ which determines the statistics of residence times on each
attractor j before the jumping to attractor i takes place. This parameter
depends on the intrinsic dynamics of a generalised (forced) intermittency
after which the individual attractors are permutated by the thalamocortical pacemaker - in the absence of externally pumped messages.

c) The problem of estimating the (composite) holding time mass
function $h(i,j,\tau_j)$ when now messages pumped from the peripheral to the
central nervous system interrupt the activity of the specific thalamic
nuclei (by amounts of time depending on the speed of peripheral
signalling), thereby causing the thalamocortical pacemaker to hold on to
a given attractor an extra amount of time.

Here we have then two intermittent mechanisms superimposed, one
intrinsic the other extrinsic and we do not yet know how to synthesise
them.

Let us finally conclude by mentioning once again the physical meaning
of the three basic concepts involving characteristics of attractors as far
as information processing is concerned. "Dimensionality": it stands for

the capacity of the dynamical memory (and its compressing or "descriptive" ability). "Transinformation": it stands for the degree of memory persistence (either, in a single attractor, or between attractors). "Basin;s width": it stands for the fault-tolerance ability of the attractor or its "predictive" ability.

REFERENCES

[1]   F.T. Arecchi and A. Califano, Phys. Lett. 101A: 443-446 (1984).
[2]   F.T. Arecchi, R. Badii and A. Politi, Phys. Rev. A32: 402-408 (1985).
[3]   F.T. Arecchi, "Noise-induced trapping at the boundary between two attractors: a source of 1/f spectra in nonlinear dynamics", preprint (1986).
[4]   B. Pompe, J. Kruscha and R.W. Leven, Z. Naturforsch. 41a: 801-808 (1986).
[5]   E.W. Montroll and M.F. Shlesinger, Proc. Natl. Acad. Sci. USA 79: 3380-3383 (1982).
[6]   A. Babloyantz and C. Nicolis, U.L.B., preprint (1985).
[7]   A. Babloyantz, in "Dimensions and Entropies in chaotic systems", G. Mayer-Kress, ed., Springer-Verlag, 241-245 (1986).
[8]   G. Nicolis and I. Prigogine, Proc. Natl. Ac. Sci. U.S.A. 78: 654-633 (1981).
[9]   J.S. Nicolis, Kybernetes 11: 123-132 (1982).
[10] J.S. Nicolis, Kybernetes 11: 269-274 (1982).
[11] D. Gikas and J.S. Nicolis, Z. Phys. B - Condensed Matter 47: 279-284 (1982).
[12] J.S. Nicolis, G. Mayer-Kress and G. Haubs, Z. Naturforsch. 38a: 1157-1169 (1983).
[13] J.S. Nicolis, J. Franklin Institute 317: 289-307 (1984).
[14] J.S. Nicolis, Kybernetes 14: 167-173 (1985).
[15] J.S. Nicolis and I. Tsuda, Bull. Math. Biology 47(3): 343-365 (1985).
[16] J.S. Nicolis, "Dynamics of hierarchical systems", Vol. 25, Synergetics, Springer-Verlag (1986).
[17] J.S. Nicolis, Rep. on Progress in Physics 49(10), 1109-1196 (1986).
[18] J.S. Nicolis, J. Milias-Argitis and C. Carabalis, Kybernetes 12: 9-20 (1983).
[19] J.S. Nicolis, E.N. Protonotarios and M. Theologou, Int. J. Man-Machine Studies 10: 343-366 (1978).
[20] J.S. Nicolis and E.N. Protonotarios, Int. J. Bio-Medical Computing 10: 417-447 (1979).
[21] John Maynard Smith, "Evolution and the theory of Games", Cambridge Univ. Press (1982).
[22] P.W. Anderson, Science 177: 393 (1972).
[23] L. Glass and R. Perez, Nature 246: 360-362 (1973).
[24] C. Grebogi, E. Ott and J. Yorke, Phys. Rev. Lett. 50: 935-938 (1983a).
[25] C. Grebogi, S. McDonald, E. Ott and J. Yorke, Phys. Lett. 99A: (1983b).
[26] G. Chaitin, Sci. Am. May: 47 (1975).
[27] J.S. Nicolis and M. Benrubi, J. Theoret. Biol. 59: 77-96 (1976).
[28] S.W. McDonald, C. Grebogi, E. Ott and J.A. Yorke, Physica 17D: 125 (1985).
[29] J.D. Farmer, J. Crutchfield, H. Froeling, N. Packard and R. Shaw, in "Non-linear Dynamics", R. Helleman, ed., Ann. of N.Y. Acad. of Sci. 357: 453-472 (1980).
[30] S.P. Layne, G. Mayer-Kress and J. Holzfuss, in "Dimensions and entropies in chaotic systems", G. Mayer-Kress, ed., Springer-Verlag, 246-256 (1986).

CHAOS IN ECOLOGY AND EPIDEMIOLOGY

W.M. Schaffer

Department of Ecology and Evolutionary Biology and
Program in Applied Mathematics
The University of Arizona
Tucson, Arizona 84721

ABSTRACT

   Methods of identifying chaos as the origin of irregularity in time
series are applied to observations and models from ecology and epidemiology.

I INTRODUCTION

   Since the mid-1970s, there has been a great deal of interest in non-
linear dynamics, particularly in the phenomenon called "chaos". Much of
the early work was motivated by considerations that were frankly biological: the growth of populations with discrete, non-overlapping generations
[23,25,15], the physiology of blood production [22] and the dynamics of
biochemical reactions [28]. Today, the study of chaos is a major enterprise [5,7], with convincing experimental evidence for its relevance to
electronics [44], optics [12], chemistry [29], and hydrodynamics [6]. A
recurrent theme is that apparently random fluctuations sometimes turn out
to be entirely deterministic. Indeed, chaos makes us rethink the very
idea of randomness.

   While the early work on chaos in theoretical biology found a receptive audience among physicists and mathematicians, biologists tended to
ignore it. One can guess the reasons: first, due to the treatment by
May [23,25], it was generally believed that chaotic behaviour can only
arise in the context of discrete dynamical systems. Second, until 1980,
no-one knew how to look for chaotic behaviour in a time series. Finally,
there was, and remains, a genuine lack of data which might reveal the
field-marks of chaos. The possibility that the irregular fluctuations
often observed in biological systems are chaotic was therefore either
rejected [17] or, more commonly, ignored. Ecologists, in particular,
continued to endorse the so-called "Balance of Nature", i.e. the idea that
populations are at or close to equilibrium. In epidemiology, the emphasis
[3] was on regularly recurrent epidemics. These were presumed to result
from external forcing in the form of seasonal variations in contact rates.
In neither case, however, are the empirical facts consistent with these

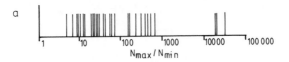

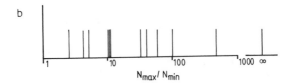

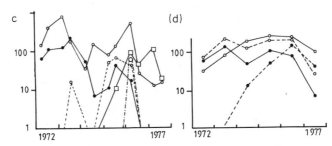

Fig. 1. Fluctuations in abundance of phytophagous insects. (a) Max/min ratios collated from the literature. The minimum number of generations or years that each population was studied is five, with a maximum of 60 and a mean of 19. (b) Max/min ratios for insects feeding on bracken (Pteridium aquilinum) at Skipwith Common, Yorkshire, over a period of 11 years. Data on six species of sawfly caterpillars were pooled. Six non-sawfly species became locally extinct during the course of the study. (c,d) Adult population densities for nine species of grassland leaf-hoppers (Auchenorrhyncha, Homoptera) studied for eight years at Silwood Park, Berkshire. (c) Bivoltine species. (d) Univoltine species. Note the pronounced changes in relative abundance, i.e. common species become rare, and rare species abundant. Reproduced from Strong et al. [42].

interpretations. Ecological populations, for example, often exhibit enormous variations [42]. Moreover, the relative as well as the absolute abundances of the different species generally fluctuate (Fig. 1). Thus, if there is a balance at all, it pertains to some overall statistical distribution [45,24] and not to the abundances of individual taxa. Similarly, while the incidence of some diseases is roughly cyclical, in other cases the pattern is more complicated (Fig. 2). One concludes either that chance perturbations are often paramount or that the deterministic component is chaotic.

III ATTRIBUTES OF CHAOS

What is chaos and how might we recognise it? We begin by enumerating the attributes of chaotic motion.

1. Determinism. The equations of motion are entirely deterministic. That is, there are no random inputs.

2. Complex dynamics. Chaotic systems exhibit sustained motion. They do not settle down to equilibria or simple cycles. Instead, one observes solutions which never repeat and are often highly irregular.

3. Sensitive dependence on initial conditions. Chaotic systems exhibit a property called sensitivity to initial conditions [31]. By this it is meant that nearby trajectories on average separate exponentially. Consequently, small differences in initial conditions are amplified, and since one can never specify a system's state with infinite precision, long term forecasting becomes impossible. Nonetheless, certain statistical properties of chaotic systems appear to be invariant. A good example is provided by the pair of difference equations first studied by Hénon [18]. Figure 3 (Top, Middle) shows the results of two simulations for which the initial conditions differed by the smallest number that the computer could distinguish from zero. After about 80 iterates, the time series diverge. Nonetheless, if the points are plotted in the X-Y phase plane, one observes essentially the same chaotic attractor for a wide range of initial conditions (Fig. 3, Bottom).

4. Road to chaos. Typically, one is interested not in single dynamical systems, but in whole families of equations which differ, the one from the other, in the value of a parameter. Often, a succession of dynamical states is observed as one varies the parameter. The transitions are called bifurcations, and one of the great triumphs of modern dynamics has been the association of "universal", i.e. system-independent, scaling laws [11] with the changes. Most familiar are one-dimensional, unimodal maps, for example, the discrete logistic equation. Here, we have

$$X_{n+1} = r X_n(1-X) \tag{1}$$

where X_{n+1} is again the next value of X_n, and r is a parameter. In Figure 4, we compare the asymptotic dynamics, i.e. after transients have died out, of Eq. (1) for a range of parameter values with those that result from a three species (one predator, two prey) continuous model (see below).

What is striking about Figure 4 is the overall similarity between the two pictures. Both bifurcation diagrams are laced with period-doublings. That is, there is the initial cascade on up to chaos, followed by the so-called "chaotic region". The latter contains truly chaotic solutions as well as periodic windows, wherein base cycles of various periods emerge and undergo their own sets of period doublings. Additionally, one sees pair-wise mergings of the dense bands, reverse bifurcations by which so-called [21] semiperiodic orbits coalesce.

The sequence of bifurcations observed in Eq. (1) is not the only route to chaos. In two-dimensional maps, i.e. difference equations with two dependent variables, chaos can develop by way of what is called an invariant circle. Here, the system hops around a closed curve which is topologically equivalent to a circle. Figure 5 gives an example for a discrete predator-prey model [20,35]. As the caloric value of the prey is increased, the invariant circle is replaced by a periodic orbit, in this case, a 7 point cycle. With further increases in the parameter, we observe period doubling until a 7 piece chaotic orbit emerges. With still further increases, the orbit is "smeared out", more or less filling the entire plane.

5. Fractal geometry. Often, chaotic motion exhibits noninteger, or fractal [10] dimension. Again, the Hénon map provides an example. Each of the apparent lines in the phase portrait, when suitably magnified, turns out to be composed of additional lines, which themselves are composed of more lines. In fact, at every length scale, one sees structure within

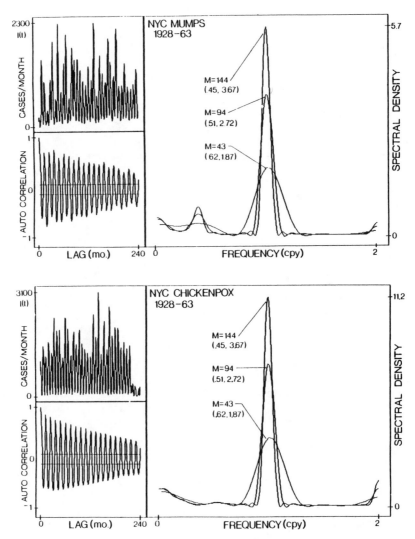

Fig. 2. Recurrent epidemics in three childhood diseases (monthly
         physicians' reports) in New York City and Baltimore, USA,
         1928-1963 [46]. Top: For chickenpox and mumps, the data
         suggest annual fluctuations. For measles, long term
         fluctuations are superimposed on the annual cycle. Time
         series shown in boxes at the upper left; autocorrelation
         functions in boxes at the lower left, and smoothed power
         spectra in main parts of the figures. Spectral smoothing
         was accomplished with a Tukey window after log transformation
         and 10% tapering of the data. For each data set, spectra
         were computed for three lag windows, M = 144, 94 and 43.
         Multiplicative confidence intervals shown in parentheses.
         The horizontal lines drawn through the autocorrelation
         functions represent 95% confidence intervals about zero.
         All time series were mean corrected. A linear trend was
         removed from the Baltimore data.

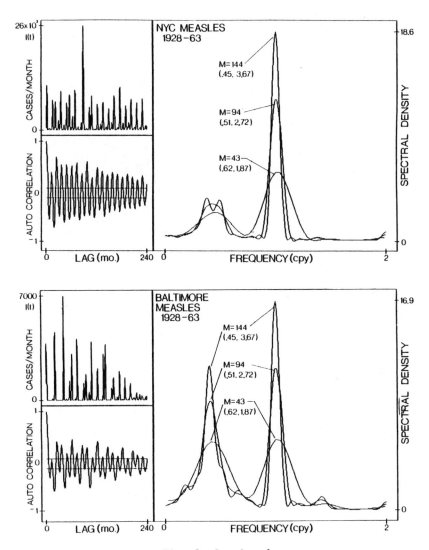

Fig. 2. Continued.

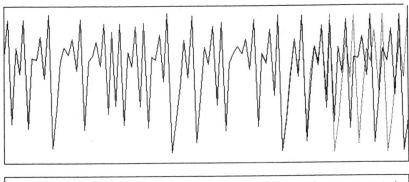

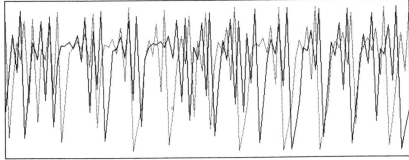

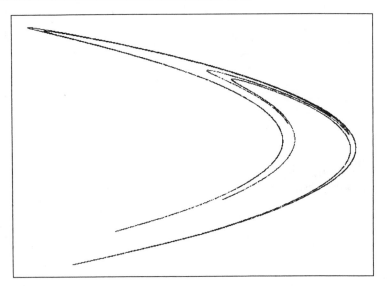

Fig. 3. Sensitive dependence on initial conditions in the Hénon Map.
Top, Middle: The time series, X(t) vs. t, for two simulations
for which the initial values of X differed by the smallest
number the computer could distinguish from zero. Bottom: The
output of one of the simulations plotted in the X-Y phase
plane. Essentially, the same picture results for a wide range
of initial conditions.

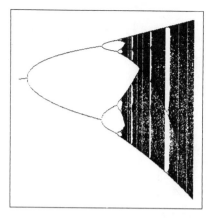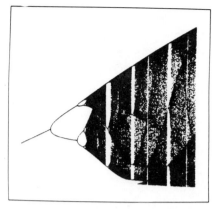

Fig. 4. Bifurcation diagrams for the logistic map (left) and Gilpin's
         [13] Equations (right), a three species Lotka-Volterra model.
         In both cases, the long term dynamics (successive maxima in
         the number of predators for Gilpin's model) are plotted
         against the value of a parameter. Note the overall similarity
         of the two pictures. As discussed in Section III, the
         continuous model exhibits a type of chaos not found in the 1-D
         map.

structure, ad infinitum [8]. Nor is this geometry accidental. Chaotic
motion typically entails the successive stretching and folding of
trajectories. For continuous systems, this produces a cross-sectional
structure resembling phyllo dough, i.e. a Cantor set. It also results
in the exponential separation of trajectories to which we have already
made reference.

III CHAOS IN ECOLOGICAL MODELS

     Figure 6 gives some examples of chaos in ecological models. Here,
we consider systems for which the dynamics are continuous. In Figure 6a,
we show the case of single species growth with a time delay. Specifically,
we have

$$dX(t)/dt = -DX(t) + B\ F(t-T) \tag{2}$$

where

$$F(t-T) = X(t-T)/[1 + X^z\ (t-T)] \tag{3}$$

Here, D, B and z are constants, and T is the time delay. The coordinates
of the phase space are $X(t)$, $X(t+T)$ and $X(t+2T)$, a choice which is
justified in the Section V. Equation (2) has been proposed [26] as a
model for the population dynamics of species with a long gestation period.
Similar equations arise in the study of red blood cell production [22,14].

     Figure 6b shows the three species, one predator, two prey model for
which we have already given a bifurcation diagram (Fig. 4). To obtain
the dynamics shown, one assumes that in the absence of predation, the
first victim species would outcompete the second. But the first victim is
also the predators' preferred food species, and when it becomes abundant,
the predators chow down. This allows the second victim to recover.
Underlying the figures are standard Lotka-Volterra equations. That is,
for each species, we write

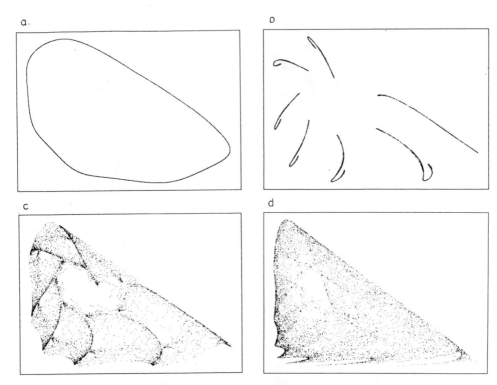

Fig. 5. Road to chaos in a discrete predator prey model. Shown are four asymptotic states obtained on increasing the bifurcation parameter. (a) Invariant circle. (b) Seven piece chaotic orbit. (c,d) Plane-filling chaotic orbits. Between (a) and (b) a seven point cycle emerges and undergoes period-doubling.

$$(1/X_i) \ (dX_i/dt) \ = \ r_i + \sum_{i=1}^{n} a_{ij} \ X_j \tag{4}$$

Here, X_i is the density of the ith species and the r's and a's are parameters. The possibility of chaotic orbits in systems of this sort was first discussed in a general way by Smale [41] who observed that given a sufficient number of species, namely 5, Eqs. (4) are consistent with arbitrary dynamics. Smale's observation was subsequently confirmed by Gilpin [13] and Arneodo et al. [4]. The latter authors also called attention to the possibility of homoclinic orbits and the relevance of a theorem due to Shil'nikov [40]. With the approach of homoclinicity, the simpler spiral chaos of Fig. 6b (left) gives way to the more complicated screw chaos (right). For details, see [37].

Figure 6c shows another example involving predation. Here, we have a two species, herbivore-plant system, for which the governing equations generate isoclines of the sort first described by Rosenzweig and MacArthur [30]. To this familiar scenario has been added [36] seasonal fluctuations in the caloric value of plant tissues. The figure itself is a so-call "time-one" or Poincaré map. It was constructed by plotting the densities of both species at time intervals equal to the period of the forcing function.

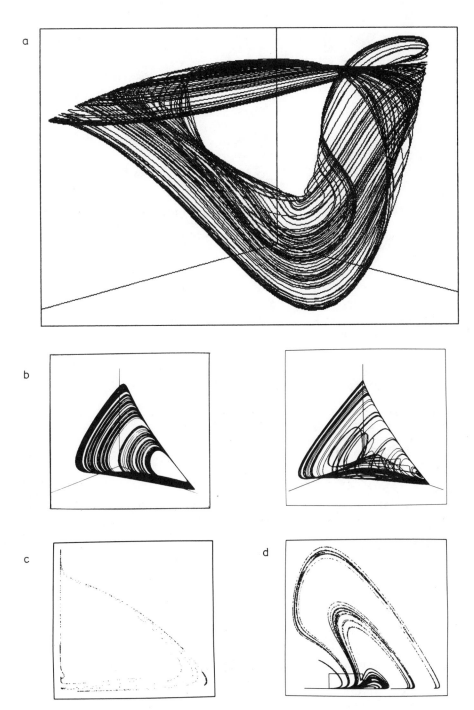

Fig. 6. Four examples of chaos in ecological models. (a) Single species delay differential equation. (b) Spiral (left) and screw chaos (right) in Gilpin's model for two prey species and one predator. (c) Two species plant-herbivore model with seasonal changes in plant quality. (d) SEIR epidemic model. In (c) and (d) the orbit is displayed as a "time-one" or "Poincaré" map.

Finally, in Fig. 6d, we show the so-called SEIR equations [3] which have been used to model the incidence of childhood diseases such as measles and chickenpox. Here, the population is divided into four classes: Susceptibles, Exposed, Infectives, and Recovered. Again, there is a seasonal factor, and the dynamics are represented as a time-one map.

IV REDUCTIONS IN DIMENSIONALITY

Often one can reduce the dimensionality of a chaotic system by slicing the orbit with a surface of lower dimension. This is called taking a Poincaré section, and the result can be informative. Consider, for example, what happens if we slice the solution to Gilpin's equations with a plane. The resulting section is indistinguishable from a one-dimensional curve and there consequently exists a mapping of the form

$$X_{n+1} = F(X_n) \tag{5}$$

relating successive points on the slice. Figure 7a shows two such 1-D maps. These correspond to the spiral and screw chaos of Fig. 6b.

The ability to extract essentially one-dimensional maps from chaotic flows is important. First, it suggests that some of the properties of one-dimensional maps, for example the bifurcation sequences shown in Fig. 4, may carry over the the more complex continuous system. Second, and despite sensitivity to initial conditions, the map can be used to make short term predictions.

Extracting one-dimensional maps from Gilpin's equations is possible because the orbit is essentially two-dimensional [8]. Of course, not all chaotic systems share this property. In this regard, the delay differential equation (2) is interesting, since it is equivalent to an ordinary differential equation of infinite order, i.e. equations such as (3) can be rewritten as an infinite set of ordinary differential equations. At the same time, it can be shown that the long term dynamics will be confined to a finite dimensional set. Modulo the ubiquitous period windows, increasing the time delay increases the orbit's dimension [8]. Nonetheless, for certain parameter values, a tolerable approximation to a one-dimensional map can still be obtained (Fig. 7b). The same trick can sometimes also be applied to partial differential equations, e.g. continuous systems with a spatial component, in general also of infinite dimension [19].

Not all low dimensional systems can be described by unimodal maps. Systems with periodic forcing, e.g. the predatory-prey model with seasonality, often have solutions which live on the surface of a torus. In this case, slicing the continuous system yields an invariant circle of the sort shown in Fig. 5. Then the appropriate one-dimensional description is a circle map. To construct the map, impose coordinate axes on the centre of mass, and for each point compute an angle, y. Then plot y_{n+1} vs. y_n, for all points. Just as the unimodal maps shown in Fig. 7 summarise the sequence of excursions in the higher dimensional system, so circle maps summarise the sequence of rotations on the invariant circle. Figure 8 shows two such maps for the discrete predator-prey model depicted in Fig. 5. In the second case, the invariant circle has broken up, and the motion is chaotic. Here the general form of the circle map persists, but it is no longer monotonic or unique.

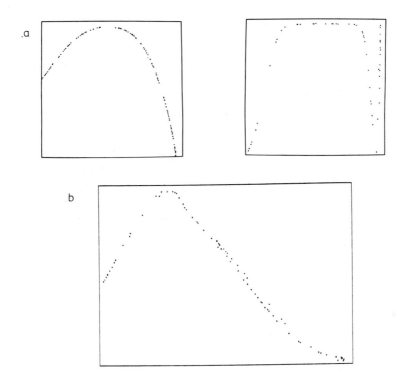

Fig. 7. Extracting one-dimensional maps from continuous systems. (a)
Gilpin equations. Left: Spiral chaos. The map has a single
extremum. Right: Screw chaos. The map has three extrema.
(b) Delay-differential equation (2), for a single species.
Although the Poincaré section has visible thickness, a
tolerable approximation to a 1-D map can still be obtained.

V DETECTING CHAOS IN NATURE

How does one go about detecting chaos in real data? The usual way
of looking for order in a time series is to compute its power spectrum.
However, in the case of chaotic systems, spectral analysis is not always
helpful. For example, Farmer et al. [9] computed spectra for a series of
well-known chaotic attractors first studied by Rossler in 1976. For each
of six parameter values, the motion was chaotic, but the spectra varied
enormously. Depending on the parameter values, everything from instru-
mentally sharp peaks to essentially featureless spectra were observed.
More recently, Anderson et al. [3] computed smoothed power spectra for
several childhood diseases. In some cases, they observed a single peak
corresponding to the annual cycle (e.g. Fig. 2). In others, there were
indications of multi-year periodicities. And in one instance, pertussis,
the qualitative form of the spectrum changed following the onset of mass
innoculations. However, a check on the SEIR equations in the chaotic
region suggests that such results be viewed with caution. Figure 9
shows results obtained from dividing a single chaotic trajectory (360
years) into 10 subsamples. Clearly, if the data sets are relatively
short, the same chaotic process can produce spectra that differ radically
in appearance. In fact, this sort of result is to be expected on the
following grounds. The onset of chaos in the SEIR equations is preceded
by an apparently infinite cascade of period doublings [1]. Thus, by the
time one reaches the chaotic region, all cycles of period 2^n, as well as
some others [33] have gone unstable. But they are still there! Hence,

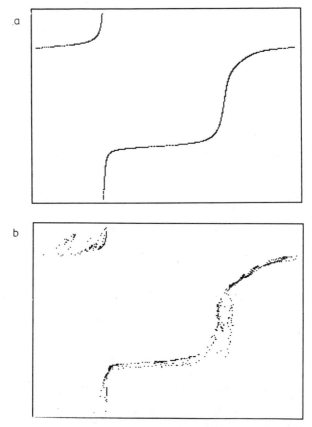

Fig. 8.   Circle maps in the discrete predator-prey model.   (a) The
motion is on an invariant circle (Fig. 5a).   (b) Chaos
following breakup of the invariant circle (Fig. 5c).

a chaotic orbit should "wander" among the former basins of attraction.
The same phenomenon, i.e. the tendency to remain in the vicinity of an
unstable orbit, produces the apparent "threads" that interlace the dense
bands in the bifurcation diagrams in Fig. 4.

For low dimensional systems, it may be more informative to plot the
motion in phase space. For experimental systems this poses a difficulty.
To construct a phase portrait, one needs to keep track of all the important variables over time. Generally, this is not possible. However, in
1981, Floris Takens [25] devised an alternative whereby one can produce
pictures like those in Fig. 4 from univariate time series. The technique
is deceptively simple. Suppose one has data for an unobservable, x,
taken at times t, t+T, t+2T, ... Suppose further that the number of
variables in the system is n. Then, for almost every variable x and
interval T the portrait constructed by plotting

$$x(t) \text{ vs. } x(t+T) \text{ vs. } \dots \text{ vs. } x[t+(m-1)T] \tag{6}$$

will have the same dynamical properties as the full system, provided

$$m > (2n+1) \tag{7}$$

Condition (7) is closely related to the Whitney embedding theorem [16].

Of course, the quantity (2n+1) can be a large number. Fortunately, Eq. (7) is merely a sufficiency condition. If trajectories are restricted to an almost two-dimensional surface, one can set m equal to three and view the motion directly.

VI CHAOS IN EPIDEMICS

Application of Takens' method has yielded convincing evidence for chaos in a variety of physical contexts [29,6]. In ecology, the results are less definitive because the available time series are relatively short. Taken together [32,33,34,35,36,38], however, they are nonetheless suggestive. By far the best evidence for chaotic dynamics comes from childhood diseases. Here, a relative abundance of information provides

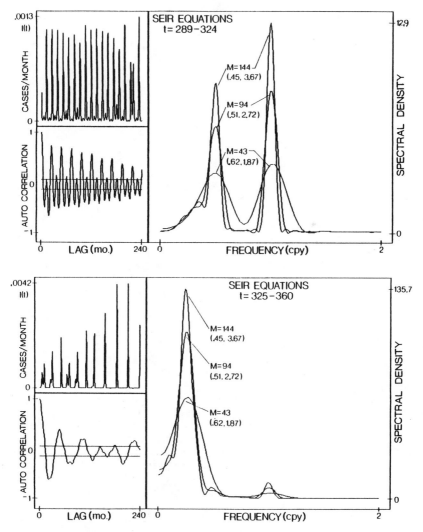

Fig. 9. Smoothed power spectra for the SEIR equations in the chaotic region. A single trajectory (360 years) was divided into 10 time series, each of which was equal in length to the data sets shown in Fig. 2. Shown here are the results of spectral analysis for two of the subsamples. Note the differences.

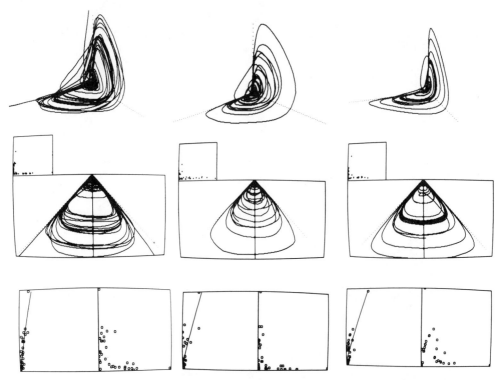

Fig. 10   Measles epidemics real and imagined. Top: Orbits reconstructed from the numbers of infective individuals. Middle:
Orbits viewed from above (main parts of figures) and sliced
with a plane (vertical line) normal to the page. Poincaré
sections shown in small boxes at upper left. Bottom: One
of the Poincaré sections magnified (left) and resulting "1-D"
map (right). Left: Data from New York. Centre: Data from
Baltimore. Right: Output from SEIR equations in the chaotic
region.

good evidence for the kind of chaos generated by the SEIR mode. In some
cases, the output of the model is virtually identical to actual epidemics.
Thus, in Fig. 10, we show three data sets. The first two are real - the
incidence of measles in New York and Baltimore. Here, the phase portraits
were constructed from the numbers of reported cases shown in Fig. 2. The
last data set is hypothetical - a solution to the SEIR equations for
parameters appropriate to measles in large cities [1]. By treating the
output of the simulation exactly as one is compelled to treat the real
world data, i.e. by lagging the numbers of infectives according to Takens,
one produces a picture which is remarkably similar to what is actually
observed. Comparable results for childhood diseases in Denmark have been
reported by Olsen [27]. One also obtains reasonable agreement (within
20%) with regard to quantitative attributes, e.g. fractal dimension,
positive Lyapunov exponent, of the motion. This concordance - between
mechanistic equations and real data - is instructive, because it
illustrates the role that theory can play. Considering just the data,
one obtains what looks like a unimodal map in the presence of noise, and
initially, the data were so interpreted [39]. There was, however, an
anomaly, namely that the fractal dimension (2.5 - 2.6) was much higher
than one would have expected. Later it was observed [33,34] that the
SEIR equations give about the right dimension (2.4 - 2.5). Moreover,

246

these equations yield the same "messy" approximation to a one-dimensional map observed in nature. Comparing the data and the theory thus accomplished two things: it established the true nature of the chaos, and it confirmed that the SEIR model accurately describes real epidemics.

ACKNOWLEDGEMENTS

This work was supported by the John Simon Guggenheim Memorial Foundation and by grants from the National Science Foundation and the National Institutes of Health. I thank L.F. Olsen for making available various manuscripts prior to their publication. As always, Mark Kot was an essential source of informed comment.

REFERENCES

[1] J.L. Aron and I.B. Schwartz, J. Theor. Biol. 110: 665-679 (1984).
[2] N.B. Abraham, J.P. Gollub and H.L. Swinney, Physica 11D: 252-264 (1983).
[3] R.M. Anderson and R.M. May, Science 215: 1053-1060 (1982).
[4] A. Arneodo, P. Coullet and C. Tresser, Phys. Lett. 79a: 259-263 (1980).
[5] H. Bai-Lin, "Chaos", World Scientific (1984).
[6] A. Brandstäter, J. Swift, H.L. Swinney, A. Wolf, J.D. Farmer and E. Jen, Phys. Rev. Lett. 51: 1442-1445 (1983).
[7] P. Cvitanovic, "Universality in Chaos", Adam Hilger Ltd. (1984).
[8] J.D. Farmer, Physica 4D: 366-393 (1982).
[9] D. Farmer, J. Crutchfield, H. Froehlin, N. Packard and R. Shaw, Ann. N.Y. Acad. Sci. 357: 453-472.
[10] J.D. Farmer, E. Ott and J.A. Yorke, Physica 7D: 153-180.
[11] M.J. Feigenbaum, J. Stat. Phys. 19: 25-52.
[12] H.M. Gibbs, F.A. Hopf, D.L. Kaplan and R.L. Shoemaker, Phys. Rev. Lett. 46: 474-477 (1981).
[13] M.E. Gilpin, Am. Nat. 113: 306-308 (1979).
[14] L. Glass and M.C. Mackey, Ann. N.Y. Acad. Sci. 356: 214-235 (1979).
[15] J. Guckenheimer, G. Oster and A. Ipaktchi, J. Math. Biol. 4: 101-147 (1976).
[16] V. Guillemin and A. Pollack, "Differential Topology", Prentice-Hall (1974).
[17] M.P. Hassell, J.H. Lawton and R.M. May, J. Anim. Ecol. 45: 471-486 (1976).
[18] M. Hénon, Commun. Math. Phys. 50: 69-78 (1976).
[19] Y. Kuramoto, in "Statistical Physics and chaos in fusion plasmas", C.W. Horton and L.E. Reichl, eds., 93-110, J. Wiley (1984).
[20] H.A. Lauwerier, in "Chaos", A.V. Holden, ed., 58-96, Manchester Univ. Press, Manchester (1986).
[21] E.N. Lorenz, Ann. N.Y. Acad. Sci.   : 282-291 (1980).
[22] M.C. Mackey and L. Glass, Science 197: 287-289 (1977).
[23] R.M. May, Science 197:
[24] R.M. May, in "Ecology and evolution of communities", M.L. Cody and J.M. Diamond, eds., 81-120, Belknap Press (1974).
[25] R.M. May, Nature 261: 459-467 (1976).
[26] R.M. May, Ann. N.Y. Acad. Sci. 357: 267-281 (1980).
[27] L.F. Olsen, Theor. Pop. Biol., in press (1986).
[28] L.F. Olsen and H. Degn, Nature 267: 177-178 (1977).
[29] J.-C. Roux, R.H. Simoyi and H.L. Swinney, Physica 8D: 257-266 (1983).
[30] M.L. Rosenzweig, M.L. and R.H. MacArthur, Am. Nat. 47: 209-223 (1963).
[31] D. Ruelle, Ann. N.Y. Acad. Sci. 316: 408-416 (1979).

[32] W.M. Schaffer, Am. Nat. 124: 798-820 (1984).
[33] W.M. Schaffer, IMA J. Math. Appl. Med. Biol. 2: 221-252 (1985a).
[34] W.M. Schaffer, Ecology 66: 93-106 (1985b).
[35] W.M. Schaffer, Trends Ecol. Evol. 1: 58-63 (1986).
[36] W.M. Schaffer, in "Evolution of life histories: theory and patterns from mammals", M. Boyce, ed., Yale Univ. Press, in press (1987).
[37] W.M. Schaffer, S.E. Ellner and M. Kot, J. Math. Biol. 24: 479-523 (1986).
[38] W.M. Schaffer and M. Kot, Biosci. 35: 342-350 (1985a).
[39] W.M. Schaffer and M. Kot, J. Theor. Biol. 112: 403-427 (1985b).
[40] L.P. Shil'nikov, Sov. Math. Dokl. 6: 163-166 (1965).
[41] S. Smale, J. Math. Biol. 2: 5-7 (1976).
[42] D.R. Strong, J.H. Lawton and R. Southwood, "Insects on Plants", Harvard University Press (1984).
[43] F. Takens, in "Dynamical systems and turbulence", D.A. Rand and L.-S. Young, eds., 366-381, Springer-Verlag (1981).
[44] J.S. Testa, J. Perez, and C. Jeffries, Phys. Rev. Lett. 48: 714-718 (1982).
[45] C.B. Williams, "Patterns in the balance of nature", A.P. (1964).
[46] J.A. Yorke and W.P. London, Am. J. Epidem. 98: 469-482 (1973).

LOW DIMENSIONAL STRANGE ATTRACTORS IN EPIDEMICS OF CHILDHOOD

DISEASES IN COPENHAGEN, DENMARK

Lars Folke Olsen

Institute of Biochemistry
Odense University
Campusvej 55
DK-5230 Odense M
Denmark.

ABSTRACT

The monthly reported cases of six childhood diseases in Copenhagen, Denmark over periods of 30-40 years were analysed by three different non-linear techniques: (1) Calculation of the correlation dimension, (2) calculation of the Lyapunov exponents from apparently 1D maps derived from Poincaré sections of the flows and (3) calculation of the maximal Lyapunov exponents from the flows. Combining the results from these three types of analysis leads to the following conclusions: three of the diseases (measles, mumps and rubella) evolve with chaotic dynamics, one disease (chicken pox) evolves with dynamics corresponding to a noise-perturbed damped oscillation and the remaining two diseases (pertussis and scarlet fever) evolve with dynamics that are completely obscured by random fluctuations.

I INTRODUCTION

The highly irregular outbreaks of some epidemic diseases have been intensively studied over decades. Until recently most theoretical descriptions of such epidemics used stochastic rather than deterministic laws of motion to explain the fluctuations in incidence rates [2,3]. Recently, however, applications of non-linear algorithms to the recorded data of some childhood diseases have resulted in an alternative explanation of these fluctuations [8-11]. It is suggested that the irregular dynamics are due to chaotic attractors and thus have a deterministic origin. For instance Schaffer and Kot [11] analysed the monthly recorded cases of measles epidemics in New York and Baltimore over a 37-year period by embedding the data in 3 dimensions and constructing Poincaré sections and return maps for the resulting flows. They showed that the return maps could be approximated by almost 1-dimensional humped maps. The Lyapunov exponents estimated for these maps were in the range 0.6 to 0.7 bit/iteration, suggesting that the dynamics of measles epidemics in the two cities were chaotic. Olsen [8] analysed the epidemics of measles, mumps, rubella and chicken pox in Copenhagen by calculating the correlation integrals of the data embedded in dimensions

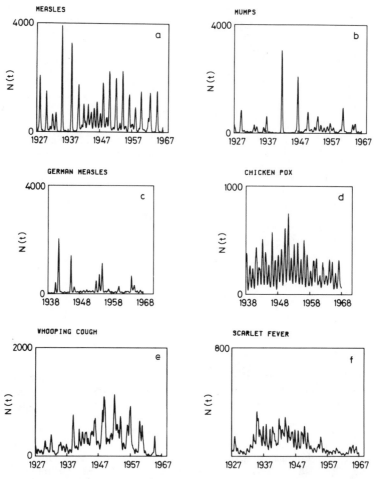

Fig. 1.   The monthly reported cases of measles, mumps, rubella,
chicken pox, pertussis and scarlet fever in Copenhagen.
Data from [7].

3-6 and found evidence for low dimensional (chaotic) attractors associated with the three former diseases. These results are in good agreement with numerical predictions obtained with epidemiological models
[1,8,9].

    In the present work the analysis of childhood diseases in Copen-
hagen [8] will be extended to include calculations of the maximal
Lyapunov exponents for the flows and the exponents estimated from 1-
dimensional maps derived from Poincaré sections of the flows. The
results support the previous suggestions that the epidemics of measles,
mumps and rubella are chaotic whilst chicken pox epidemics correspond
to a periodic oscillation perturbed by noise. The analyses of two
additional diseases - pertussis (whooping cough) and scarlet fever -
failed to provide evidence for deterministic behaviour due to too large
contributions from stochastic factors.

II   THE DATA

    Fig. 1 shows the monthly reported cases of the six childhood
diseases: measles, mumps, rubella (German measles), chicken pox,
pertussis (whooping cough) and scarlet fever over the periods 1927-1967
or 1938-1968 [7].   In order to analyse these data we need to know how
reliable they are.   The data were obtained from the City of Copenhagen
with a population size of 600,000.   The population size, the age structure and the birth rate were essentially constant over the 40-year
period of investigation.   The population size changed by ±10% and the
relative birth rate changed by ±2% in this period [7].   The reporting
rates of the diseases are high compared to the reporting rates of
measles, mumps and chicken pox in New York and Baltimore.   If we assume
that by the age of 20 every individual in the population has acquired
measles, we can estimate the reporting rate for measles by dividing the
accumulated number of cases by the accumulated number of live births
during a 20-year period.   We then obtain a reporting rate of 50%.   For
the other diseases infectibility is not 100%.   If we assume infectibil-
ities between 50 and 100% for these diseases we obtain reporting rates of
no less than 30%.   These reporting rates should be compared with the
rates of 12.5% and 25-30% for measles in New York and Baltimore respectively and 8-10% for chicken pox and mumps in New York [6].

III   CORRELATION EXPONENTS

    The calculation of correlation integrals and estimation of correla-
tion exponents will only be briefly discussed here.   The data were
embedded in dimensions 3-6 as $N(t)$, $N(t+h)$, $N(t+2h)$,..., $N(t+(m-1)h)$ [12]
where $N(t)$ is the number of cases in month t, h is a fixed time delay
(usually 2-5 months) and m is the embedding dimension.   The correlation
integrals were then calculated as described by Grassberger and
Procaccia [5].   For three of the diseases the log-log plots of correlation integral versus length scale defined straight lines at medium
length scales with slopes that eventually became independent of the
embedding dimension (for m>4).   For the remaining three diseases the
plots of correlation integrals versus length scale did not define
straight lines and the slope of the curves approached the embedding
dimension.   The results are summarized in Table 1.

Table 1.   Correlation exponents calculated from the data displayed
           in Fig. 1.

Disease	No. of points	υ
Measles	480	1.9 ± 0.2
Mumps	480	1.6 ± 0.1
Rubella	360	3.2 ± 0.1
Chicken pox	360	−
Pertussis	480	−
Scarlet fever	480	−

- indicates that the slope of the log-log plot of correlation
integral vs length scale did not converge when the embedding
dimension was increased.

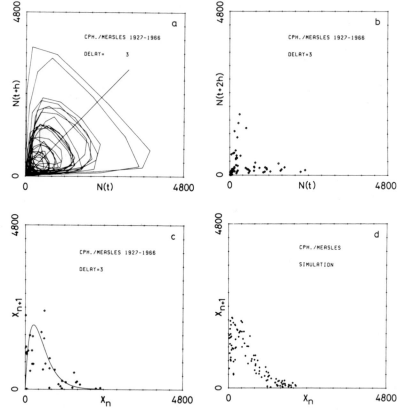

Fig. 2. Flow, Poincaré section and return maps of the measles data
constructed from embedding the data in 3 dimensions using a
time delay of 3 months. a, Projection of the flow in the
N(t), N(t+h) plane. b, Poincaré section of the flow using a
plane normal to the N(t), N(t+h) plane. The plane is indicated by the solid line in a. c, Return map constructed using
the points on the lower branch of the Poincaré section. The
curve represents the best fit of a 1D map to the points.
d, Simulation of the 1D map from c in the presence of 20%
noise (uniformly distributed).

IV FLOWS, POINCARÉ SECTIONS AND RETURN MAPS

The values of the correlation exponents indicate that an embedding
dimension of 3 may be sufficient to describe the flows. It could be
shown that for measles, mumps, rubella and chicken pox the flow of N(t),
N(t+h), N(t+2h) is almost 2-dimensional. We shall illustrate this for
the measles data. Fig. 2a shows the resulting flow of measles epidemics
projected in the N(t), N(t+h) plane. The straight line indicates the
plane (normal to the page) used for the construction of the Poincaré
section, which is shown in Fig. 2b. The Poincaré section is a 'V'
shaped curve indicating that the flow is confined to an almost 2-dimen-
sional conical surface. Fig. 2c shows a return map constructed from one
(the lower) of the branches of the Poincaré section. The return map may
be approximated by a 1D map of the type

$$X_{n+1} = aX_n \exp(-bX_n)$$

252

where a and b are constants. Such a map is indicated in Fig. 2c. We assume here that the deviations from the map are due to stochastic perturbations. Fig. 2d shows an iteration of the theoretical map in Fig. 2c following addition of noise to the values of X_n and X_{n+1}. The data for the other diseases were analysed in a similar way with the following results: the flows of mumps and rubella were almost 2-dimensional and the return maps could be approximated by 1D maps similar to the one obtained for measles. The flows obtained with scarlet fever and pertussis were far from being almost 2-dimensional and the return maps were randomly scattered points. The flow of the chicken pox data was almost 2-dimensional, but the return map was a random scatter of points. For those diseases that yielded almost 1D return maps the Lyapunov exponents were calculated from iterations of the maps. The exponents are listed in Table 2.

V LYAPUNOV EXPONENTS DERIVED FROM THE FLOWS

Recently Wolf et al. [13] have described an algorithm for estimation of the maximal Lyapunov exponent from the flow of a time series. The method was applied to the present data and the results are included in Table 2. We observe that the values of λ_1 obtained for measles, mumps and rubella are very close to those obtained from iterations of the return maps. The very low value of λ_1 obtained for chicken pox suggests that in the absence of perturbations this value should be zero. Although an apparently stable Lyapunov exponent could be estimated for pertussis it is not possible to draw any conclusions about the meaning of this value in the absence of any other evidence for a deterministic mechanism underlying the flow of the disease.

VI DISCUSSION

The combined results of the three non-linear techniques applied to the Copenhagen childhood disease data suggest that three of the diseases - measles, mumps and rubella - evolve with dynamics corresponding to low dimensional chaotic attractors. As for chicken pox the low value of λ_1 is consistent with dynamics corresponding to a noise perturbed periodic oscillation. However, the dynamics of the two remaining diseases - pertussis and scarlet fever - cannot be resolved with the present techniques. It is worth mentioning here that numerical simulations of an

Table 2. Lyapunov exponents estimated from 1D maps (λ) and from the flows (λ_1) of the six childhood diseases.

Disease	λ (bit/iteration)	λ_1 (bit/cycle)
Measles	0.67 ± 0.1	0.8 ± 0.2
Mumps	0.7 ± 0.1	0.7 ± 0.15
Rubella	0.45 ± 0.1	0.4 ± 0.1
Chicken pox	–	0.15 ± 0.1
Pertussis	–	1.2 ± 0.2
Scarlet fever		

epidemiological model, the so-called SEIR model [4] with seasonal variations in contact rates, using parameters corresponding to measles and
chicken pox respectively, reveal the same types of dynamics as those
observed for the real diseases when comparing correlation exponents,
flows and return maps as well as Lyapunov exponents (W.M. Schaffer and
L.F. Olsen, in preparation). It is interesting to note that the numerical simulation with parameters corresponding to measles showed chaotic
dynamics whilst the simulation with parameters corresponding to chicken
pox showed a periodic oscillation.

When comparing the present results with those from previous studies
of measles, mumps and chicken pox epidemics in New York and Baltimore
[9-11] we note that measles and chicken pox have the same dynamics in
Copenhagen as in the two North American cities whilst the dynamics of
mumps in North America apparently differ from the dynamics in Copenhagen.
However, recent numerical studies of the SEIR model using parameters
corresponding to mumps indicate that this disease may evolve with
dynamics corresponding to a time dependent alternation between periodic
and chaotic attractors (W.M. Schaffer and L.F. Olsen, in preparation).

ACKNOWLEDGEMENTS

I should like to thank the Danish Natural Science Research Council
for financial support and Dr. W.M. Schaffer for stimulating discussions.

REFERENCES

[1]   J.L. Aron and I.B. Schwartz, J. Theor. Biol. 110:665-679 (1984).
[2]   N.T.J. Bailey, "The Mathematical Theory of Infectious Diseases and
        its Applications", Charles Griffin & Co., London (1975).
[3]   M.S. Bartlett, "Stochastic Population Models in Ecology and
        Epidemiology", Methuen & Co., London (1960).
[4]   K. Dietz, Lect. Notes Biomath. 11:1-15 (1976).
[5]   P. Grassberger and I. Procaccia, Physica 9D:189-208 (1983).
[6]   W.P. London and J.A. Yorke, Am. J. Epidemiol. 98:453-468 (1973).
[7]   Medical Report for the Kingdom of Denmark 1927-1968, The National
        Health Service of Denmark, Copenhagen.
[8]   L.F. Olsen, in "Proceedings of Workshop on Dynamical Systems and
        Environmental Models", Wartburg Castle, Eisenach, GDR, March
        16-21, 1986, Akademie Verlag, Berlin (in press).
[9]   W.M. Schaffer, IMA J.Math. Appl. Med. & Biol. 2:221-252 (1985).
[10] W.M. Schaffer, this volume (1987).
[11] W.M. Schaffer and M. Kot, J. Theor. Biol. 112:403-427 (1985).
[12] F. Takens, Lect. Notes Math. 898:366-381 (1981).
[13] A. Wolf, J.B. Swift, H.L. Swinney and J.A. Vastano, Physica 16D:
        285-317 (1985).

PERIODICITY AND CHAOS IN BIOLOGICAL SYSTEMS: NEW TOOLS FOR THE STUDY

OF ATTRACTORS

J. Demongeot[1], C. Jacob[2] and P. Cinquin[1]

[1]University of Grenoble I   [2]INRA-CNRZ
TIM3-IMAG BP 68   Laboratoire de biométrie
38 402 St. Martin d'Hères   78 350 Jouy-en-Josas
Cédex   France
France

ABSTRACT

In many biological systems, the localisation of attractors of the dynamics and the quantification of the degree of synchrony is very difficult to obtain from experimental data. We propose here new notions of confiners and entropy to solve this problem and we give some examples of application (an "academic example" concerning the noised van der Pol oscillator and two realistic possible fields of application, namely respiratory physiology and population dynamics).

I INTRODUCTION

From enzymatic systems to ecosystems, we can meet many situations where the behaviour of the biological system shows noised oscillations (cf. [7], [15], [22], [23], [33]). Many authors have pointed out the fact that, because of the intrinsic random character of the observed phenomenon, the fit to a deterministic model has less explanatory power than a stochastic one ([2], [3], [27], [30]). In fact, the larger the variance of the noise, the more the behaviour of the noised system is different from that of the underlying deterministic system. For example, we can obtain random oscillations from a weakly attractant stable focus [23]. Then it has been proposed to define random oscillations systematically with respect to an underlying deterministic attractor, in general a limit cycle [33]; this approach needs the generalisation of the classical deterministic notions of stability and attraction basin [2,16,25]. The notion of a stochastic limit cycle has been introduced by C. Jacob [20,21,22], from the notion of a Markovian cycle defined by J.L. Doob [14] and S. Orey [29]. This notion has since been recognised as a particular case of the concept of confiner, the stochastic analogue of the concept of attractor (cf. [6], [8], [9], [11], [12], [13]). In this paper, we will define in Section II first attractors and then confiners; in Section III we will give some indications about the spatial estimation of a confiner and the temporal estimation of its period and entropy. Finally, in the last Section examples in respiratory physiology and in population dynamics will be proposed.

II DEFINITION OF ATTRACTORS AND CONFINERS

The definition of an attractor is based on the intuitive property of invariance in two reciprocal operations consisting firstly of searching its basin and after of taking the limit of this basin, when the time tends to infinity. We demand also the Ruelle-Bowen-connexity from an attractor, and we call c-connexity this property [8,9,11,13].

Now we can define on a dynamical system (X,f_t), where X is a compact state space and f_t a continuous flow on the product space $X \times T$ (T being the time space equal to N or R_+), the limit set $L(x)$ of the trajectory beginning in x as its classical ω_+-limit set. Then, for any subset A of X, we have: $L(A) = \underset{x \in A}{\cup} L(x)$ and $B(A) = \{x \in X \backslash A; L(x) \subset A \text{ and } \nexists y/x \in L(y)\}$; $L(A)$ is called the limit of A and $B(A)$ its basin. Then we can define an attractor A by:

(i) $LoB(A) = A$
(ii)   for each c-connex component C of A, there is no A' containing strictly C such that A' verifies (i)
(iii) it does not exist A" strictly contained in A such that A" verifies (i) and (ii).

We can easily prove that attractors are exactly the c-connex components of the set $L(X \backslash L(X))$; therefore, it suffices that $X \neq L(X)$ to ensure the existence of at least one attractor.

Let us now generalise to the notion of confiner; we have to define the stochastic analogues of the operator's limit and basin: let us consider a stochastic dynamical system $(X^T, \alpha^{\boxtimes T}, P_\mu)$, where X^T is the set of all the trajectories ω on X supposed to be metric, locally compact and separable, $\alpha^{\boxtimes T}$ is the product σ-field completed for the canonical probability P_μ induced by the initial measure μ; for each trajectory ω, $L(\omega)$ is a closed subset of X and we can prove that L is a random variable from $(X^T, \alpha^{\boxtimes T})$ to (F, B_F), where F is the set of all closed subsets of X and B_F the Matheron's σ-field on F (cf. Demongeot & Jacob (to appear)). The stochastic basin $B_\theta^\varepsilon(A)$ is defined by:
$B_\theta^\varepsilon(A) = \{x \in X \backslash A; P_x(E) f(x) > \theta\}$, where μ is supposed to have a density function f with respect to a reference measure on x denoted by ν, E being
$-\{L(\omega) \in A$, if this event is not trivial (in the case of a diffusion system for example, $L(\omega)$ is often P_μ-a.s. equal to X)
$-\{o_\omega(L(\omega) \backslash A) \in (\varepsilon, 2\varepsilon)\}$, if the staying time measure of Takens o_ω exists P_μ-a.s. (let us recall that o_ω is defined by:

$\forall C \in \alpha$, $o_\omega(C) = \underset{\notin}{\lim} \underset{t \to +\infty}{\underset{U \text{ open} \downarrow C}{\lim}} \lambda (s \leq t; \omega(s) \in U)/t$, where λ is the Lebesque measure, if $T = R_+$ and the counting measure, if $T = N$. If there exists a unique invariant measure, o_ω is P_μ-a.s. constant and equal to this measure). Let us remark that $P_x(E) f(x) > \theta$ expresses in any case the fact that $L(\omega)$ is "almost" contained in A.

Then let us denote by $L(B_\theta^\varepsilon(A))$, the union of all limit sets $L(\omega)$ of the trajectories ω beginning in the basin $B_\theta^\varepsilon(A)$. We express that A is approximately equal to this set, by assuming:
there exists a positive constant η such that, for any subset D of trajectories beginning in $B_\theta^\varepsilon(A)$ whose difference to the set of all trajectories beginning in $B_\theta^\varepsilon(A)$ has a P_μ-measure less than $\eta.\varepsilon$, we have:

$\nu(A \backslash L(D)) \leq \varepsilon$

We shall denote by: $L(B_\theta^\varepsilon(A) \overset{\varepsilon}{=} A$ this quasi-invariance principle. Finally, after a natural extension of the c-connexity (cf. Demongeot & Jacob (to appear)) in the stochastic c-connexity, we can define a

confiner A as a measureable subset of X verifying the conditions below
(such a confiner will be called a θ,ε-confiner, because of the presence
of the 2 degrees of freedom θ and ε):

(i) $L(B_\theta^\varepsilon(A)) = A$
(ii)   for each s-c-connex component C of A, there is not A' s-c-connex
       containing C such that $\nu(A')>\nu(C)$ and $A'\cup A$ verifies (i)
(iii)   there does not exist A" such as: $A"\neq A$, $A"\cap A\neq\emptyset$, $B_\theta^\varepsilon(A")\cap B_\theta^\varepsilon(A)\neq\emptyset$,
       $\nu(A")<\nu(A)$ and A" verifies (i) and (ii).

Examples:
a) if the stochastic system is a deterministic one having only its initial conditions random, following the distribution μ, if $supp\mu = X$ and
$E=\{L(\omega)\setminus A\subset N\}$, where $\nu(N)=0$, any $\theta,0$-confiner is ν-equal to an attractor;
b) if A is closed, ν-recurrent, super-absorbing (i.e. there exists an
open neighbourhood V(A) verifying:
$\forall\ x\epsilon A^c$, $P_x(L(\omega)\cap V(A)\setminus A\neq\emptyset = 0$ and $\forall\ x\epsilon V(A)\setminus A$, $P_x(L(\omega)\cap A^c\neq\emptyset) = 0$, then A is
a $\theta,0$-confiner, if $V(A)\subset supp\nu$, $\theta<\inf_{x\epsilon V(A)\setminus A} f(x)$, A non s-c-connected to A^c.

    For these two examples above, the intuition was waiting the results
obtained from the definition of a confiner.

    Let us now generalise the notion of bifurcation: if we suppose the
distribution P_μ depending on a parameter β, if A is a θ,ε-confident
obtained for the value β_0 and if $C(\delta)$ is the set of non-disjoint of A
confiners verifying the minimal weight for ν of the condition (iii), for
the value $\beta_0+\delta$ of the parameter, then β_0 is a geometric bifurcation value,
if there is in $C(\delta)$ a confiner having an homotopy type different from A
(Godbillon, 1971), for each δ sufficiently larger (resp. smaller) than
zero.

    If we consider now the temporal subsets T_ϕ^τ, made from all times
$\phi+k\tau$, equal to ϕ modulo an assay period τ, let us denote by C_ϕ^τ, the sub-
confiner obtained for the subdynamics corresponding to T_ϕ^τ, the origin of
the times being, for a trajectory ω, the entry time in a subset C_0 chosen
as reference in A; then, the η-period of A is defined by:

$$d=\inf\{\tau\epsilon\bar{T};\nu(A\triangle C_\phi^\tau)\leq\eta \text{ and } \not\exists\tau'\neq\tau;\inf_{\phi<\tau'}\nu(C_\phi^{\tau'})<\inf_{\phi<\tau}\nu(C_\phi^\tau),\ \sup_{\phi<\tau'}\nu(C_\phi^{\tau'})<\sup_{\phi<\tau}\nu(C_\phi^\tau)+\eta\}$$

For the convenient period d, the sub-confiners recover A and are approximately disjoint. We can consider 3 types of period: singular, if d=0,
period, if $0<d<+\infty$, and chaotic, if $d=+\infty$. If we change the type of period
at the place of type of homotopy, we call the bifurcation temporal.

III ESTIMATION

    We shall give indications about the estimation procedures for the
localisation and the period of a confiner, in the academic case of the
van der Pol oscillator. The ideal estimate is obtained by considering the
contour-lines of the staying time measure σ_ω, when we observe only one
trajectory ω. The contour-line having $1-\varepsilon\%$ of the empirical accumulation
points inside is then a good estimate for the localisation of a θ,ε-
confiner; more generally, if we have more than one trajectory beginning
in x, we can search for the minimal area subset A of X containing $M\theta\%$ of
such $1-\varepsilon\%$ contour-lines (μ being taken equal to the empirical measure of
the M initial points).

    A faster procedure can be used, if an a priori knowledge is in
favour of a localisation on a closed sub-manifold of X:

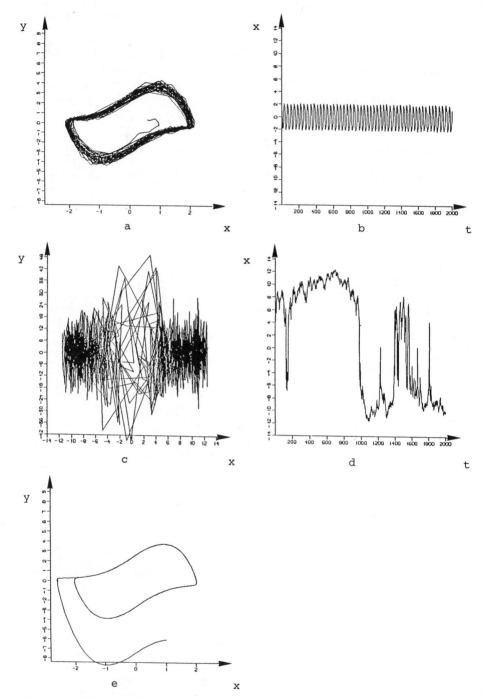

Fig. 1. Noised van der Pol system $dx/dt = y$, $dy/dt = -x+2(1-x^2)y+W(t)$,
where $W(t)$ is Gaussian centred with variance .1 (a,b) and
10,000 (c,d). (e) shows the deterministic underlying system;
let us note that the noise system remains a long time in the
"beak" regions of the limit cycle, when the variance of the
noise is increasing; the system has then the same behaviour
as a random bistable system.

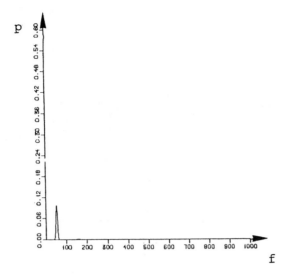

Fig. 2. Fourier power spectrum for the signal (b) of the Figure 1;
         after complex demodulation, the period has been found equal
         to 38.8.

- we determine the empirical accumulation points (sufficiently recurrent
with the help of any criterion of empirical recurrence);
- we determine the closed cubic spline curve minimising the least squares
criterion for the distances to the empirical accumulation points (if the
manifold has a dimension greater than 1, we determine after the closed
spline surface having boundary conditions of closure on the preceding
closed curve) and we keep 1-ε% of the nearest points.

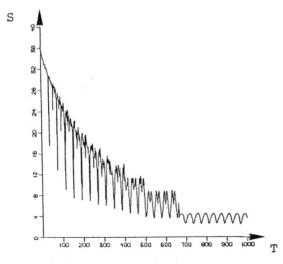

Fig. 3. S(T) represents the mean area of the Markovian cycle related
         to the signal (b) (Jacob, 1985). It shows a local minimum for
         each multiple of the period (here 40).

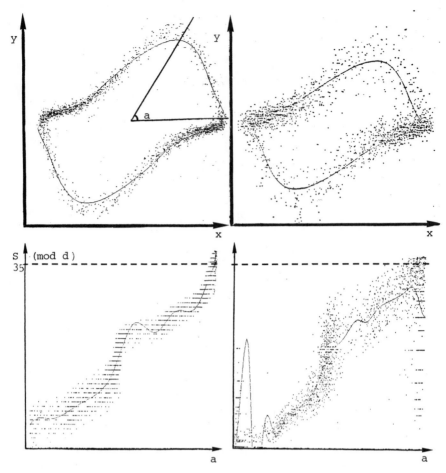

Fig. 4. On the left-hand side we have above the 5%-confiner and below
the spline estimating the period (=38 for the system (a); on
the right-hand side, we have the same pictures for a noised
van der Pol system with variance 1: let us remark that the
criterion for the spline below is worse than in the case of
variance .1; S(a) represent the times module d of the points
having a as polar angle.

The convergence of these estimates to a theoretical confiner, when
the number of observed points increases is related to the convergence of
the empirical measure of the accumulation points to the invariant measure,
whose Fokker-Planck equations can be obtained in certain cases [15].

The estimation of the period can be made by considering the sub-
confiners (Figure 5) or by searching the period d minimising the least
squares criterion of the cubic spline curve approximating the times S(a)
modulo d of the points whose polar angle with the barycentre of the
confiner above is a (Figure 4). If the criterion is too large, we can
use the staying time measure of Takens to calculate the Grassberger
fractal dimension; the calculation of this dimension and of the other
fractal dimensions (Renyi, Haussdorff-Besicovitch) could be made at the
beginning to define the dimension of the spline used in the step of
localisation.

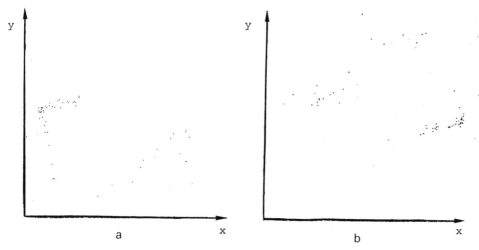

Fig. 5. Sub-confiners of phase 0 (a) and 19 (b) (d = 38).

The sub-confiners corresponding to the phase $\phi=0$ and to the phase $\phi=19$ (for an estimated period d=38) can be plotted for the system (a) (cf. Figure 5); the consideration of such sub-confiners has three advantages:
- their overlapping permits appreciation of the degree of variability of the periodic description for the whole confiner;
- by searching for the basin of a confiner of phase ϕ for the subdynamics T_ϕ^d, we define in fact the empirical isochron of phase ϕ; the isochronal landscape is particularly important to define the regions in which a biological system can be perturbed: if one sends the confiner in a domain where the isochrons are very spread, the system returns after perturbation to the same sub-confiner; it is an excellent way to synchronise for example a population of cells describing the same confiner, with the same period, but with different phases;
- if we consider a new trajectory, we can describe it backward by giving the genealogy (i.e. the succession) of the different isochrons in which it has been in the past (the assignment of an isochron being made for example by minimising the sum of the distances to the N nearest points of the isochron (N suitable to each observation)). Then we can define the Kolmogorov-Sinai entropy of this sequence (cf. for example Demetrius & Demongeot (submitted)). It characterises the degree of variability in the periodic description near the confiner and hence can be a good para-meter of synchronisability for the stochastic system.

One of the main interests of the notion of confiner lies in fact in the possibility of considering sub-confiners, their basins and the vari-ability in the transitions between sub-confiners.

IV CONCLUSION: POSSIBLE FIELDS OF APPLICATION

Many phenomena in biology present noised oscillatory or chaotic behaviours:
- in respiratory physiology, entrainment experiments can lead to periodic (Figure 7a) or chaotic (Figure 7b) recurrences, noised in the experimental data [4].
- in demography, the classical data relative to the Canadian lynx show a confinement in an annulus (cf. Figure 8 and [33]). In Figure 6 we observe

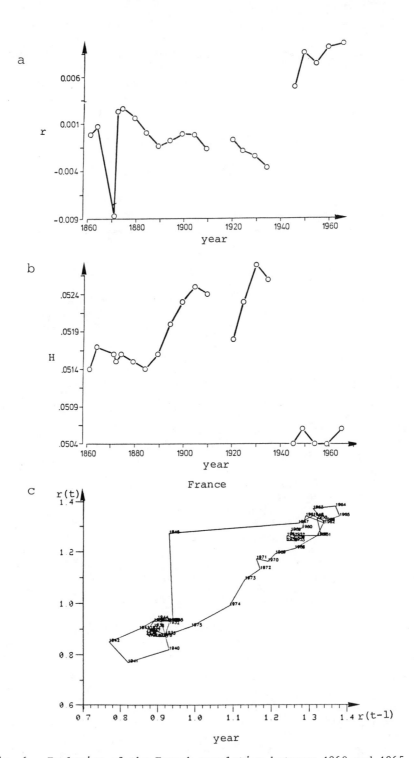

Fig. 6. Evolution of the French population between 1860 and 1965.
(a) represents the evolution of the Malthusian parameter r.
(b) represents the evolution of the population entropy H.
(c) represents the net fertility rate r(t) versus r(t-1).
(cf. Demetrius & Demongeot (submitted) and Bonneuil (personal
communication)).

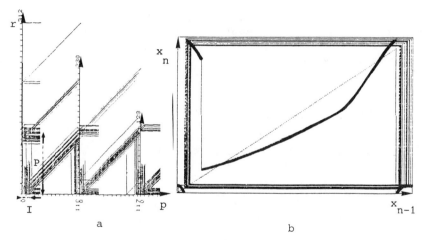

a

b

Fig. 7. (a) represents phase locked respiratory cycles (experimental
data in the rabbit); r (resp. p) corresponds to the respiratory time (resp. to the time relative to the external pump
forcing the respiratory system. P denotes the occurrence time
of the beginning of the inflation of the pump in the respiratory cycle; I denotes the occurrence time of the beginning of
the inspiration in the pump cycle. The experiment has been
processed for three different values of the entrainment
period. A big variability with respect to a unique ideal
cycle is observed. (b) represents the succession of the values
I for successive cycles (theoretical data); we can show that
the value x_n at time n is depending on the value x_{n-1} through
a discontinuous bimodal function of the interval, whose iterations show here a chaos (box counting dimension = .99,
information dimension = .89, correlation dimension = .79).

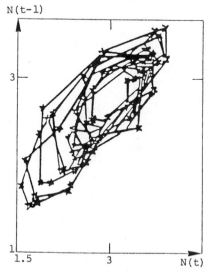

Fig. 8. Evolution of the size N(t-1) versus N(t) of the population of
Canadian lynx (logarithmic coordinates in years) (experimental
data).

two confiners (comparable to the case of the noised van der Pol system
with large variance), for the evolution of the net fertility rate: note
that the transient phase between these two confiners corresponds to a
local increase of the population entropy H (Demetrius & Demongeot (submitted)), and to an increase or decrease of the Malthusian parameter r
(depending on the sense of the transition from the low rate confiner to
the high or reciprocally from the high to the low one).
- in chemical or biochemical kinetics [17].

In these fields of application, the systematic use of the notions of
confiners, sub-confiners, stochastic isochrons and entropy can give an
alternative to the classical Fourier procedures.

ACKNOWLEDGEMENTS

This work has been written during a stay at the Centre for Mathematical Biology (Oxford University). We thank Prof. J.D. Murray, F.R.S. for
this opportunity.

REFERENCES

[1]   P. Baconnier, G. Benchetrit, J. Demongeot and T. Pham Dinh, Lect.
      Notes in Biomaths 49: 2-16 (1983).
[2]   M. Barra, C. Bruni and G. Koch, Lect. Notes in Biomaths 32: 140-154
      (1979).
[3]   M.S. Bartlett, Proc. Third Berkeley Symp. Math. Stat. & Prob. 4:
      81-109 (1956).
[4]   G. Benchetrit, P. Baconnier and J. Demongeot, "Concepts and formal-
      ization of breathing", Manchester Univ. Press (1987).
[5]   A.B. Budgor, K. Lindenberg and K.E. Shuler, J. Stat. Phys. 15: 375-
      391 (1976).
[6]   P. Cinquin, M. Cosnard, J. Demongeot and C. Jacob, in "Analyse
      numérique des attracteurs étranges", eds. M. Cosnard and C. Mira,
      Editions du CNRS, Paris (1987).
[7]   M. Cosnard, J. Demongeot and A. Le Breton, Lect. Notes in Biomaths
      49, Springer Verlag, New York (1983).
[8]   M.Cosnard and J. Demongeot, C.R. Acad. Sci. 300: 551-556 (1985).
[9]   M. Cosnard and J. Demongeot, Lect. Notes in Maths 1163: 23-32
      (1985).
[10] L. Demetrius and J. Demongeot, "Etude démographique de la France
      entre 1860 et 1965", submitted.
[11] J. Demongeot, M. Cosnard and C. Jacob, in "Dynamical systems - a
      renewal of mechanism", eds. Diner et al., World Sc. Publ.,
      Singapore (1986).
[12] J. Demongeot and C. Jacob, "Confineurs: une approche stochastique",
      C.R. Acad Sc., to appear.
[13] J. Demongeot, C. Jacob and P. Cinquin, in "Formalisation en biologie
      et en économie", eds. J. Demongeot and P. Malgrange, Presses Un.
      de Dijon (1987).
[14] J.L. Doob, "Stochastic processes", J. Wiley, New York 18953).
[15] W. Ebeling and H. Engel-Hebert, Physica 104A: 378-396 (1980).
[16] A. Friedman, "Stochastic differential equations and applications",
      Academic Press, New York (1976).
[17] D.T. Gillespie, J. Phys. Chem. 81: 2340-2361 (1977).
[18] C. Godbillon, "Eléments de topologie algébrique", Hermann, Paris
      (1971).
[19] D. Henry, Lect. Notes in Maths 840, Springer Verlag, New York (1981).
[20] C. Jacob, in "Time series analysis: theory and practice",
      ed. O.D. Anderson, North Holland, Amsterdam (1985).

[21] C. Jacob, "Stochastic limit cycles and confiners: definitions and comparative study in the case of Markovian processes", Lect. Notes in Maths, to appear.

[22] W. Jager and J.D. Murray, Lect. Notes in Biomaths 55, Springer Verlag, Berling (1984).

[23] C. Jeffries, Lect. Notes in Biomaths 2: 123-131 (1974).

[24] M.A. Krasnosel'skii, Translation operator along trajectories of differential equations. Transl. of Math. Monographs, AMS, Providence (1968).

[25] H.J. Kushner, "Stochastic stability and control," Ac. Press, London (1967).

[26] G. Matheron, "Random sets and integral geometry", J. Wiley, New York (1975).

[27] R.M. May, "Stability and complexity in model ecosystems", Princeton Un. Press, Princeton (1974).

[28] J. Neveu, "Bases Mathématiques cu calcul des probabilités", Masson, Paris (1970).

[29] S. Orey, "Lecture notes on limit theorems for Markov chain transition probabilities", van Nortrand Rheinhold Math. Studies, London (1971).

[30] R.A. Parker, Lect. Notes in Biomaths 2: 174-183 (1974).

[31] T. Pham Dinh, J. Demongeot, P. Baconnier and G. Benchetrit, J. Theor. Biol. 103: 113-132 (1983).

[32] W.M. Schaffer, S. Ellner and M. Kot, "Effects of noise on some dynamical models in ecology", J. Math. Biol., to appear.

[33] H. Tong, Lect. Notes in Stats. 21, Springer, Berlin (1983).

POPULATIONS UNDER PERIODICALLY AND RANDOMLY VARYING GROWTH CONDITIONS

M. Markus[1], B. Hess[1], J. Rössler[2] and M. Kiwi[3]

[1]Max-Planck Institut für   [2]Depto. de Fisica
Ernährungsphysiologie   Facultad de Ciencias
Rheinlanddam 201   Universidad de Chile
D-4600 Dortmund 1, FRG   Casilla 653
                             Santiago, Chile
[3]Facultad de Fisica
Universidad Catolica
Casilla 6177
Santiago, Chile

I INTRODUCTION

A large number of natural populations result from single generations that do not overlap, so that population growth occurs in discrete steps. The growth of a single species can then be described by an equation of the type

$$X_{t+1} = f(X_t) \tag{1}$$

where we consider the time lapse between two generations as time unity. Examples of this type of poulation are many temperate zone arthropod species with one short-lived adult generation per year [15], bivoltine insects (i.e. insects having a summer and a winter generation [8]) and cicadas with adults emerging every 13 years [13].

The functions f in eq. (1) that have been proposed in the literature all share the following features: (a) $f(0)=0$, i.e. if the population vanishes, it will remain zero forever after; (b) $f(N)$ increases mono-tonically for small values of N; (c) $f(N)$ decreases monotonically for large values of N because of growth restrictions imposed by the environ-ment. The simplest model which contains these features is the well-known logistic equation

$$X_{t+1} = r X_t (1-X_t) \tag{2}$$

In spite of its simplicity, this equation leads to phenomena that are qualitatively of the same kind as those obtained from more complex recurrence equations [14]. Its disadvantages are: (a) it reflects only crudely the ecological observations; and (b) for $X_t>1$ it is not biologic-ally acceptable since it leads to negative populations.

A biological more realistic ansatz is the Ricker equation [18]:

$$X_{t+1} = X_t \exp [r(1-X_t)] \tag{3}$$

It has been recognised [2,10,13] that this recurrence equation has properties closely related to the logistic differential equation

$$\frac{dN}{dt} = r\, N(1-N) \tag{4}$$

which in turn was found to describe the growth of a variety of species: the protozoan Paramecium, yeast cells, E. coli, Drosophila, and various flour beetles [9,15].

Ricker introduced equation (3) to describe salmon populations from the Pacific coast of Canada. Equation (3) has also been found to describe epidemiological systems. In fact, this equation fits to the records of incidences of measles, mumps and german measles in Copenhagen [17]. In addition, similar recurrence maps were obtained for the incidence of measles in New York City and Baltimore [21].

It has been shown [14] that equations (2) and (3) may lead to constant populations as well as to periodic and chaotic oscillations. Such oscillations have also been found in nature. For a review on periodic population oscillations one may consult reference [15]. Especially impressive examples in what concerns amplitude and predictability are the periodic oscillations of the Colorado potato beetle Leptinotarsa [7] and the almost legendary lemming [5]. Chaotic oscillations are exemplified in ecology by the Canadian lynx [19], the Australian insect pest Thrips imaginis [4,22], and the devastating "gipsy moth" Lymantria dispar [1]. A review table of chaos in ecological and epidemiological systems is given in [20]

As written in equations (2) and (3), r is independent of t. Populations, however, are subject to more or less unpredictable changes in environmental conditions, such as temperature, rainfall and human intervention. These changes influence survival and fecundity rates and we incorporate them by transforming equations (2) and (3) into:

$$X_{t+1} = r_t\, X_t\, (1-X_t) \tag{5}$$

$$X_{t+1} = X_t \exp [r_t(1-X_t)] \tag{6}$$

Here, r_t may assume different values in a periodic or random sequence. For simplification, we make the dichotomic hypothesis that r_t can assume only two values A and B ("good" and "bad" environmental conditions). In the special case A=B, i.e. r_t = const, equation (5) and (6) reduce to equations (2) and (3).

For a convenient representation of conditional probabilities, the Cowley parameter γ [3] is defined as

$$\gamma = 1 - \frac{P_{B|A}}{P_A} \tag{7}$$

Since $P_A\, P_{A|B} = P_B\, P_{B|A}$, $P_A + P_B = 1$, $P_{A|B} + P_{A|A} = 1$ and $P_{B|A} + P_{B|B} = 1$, we can write the conditional probabilities as functions of P_A and γ: $P_{A|A} = \gamma + P_A(1-\gamma)$, $P_{B|B} = 1 - P_A(1-\gamma)$, $P_{A|B} = (1-P_A)(1-\gamma)$ and $P_{B|A} = P_A(1-\gamma)$. In the present work, we restrict to cases where $P_A = P_B = 0.5$. Different "degrees of randomness" are obtained by different $\gamma \epsilon [-1,1]$. Special cases are:
(a) $\gamma \to 1$, meaning $\{r_t | t=1,2,3, \dots\} = \{\dots AAAAABBBBB\dots\}$.

(b) γ=C which describes the truly random situation without short range correlations. Analytically this case is defined by $p_{B|A}=p_A$.
(c) γ= -1, meaning $\{r_t|t=1,2,3,...\} = \{ABABABAB...\}$ or $\{BABABABA...\}$, i.e. A and B alternate periodically. This case has already been studied to some extent [8,16]. For bivoltine insects, the condition γ = -1 simulates the alternation of spring and fall.

In what follows, we study equations (5) and (6) for different A, B and γ. In particular, we are interested in computing the Lyapunov exponent λ, as an indicator of order (λ<0) or chaos (λ>0) (see e.g. [23]). The value of λ tells us how well we can forecast the system behaviour. For λ<0, uncertainties in the initial conditions are damped out. For λ>0, any uncertainty in the initial conditions will grow exponentially, thus making long-term predictions impossible, even if all details of the mechanism were known. The magnitude of a positive λ is a measure of the inability to forecast "outbreaks" and "crashes" in agroeconomical pests or epidemic diseases.

II METHODS OF ANALYSIS AND DEFINITION OF "COUPLING PROCESSES"

Starting at a "seed" $X_t=0.35$, equations (5) and (6) were first iterated 600 times to allow for transients to die away. Then, N iterations were performed to compute the Lyapunov exponent λ using the equation

$$\lambda = \frac{1}{N} \sum_{t=1}^{N} \log_2 \left| \frac{dX_{t+1}}{dX_t} \right| \tag{8}$$

We set $N=4 \times 10^4$ (except for Fig. 6, where $N=2 \times 10^5$). At each iteration, r_t was set equal to A or to B, according to a random number generator (subroutine FA01AS from [6]) under consideration of the given γ. All operations were performed using double-precision arithmetics.

We introduce here a nomenclature to describe the effect of A and B on the behaviour of the system. We write the "coupling process" $\xi + \eta \rightarrow \zeta$, where ξ, η and ζ can be C, 0 or ∞. "C" means chaotic behaviour (λ>0), "0" means ordered behaviour (λ>0) and "∞" means divergence of X_t. The latter case is obtained for example with the logistic equation (5) for a constant $r_t>4$, for which X_t diverges to -∞. Given A and B, ξ describes the behaviour for the sequence {AAAAAA...}, η the behaviour for {BBBBBB...}, and ζ the behaviour for variation between A and B according to a specified γ. We give some examples:
(a) 0 + 0 → C means: A alone leads to order, B alone leads to order, and variation between A and B to chaos;
(b) 0 + C → 0 means: A alone leads to order, B alone leads to chaos, and variation between A and B to order;
(c) ∞ + 0 → C means: A alone leads to divergence, B alone to order, and variation between A and B to chaos.

III RESULTS

The sign of the Lyapunov exponent λ in dependence of A and B is given in Figs. 1 to 5 and 7 to 8. A black point on these figures indicates λ>0 (chaos), and a white point λ<0 (order). γ was held constant for each figure. Figures 1 to 5 were calculated with the logistic equation (5), Figs. 7 to 8 with the Ricker equation (6). The diagonal indicated by the letter D in the figures corresponds to the condition r_t=const, i.e. to equations (2) or (3).

269

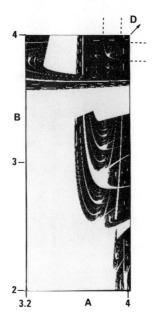

Fig. 1. Sign of Lyapunov exponent (white: negative, black: positive)
for the logistic equation (5) with r_t alternating periodically
between A and B (starting with A). Step length for A and B:
3.2×10^{-3}.

Figure 1 corresponds to the case of periodically alternating A and B
($\gamma = -1$). This figure is asymmetric with respect to the diagonal D,
since the sequence {ABABAB...} gives, in principle, a different result
than the sequence {BABABA...}. This occurs because the composed mapping
f_{AB} resulting from the two functions $f_A(X_t)=AX_t(1-X_t)$ and $f_B(X_t)=BX_t(1-X_t)$
has at least two attractors in some regions of the (A,B)-plane. Note that
in contrast to Fig. 1, Figs. 3 to 5 and 7 to 8 are all symmetric with
respect to the diagonal D, owing to randomness. In Fig. 1, we left out
the cases where the logistic equation diverges, setting A,B≤4. Note that
all cases of coupling processes (except those involving ∞) occur in
Fig. 1:   0+0→0, 0+0→C, 0+C→0, 0+C→C, C+C→0 and C+C→C.

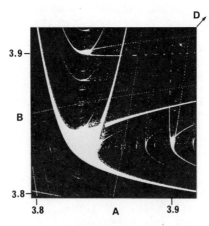

Fig. 2. Magnification of the square delimited by dashed lines in the
upper right of Fig. 1 (neighbourhood of the window with
period 3 obtained with the logistic equation). Step length
for A and B: 2.5×10^{-4}.

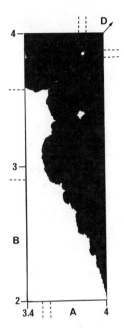

Fig. 3.   As Fig. 1, but with random variation between A and B ($\gamma=0$).
Step length for A and B: 3.2×10^{-3}.

To locate these cases, one just needs to compare the behaviour at
the point (A,B) with the behaviour at the two points (A,A) and (B,B) on
the diagonal D.

Figure 2 shows an enlargement of the square delimited by the two
pairs of dashed lines at the (upper and right) edges of Fig. 1.   In this
enlargement, we see the neighbourhood of the "window" with period 3 that
is obtained with equation (2) within the chaotic region (see e.g. [14]).
The shapes in Fig. 2 are closely related to curves on the (A,B)-plane
where the composed mapping f_{AB} is superstable [11].

When leaving an ordered (white) region in Fig. 2, the route to chaos

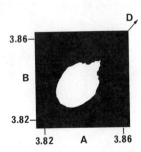

Fig. 4.   Magnification of the square delimited by dashed lines in
the upper right part of Fig. 3 (neighbourhood of the window
with period 3 obtained with the logistic equation).   Step
length for A and B: 10^{-4}.

depends on the way one proceeds. Some transitions occur through intermittency and others through a cascade of bifurcations. However, there is
no clear cut boundary between intermittency and bifurcations, and moreover, there are self-replicating copies throughout. A detailed study
shows that fractal structures appear at the edges of the "tails" in Fig.
2, leading to a consistent understanding which will be made available
shortly [11].

Figure 3 shows an excerpt of the (A,B)-plane with a random sequence
of A and B ($\gamma=0$). A comparison with Fig. 1 shows that most periodic
"islands" in the chaotic region are "smeared out" due to randomness.
Moreover, the case $0+0 \rightarrow C$ appears for values A and B below the "chaotic
threshold" r=3.57 of equation (2). Thus, the random character of the
process can "accelerate" the onset of chaos. The region where this phenomenon takes place is delimited by the two pairs of dashed lines at the
lower and left edges of Fig. 3. The region around the period 3 window
for $\gamma=0$ is shown in Fig. 4. As compared to Fig. 2, Fig. 4 shows once more
how randomness "smears out" the ordered islands. However, randomness does
not destroy the order in the immediate neighbourhood of the period 3
window, that is if A and B are close enough to each other.

Figure 5a shows the neighbourhood of this window for variation
between A and B with a low degree of randomness ($\gamma = -0.98$). We see
here the remarkable case $C+C \rightarrow 0$ ("chaos and chaos produces order") in spite
of the superposed randomness. This effect is seen more clearly in the
enlargement given in Fig. 5b which corresponds to the square delimited by
the two pairs of dashed lines at the (upper and right) edges of Fig. 5a.
In fact, there are white (ordered) points (A,B) such that the diagonal
points (A,A) and (B,B) are black (chaotic). These white points are inside
the rectangle delimited by dashed lines at the lower right of Fig. 5b as
well as on the symmetrically opposed rectangle in the upper left. However,
points obeying $C+C \rightarrow 0$ are very difficult to detect for $\gamma > -0.9$.

Table 1 shows examples for the different cases of coupling processes
obtained with the logistic equation (5). It can be seen in this table
that all possibilities for $\xi + \eta \rightarrow \zeta$ containing 0, C and ∞ occur, except
that ζ can only be ∞ if ξ or η are ∞. Note that sequences where ξ or η
are divergent (∞) will diverge if $\gamma > -1$. The reason is that any randomness allows X_t to escape to $-\infty$ with a finite probability. All examples

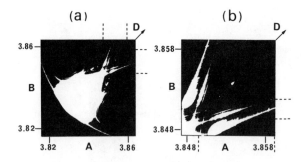

Fig. 5. (a) Neighbourhood of the window with period 3 obtained with
the logistic equation. A and B change randomly with $\gamma = -0.98$.
Step length for A and B: 10^{-4}. (b) Magnification of square
delimited by dashed lines in upper right part of (a). Step
length for A and B: 2.5×10^{-5}.

Table 1: Examples of coupling processes for the logistic equation (5). The degree of randomness of r_t is determined by the Cowley parameter γ. "O" means order, "C" means chaos and "∞" means divergence. The sets Ω, Γ, Δ, P_3 and P_6 are defined for r_t = const: Δ =]4,∞] (divergence), Ω = [0,0.357] (order), $\Gamma \subset$ [3.57,4] (chaos), P_3 (period 3 window) and P_6 (period 6 window).

Coupling process	A	B	γ	Remark	
O+O→O	3.4	2.8	[-1,0.99]	A,B∈Ω	
	3.39	3.83	[-1,0.99]	A∈Ω	B∈P₃
C+C→C	3.6	3.9	[-1,0.99]	A,B∈Γ	
C+O→O	2.3	3.7	[-1,0.99]	A∈Ω	B∈Γ
	3.828	3.834	[-1,0.99]	A∈Γ	B∈
C+O→C	2.9	3.9	[-1,0.99]	A∈Ω	B∈Γ
	3.6	3.83	[-1,0.99]	A∈Γ	B∈P₃
O+O→C	3.289	3.848	[-1,0.95]	A,B∈P₃	
	3.025	3.569	[-1,0.9]	A,B∈Ω	
	3.08	3.52	[-0.77,0.2]	A,B∈Ω	
C+C→O	3.85232	3.84971	[-1,-0.89]	A,B∈Γ and close to P₃	
	3.571	3.823	[-1,-0.88]	A,B∈Γ	
	3.58	3.98	-1	A,B∈Γ	
C+∞→O	3.64	4.2	-1	A∈Γ	B∈Δ
C+∞→C	3.64	4.25	-1	A∈Γ	B∈Δ
O+∞→O	3.63	4.3	-1	A∈P₆	B∈Δ
	2.6	4.07	-1	A∈Ω	B∈Δ
O+∞→C	3.63	4.2	-1	A∈P₆	B∈Δ
	2.58	4.1	-1	A∈Ω	B∈Δ
C+∞→∞	3.9	4.1	[-1,1]	A∈Γ	B∈Δ
O+∞→∞	3.2	4.1	[-1,1]	A∈Ω	B∈Δ
∞+∞→∞	4.1	4.1	[-1,1]	A∈Δ	B∈Δ

for $\gamma = -1$ in Table 1 were chosen such that they are valid for the sequence {ABABAB...} as well as for the sequence {BABABA...}. Note that the case C+C→0 may occur not only for A and B in the vicinity of the period 3 window, as illustrated in Fig. 5, but also in "ordered islands" on the (A,B)-plane occuring for large differences between A and B (last two cases for C+C→0 in Table 1).

We would like to point to the interesting result that a system which is chaotic under periodic environmental variations ($\gamma = -1$) may become ordered under random environmental variations ($\gamma > -1$). In a different system, namely in a model for the Belousov-Zhabotinskii reaction with added noise, this phenomenon has already been reported and was called "noise-induced order" [12]. In the present work, we obtain noise-induced order with the logistic equation (5) for example in the neighbourhood of the point (A,B) = (3.34,3.6). The effect is illustrated in Fig. 6, which shows λ as a function of γ. For $\gamma = -1$ (periodic alternation of A and B) the system is chaotic ($\lambda>0$). The value of λ obtained for $\gamma = -1$, in the case {ABABAB...} is the same as in the case {BABABA...}. The function $\lambda(\gamma)$ is discontinuous for $\gamma = -1$, jumping to a negative value for $\gamma > -1$. In other words, the slightest randomness in the sequence of A and B orders the system. For any $\gamma\epsilon]-1, 1[$, $\lambda<0$. The situation can be different if another "seed" is chosen as initial condition for X_t. In fact, if the seed is set to 0.5 instead of 0.35, two values of λ are obtained for $\gamma = -1$, a positive and a negative one, depending on the A-B-sequence ({ABABAB ...} or {BABABA...}). In such a case, noise-induced order occurs only for one A-B-sequence.

For Rickers equation (6) the same type of phenomena are obtained, although for quantitatively different parameter values. Figure 7 shows the case $\gamma=0$. As for the logistic equation (Fig. 3), the case 0+0→C is obtained for A and B below the chaotic threshold, which for Rickers equation (3) is given by r = 2.6924. This case holds for the black points in the "appendix"-shaped region and within the rectangle delimited by dashed lines at the (lower and left) edges of Fig. 7. For Rickers equation we found this effect to be most pronounced for values of γ around

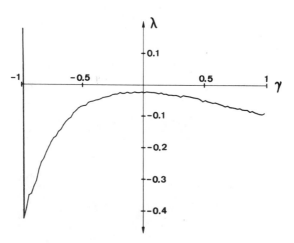

Fig. 6. Dependence of the Lyapunov exponent λ on the Cowley parameter γ. $\gamma = -1$ means periodic alternation of A and B. In the presence of randomness ($\gamma < -1$), λ becomes negative (noise induced order). A = 3.34, B = 3.6. Initial X_t: 0.35. Step length for γ: 0.01.

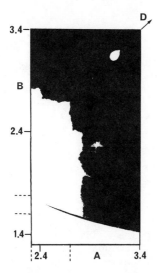

Fig. 7. As Fig. 1, but for Rickers equation (6) and γ = 0. Step length for A and B: 3.5 x 10^{-3}.

γ = 0, while for the logistic equation it was most pronounced around γ = −0.5. In Fig. 8 we show the neighbourhood of the period 3 window for γ = −0.96. As in the case of the logistic equation (see Fig. 5), we obtain here a region where the coupling process C+C→O is fulfilled in spite of randomness. This region is given by white points within the rectangle delimited by dashed lines at the lower right, and by the symmetrically opposed region at the upper left. We found that the coupling process C+C→O is more robust to resist randomness for Rickers equation than for the logistic equation. In fact, for Rickers equation, we could easily detect this effect for −1 ⩽ γ ⩽ −0.8, while for the logistic equation it was difficult to detect for γ > −0.9. All cases exemplified on Table 1 are also obtained for Rickers equation although for different values of A and B. We give here just two interesting numerical examples:
1) C+C→O for A = 3.202, B = 3.0895 and $γ_ε$[−1,−0.8]
2) O+O→C for A = 1.66, B = 2.54 and γ = 0.

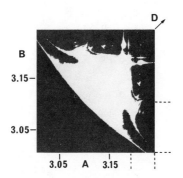

Fig. 8. Neighbourhood of the window with period 3 obtained with Rickers equation. A and B change randomly with γ = −0.96. Step length for A and B: 5 x 10^{-4}.

IV DISCUSSION

We found here that periodically or randomly changing environmental conditions may induce a variety of unexpected dynamic behaviours. In particular, a system that is ordered under constant conditions may become chaotic if the conditions change periodically or randomly (O+O→C). This effect lowers the threshold for chaos and may explain the fact that parameters leading to chaos in calculations are higher than those observed in nature [22].

It is remarkable that also the reverse effect of O+O→C occurs: a system that is chaotic under constant conditions may become ordered if the conditions change periodically or randomly (C+C→O).

Furthermore, we found that a system that is chaotic for a periodic alternation of A and B ($\gamma = -1$), can become ordered when the change between A and B is random (noise-induced order). Matsumoto and Tsuda [12] investigated this type of phenomenon. However, they did not detect such an effect for the logistic equation as they considered an additive random term in equation (2) instead of a random variation of the parameter r.

The distributions of the regions of order and chaos on the plane allow no simple generalization. The one sentence summary given in [15] for continuous systems, namely that "temporal variations in the environment are a destabilizing influence" is much too simple in view of the diversity of coupling processes found in this work.

The applicability of the present results to natural systems is of course restricted because of our assumption of a single species and our hypothesis that growth conditions can be reduced to only two states A and B with equal probabilities. However, the present work shows the existence of phenomena that may, in principle, occur in nature.

ACKNOWLEDGEMENTS

We thank Mr. Jürgen Marsch for valuable computing assistance, and Mrs. Angelika Rohde for efficient typing of the manuscript. One of us (M.M.) gratefully acknowledges an invitation to Chile, financed by UNDP-UNESCO, Universidad de Chile and the Stiftung Volkswagenwerk.

REFERENCES

[1] R. Breuer, Geo. 7: 36-54 (1985).

[2] L.M. Cook, Nature 207: 316 (1965).

[3] J.M. Cowley, Phys. Rev. 77: 669 (1950).

[4] J. Davidson and H.F. Andrewartha, J. Anim. Ecol. 17: 193-199 (1948).

[5] C. Elton, "Voles, Mice and Lemmings: problems in population dynamics", Oxford Univ. Press, Oxford (1942).

[6] Harwell Subroutine Library, A Catalogue of Subroutines (Theoretical Physics Division, A.E.R.E., Harwell, Great Britain) (1973).

[7] M.P. Hassell, J.H. Lawton and R.M. May, J. Anim. Ecol. 45: 471-486 (1976).

[8] M. Kot and W.M. Schaffer, Theor. Popul. Biol. 26: 340-360 (1984).

[9] C.J. Krebs, "Ecology: the experimental analysis of distribution and abundance", 190-200, Harper and Row, New York (1972).

[10] A. MacFadyen, "Animal Ecology: Aims and Methods", 2nd ed., Pitman, London (1963).

[11] M. Markus, B. Hess, J. Rössler and M. Kiwi, to be published.

[12] K. Matsumoto and I. Tsuda, J. Stat. Phys. 31: 87-106 (1983).
[13] R.M. May, Science 186: 645-647 (1974).
[14] R.M. May, Nature 261: 459-467 (1976).
[15] R.M. May, "Theoretical Ecology. Principles and Applications",
         Chapter 2, Blackwell, Oxford (1976).
[16] R.M. May and G.F. Oster, Am. Nat. 110: 573-599 (1976).
[17] L.F. Olsen, this workshop (1987).
[18] W.E. Ricker, J. Fish. Res. Board Canada 11: 559-623 (1954).
[19] W.M. Schaffer, Am. Nat. 124: 798-820 (1984).
[20] W.M. Schaffer, IMA J. of Math. Applied in Med. and Biol. 2: 221-252
         (1985).
[21] W.M. Schaffer and M. Kot, J. Theor. Biol. 112: 403-427 (1985).
[22] W.M. Schaffer and M. Kot, Bioscience 35: 342-350 (1985).
[23] R. Shaw, Z. Naturforsch 36a: 80-112 (1981).

BI-FRACTAL BASIN BOUNDARIES IN INVERTIBLE SYSTEMS

O.E. Rössler[1], C. Kahlert[1] and J.L. Hudson[2]

[1]Institute for Physical
and Theoretical Chemistry
University of Tubingen
7400 Tubingen
West Germany

[2]Department of Chemical
Engineering
University of Virginia
Charlottesville
VA 22901, U.S.A.

ABSTRACT

A Lauwerier-type limiting case of a folded-towel diffeomorphism is considered. Its existence confirms the conjecture made previously that differentiable dynamical systems may possess bifractal (self-similar, in two directions) basin boundaries. A procedure how to search for such boundaries in realistic systems is indicated.

I   INTRODUCTION

Fractal basin boundaries exist in both non-invertible and invertible systems as is well known (cf. [1] for a review). However, in invertible systems, so far only 'striated' fractals (that are locally the product of a Cantor set and a line and hence are not self-similar in two directions) have been observed [1,2]. This is somewhat unexpected since a number of bifractal (fully self-similar, in 2 D) basin boundaries have been observed in noninvertible 2-D real maps which are 'close' to related invertible 3-D maps. See [3,4] for examples of 2-D endomorphisms of continuous type, and [5] for time-inverted, discontinuous (piecewise-linear) examples. Most tellingly, perhaps, the map of ref. [3] (two uncoupled logistic maps) is identical in type to a 2-D map found as an explicit cross-section for the non-invertible (3-variable) singular perturbation limit of an explicit 4-variable ODE of the hyperchaotic type [6] (see [7] for pictures). Bifractal basin boundaries therefore exist in certain limiting flows at least.

On the other hand, the non-invertible 2-D map of ref. [3], with its self-similar checkerboard pattern, is at the same time also the limiting case to an explicit 3-D invertible map [8]. Again, the bifractal pattern necessarily survives numerically in the invertible case close to the non-invertible limit. The question that naturally poses itself is whether or not this numerical survival should be taken seriously mathematically. The answer proposed in ref. [3] was that Smale's [9] global-analytic theory of differentiable dynamical systems may be used to show the persistence of these boundaries.

In the following, further arguments in support of this conjecture will be presented.

II A LOWER-DIMENSIONAL ANALOGUE

Recently, Lauwerier [10] indicated an explicit example of a 1-D non-invertible map that by adding a passively coupled second variable could be changed into an almost everywhere diffeomorphic 2-D map that neverthe-less retained the behaviour of the former single variable, for that variable. An example of the same kind (in fact a closely related example) can be derived from the following explicit 'walking-stick' map [11]

$$x_{n+1} = a\, x_n(1-x_n)-y_n$$

$$y_{n+1} = (b\, y_n-c)(1-2x_n). \tag{1}$$

This map is equivalent, under the transformation $x \rightarrow x$, $y \rightarrow y/c$, to the following map

$$x_{n+1} = a\, x_n(1-x_n)-c\, y_n$$

$$y_{n+1} = (b\, y_n-1)(1-2x_n), \tag{2}$$

as can be verified by insertion. The second version, Eq. (2), in the limit $c \rightarrow 0$ obviously is an example in the Lauwerier class (since y stays bounded if as fitting $0 > b > 1$). A numerical calculation (taken from [12]) which verifies this claim is shown in Fig. 1. Note that a single vertical line in the original box (at $x = 1/2$) gets contracted to a point (the welding-together point of the first iterate, on the left in Fig. 1a)

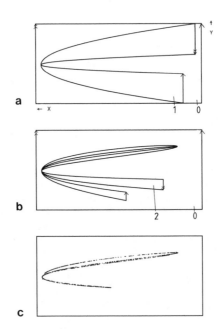

Fig. 1. Numerical calculation of Eq. (2). Parameters: a = 3.8, b = 0.4, c = 2.10^{-52}. Initial box: 0.04 to 0.98 for x, and -1.47 to 1.53 for y. a: Zeroth and first iterate. b: Zeroth and second iterate. c: Chaotic attractor; 1200 iterates of the lower right-hand corner, at 12-digit accuracy.

while all other vertical lines remain vertical, either with unchanged or with changed orientation depending on whether their x-coordinate is less or greater than 1/2.

To make the connection to one of the uncoupled logistic maps in ref. [3], where a was larger than 4, a value of a larger than 4 may also be assumed now in Eq. (1) or (2), in the limit $c \to 0$ considered here. The present map then no longer generates a chaotic attractor (as it does in Fig. 1c, and as it still does if a = 4 [10]). Instead, the upper 'tongue' of the first image then crosses the vertical stable manifold of the saddle point located at 0, −1.66, so that no attracting box remains. At the same time the walking-stick map becomes a horseshoe map in the sense of Smale [9], with the consequence that only a Smale basic set (a Cantor set of unstable periodic solutions) survives inside any initial box. Smale, incidentally, in the original construction of his set [9], also made use of a but passively coupled second variable (while the first variable was piecewise linear).

III THE FULL CASE

The very same construction that was applied above to a 1-D map and its canonic extension, can also be applied to an analogous 2-D map and its extension. A straightforward analogue to the 2-D map of Eq. (2) is the following 3-D map [8]

$$x_{n+1} = a\, x_n(1-x_n) - \frac{c}{4}(z_n+0.35)(1-2y_n)$$

$$y_{n+1} = a'y_n(1-y_n)+c\,z_n$$

$$z_{n+1} = 0.1[(z_n+0.35)(1-2y_n)-1](1-2x_n). \tag{3}$$

This map does not generate a walking-stick but rather a hyper-walking-stick - a folded towel. Recently, Stefanski [13] found a much simpler example (with just two quadratic terms and a constant Jacobian determinant), but this is not the point here. The point is that Eq. (3) is built in analogy to Eq. (2). By letting the constant c again approach zero, an endomorphism is obtained once more (in the first two variables this time). This endomorphism again becomes embedded, via the passively coupled additional variable, into an almost everywhere diffeomorphic higher-dimensional map.

Moreover, by letting the constants a and a' both exceed 4, again a Smale basic set - although of a 'higher-dimensional' type - is generated, in the form of a product of two Cantor sets of periodic solutions. Interestingly, this time two 'explosions' are involved - one toward negative values of x, one toward negative values of y. If one assumes (as this was done in ref. [3]) that a different attractor at minus infinity applies in each case, then each point of the basic set separates two basins. Moreover, a continuous separatrix is bound to connect the elements of the basic set.

Thus, the self-similar basin boundary that was found to apply to the first two lines of Eq. (3), with c = 0 and a = a' = 4.04, in ref. [3], is retained in the present almost everywhere differentiable system.

What is still needed is an argument of topological type (like that adduced by Smale [9] in the lower-dimensional situation) to show that the behaviour found is 'stable' under sufficiently small distortions - as when c is given a small non-zero value. Whether or not such a proof can be given, or whether fundamental new obstacles (there may be no analogue to the 'no cycle' assumption of ref. [9]) will be encountered along the way, is presently open.

IV DISCUSSION

Bifractal basin boundaries have apparently never been found so far in invertible systems like 4-variable ordinary differential equations. The explicit map presented above shows, however, that systems possessing a 'hyper-horseshoe' (derived from a folded-towel map) as a cross-section in their phase spaces are, in principle, capable of possessing such a boundary.

The simplest 4-variable ODE generating a folded-towel attractor possesses but a single nonlinear (quadratic) term on its right-hand side [8]. Recently a reaction-kinetic analogue to that equation, producing the same qualitative behaviour (two positive Lyapunov characteristic exponents [14]), was described [15]. Many more simple systems with the same type of qualitative behaviour (cf. [16]) are bound to exist. It therefore makes sense to search for bifractal basin boundaries, both in the above-mentioned formal systems [6,7,8,15] and in natural systems showing hyperchaos - like low-temperature semiconductors [17].

As it turns out, performing a numerical search will prove rather time-consuming. A plane of initial conditions that cuts across the folded-towel region in phase space has to be scanned point by point with a certain resolution, whereby each trajectory has to be followed up for a potentially rather long time - namely, until it has entered the ('quiet') last domain of its corresponding basin of attraction. (In this last portion of the funnel, neighbours over a finite distance are all headed toward the same attractor). Since each point in parameter space in general determines a different location for the folded-towel region, in general several planes have to be searched per point. The paths followed in parameter space are such that several paths (arcs), each leading to a different 'explosion' of the folded towel, have to be followed up - until a region in parameter space is found in which several such explosions co-exist. Of course, instead of an 'external explosion' (toward a neigh-bouring external basin), 'internal explosions' (where a periodic attractor inside the folded towel sucks up part of the initial conditions) are just as appropriate in principle.

At any rate - having access to a very fast computer in an interactive fashion should greatly facilitate the search.

Apart from fractal separatrices, other similarly shaped objects may exist in continuous dynamical systems. One possibility is bifractal attractors (cf. [18]) where too no explicit example is available so far. Another possibility is fractal objects in parameter space (cf. [19]). Here even experimental systems [16,17] may offer a chance. Recently, a fractal structure was found in the parameter space of the Rashevsky-Turing system (cf. [20]) in a neighbourhood to a 'bi-chaotic' [21] regime - if the system's trajectory was observed in a certain nonlinear fashion [22].

At the time being, it appears that all 'bifractal observables' of differentiable dynamical systems deserve the same theoretical interest. This is because there is still this major undecided question waiting to be answered: Is Nature fractal [23] because, or in spite, of the existence of differentiable systems?

ACKNOWLEDGEMENTS

We thank Lars Olsen for discussions. Work supported in part by the DFG and the NSF.

REFERENCES

[1]   S.W. McDonald, C. Grebogi, E. Ott and J.A. Yorke, Physica 17D:125
      (1985).
[2]   C. Grebogi, personal communication, August (1986).
[3]   O.E. Rössler, C. Kahlert, J. Parisi, J. Peinke and B. Röhricht,
      Z. Naturforsch. 41a:819 (1986).
[4]   C. Kahlert and O.E. Rössler, Analogues to a Julia boundary away
      from analyticity, Z. Naturforsch. 42a:324 (1987).
[5]   M.F. Barnsley and A.N. Harrington, Physica 15D:421 (1985).
[6]   O.E. Rössler and C. Mira, Higher-order chaos in a constrained
      differential equation with an explicit cross section, Topologie:
      Spezialtagung: Dynamical Systems, September 13-19, 1981,
      Mathematisches Forschungsinstitut Oberwolfach Tagungsberichte
      (1981).
[7]   O.E. Rössler, Z. Naturforsch. 38a:788 (1986).
[8]   O.E. Rössler, Phys. Lett. 71A:155 (1979).
[9]   S. Smale, Bull. Amer. Math. Soc. 73:747 (1967).
[10] H.A. Lauwerier, Physica 21D:146 (1986).
[11] O.E. Rössler, Chaos, in "Structural Stability in Physics" W.
      Güttinger and H. Eikermeier, eds., pp 290-309, Springer-Verlag,
      Berlin (1979).
[12] O.E. Rössler, "Chaos, The World of Nonperiodic Oscillations,"
      Unpublished manuscript (1981).
[13] K. Stefański, preprint, Nicholas Copernicus University (1983).
[14] H. Froehling, J.P. Crutchfield, D. Farmer, N.H. Packard and R.Shaw,
      Physica 3D:605 (1981).
[15] J.L. Hudson, H. Killory and O.E. Rössler, Hyperchaos in an explicit
      reaction system, preprint, September (1986).
[16] R. Shaw, "The Dripping Faucet as a Model Chaotic System", Aerial
      Press, Santa Cruz, Calif. (1985).
[17] J. Peinke, B. Röhricht, A. Mühlbach, J. Parisi, C. Nöldeke, R.P.
      Huebener and O.E. Rössler, Z. Naturforsch. 40a:562 (1985).
[18] O.E. Rössler, J.L. Hudson and J.A. Yorke, Z. Naturforsch. 41a:979
      (1986).
[19] J. Peinke, J. Parisi, B. Röhricht and O.E. Rössler, Instability of
      the Mandelbrot set, Z. Naturforsch. 42a:263 (1987).
[20] B. Röhricht, J. Parisi, J. Peinke and O.E. Rössler, Z. Phys. B. -
      Condensed Matter 65:259 (1986).
[21] O.E. Rössler, Z. Naturforsch. 31a:1168 (1976).
[22] B. Röhricht et al., Self-similar fractal in the parameter space of
      a Rashevsky-Turing morphogenetic system, in preparation.
[23] B. Mandelbrot, "The Fractal Geometry of Nature", Freeman, San
      Francisco (1982).

HOMOCLINIC BIFURCATIONS IN ORDINARY DIFFERENTIAL EQUATIONS

Paul Glendinning

Mathematics Institute
University of Warwick
Coventry CV4 7AL
U.K.

ABSTRACT

Some global bifurcations of low codimension are described, and the results are used to give scaling properties of travelling and solitary waves in the FitzHugh-Nagumo equations.

I   MOTIVATION

Bifurcation theory describes the way qualitative changes in behaviour occurs as some control parameter is varied. Often the analysis of such changes can be done by considering a local normal form, as in the saddle-node and period-doubling bifurcations. Such bifurcations are called local bifurcations. Figure 1 shows a schematic view of a global bifurcation. As the control parameter, μ, passes through zero a homoclinic orbit is formed (this is a trajectory bi-asymptotic to an unstable stationary point); the sketch shows that the homoclinic bifurcation destroys a periodic orbit. Such bifurcations cannot be described algebraically by a simple linearisation argument near the periodic orbit or stationary point alone, yet these bifurcations can be observed in many systems of equations which exhibit exotic behaviour, and can be used to understand many complicated sequences of local bifurcations. In some sense, we could say that local bifurcations are algebraic: they can be described by unfolding some algebraic degeneracy, whilst global bifurcations are geometric in that they are analysed by concentrating on a geometric object, viz. the homoclinic orbit.

To fix ideas, and to show that the results presented below are useful in contexts other than families of ordinary differential equations, consider the FitzHugh-Nagumo equations in an infinite axon

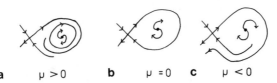

a $\mu > 0$ b $\mu = 0$ c $\mu < 0$

Fig. 1. Phase portraits for a homoclinic bifurcation in R^2.

$$u(x,t)_t = u_{xx} + f(u) - w$$
$$w(x,t)_t = \sigma u - \gamma w \qquad (1)$$

where subscripts denote partial differentiation, $\sigma, \gamma > 0, f(u) = u(1-u)(u-a)$ and $a \in (0, 1/2)$. Looking for constant stationary solutions we find that $u = w = 0$ is a solution, and there are two other constant solutions provided

$$\sigma/\gamma < (1-a)^2/4. \qquad (2)$$

Travelling waves can be investigated by looking for solutions of the form

$$u(x,t) = U(x+ct), \quad w(x,t) = W(x+ct) \qquad (3)$$

which leads to the third order ordinary differential equation

$$cU' = U'' + f(U) - W$$
$$cW' = \sigma U - \gamma W \qquad (4)$$

where the primes denote differentiation with respect to $z = x+ct$. Note that the velocity of the wave, c, now becomes a parameter and that travelling waves of (1) correspond to periodic orbits of (4) and solitary waves (or impulses) correspond to homoclinic orbits of (4). Thus homoclinic orbits also have an appealing physical interpretation, and the existence of travelling and solitary waves in reaction-diffusion equations such as (1) is equivalent (in infinite media) to the existence of periodic and homoclinic orbits in a related family of ordinary differential equations. The FitzHugh-Nagumo equations have been studied from this point of view in [6, 7, 19] and we shall return to this problem in the section IV, where we shall show how the general results of section II can be used to reproduce and extend some known results.

II HOMOCLINIC BIFURCATIONS: CODIMENSION ONE

The existence of a homoclinic orbit is structurally unstable: a typical small perturbation of the homoclinic system no longer has such an orbit (c.f. Fig. 1). By considering trajectories near the homoclinic orbit we shall describe the periodic orbits and bifurcations in one parameter families of ordinary differential equations

$$x' = f(x;\mu) \qquad (5)$$

where $f(0;\mu) = 0, x \in R^n$, $n > 2$ and $\mu \in R$ is the control parameter. Assume that for $\mu = 0$ there is a homoclinic orbit, Γ, biasymptotic to the stationary point at the origin, 0. Trajectories near the homoclinic orbit spend long periods of time in a neighbourhood of the origin, so the behaviour near 0 dominates the analysis of these solutions. The method used to describe the orbit structure is thus based on a simple observation: the flow near Γ can be split into two parts, a region near 0 dominated by the linear part of the flow at 0, and a global reinjection which takes trajectories near Γ out of the linear region and then back into it. This global reijnection can be taken to be affine (constant plus linear) for trajectories near Γ, and so a return map can be derived on a suitable surface near 0 by taking a composition of the linear map followed by the affine global reinjection.

The flow near 0 is determined by the Jacobian matrix $Df(0;0) = \partial f./\partial x.(0;0)$. Provided 0 is hyperbolic this matrix will have k_u (resp. k_s) eigenvalues $\lambda_i, 1 \leq i \leq k_u$ (resp. $\gamma_j, 1 \leq j \leq k_s$) with real part greater than (resp. less than) zero, and $k_u + k_s = n$. We choose to label these eigenvalues so that $0 < re\lambda_1 \leq re\lambda_2 \leq \ldots$ and $0 > re\gamma_1 \geq re\gamma_2 \geq \ldots$ Almost all trajectories in the unstable manifold of $0, W^u(0) = \{x$ the trajectory through x tends to 0 as $t \to -\infty\}$, tend to 0 tangent to the eigenspace associated with eigenvalues having $re \lambda_i = re\lambda_1$ as $t \to -\infty$. Similarly almost all trajectories which lie on the stable manifold of 0 tend to 0 tangent to the eigenspace associated with eigenvalues having $re\gamma_j = re\gamma_1$ as $t \to \infty$. Thus when $Df(0;0)$ is non-degenerate these eigenspaces are either one-dimensional (if λ_1 or γ_1 is real) or two-dimensional otherwise, in which case trajectories spiral out of (or into) 0. Shil'nikov [23,24,25,26] has described the dynamics near these homoclinic orbits. We give two results which determine conditions on the dominant eigenvalues for the existence and non-existence of chaotic motion. Note that in theorem 2, which describes the chaotic case, case. the complicated motions are not stable, and hence that the theorem is not a great help when applied directly to observable (i.e. attracting) chaos.

Theorem 1 [25]

Consider (5) with f sufficiently smooth. Suppose the origin 0 is a hyperbolic stationary point with eigenvalues as described above and that $\mu = 0$ there is an orbit, Γ, biasymptotic to 0, which tends to 0 as $t \to \pm\infty$ tangent to the eigenspaces associated with the dominant eigenvalues. Then if $\lambda_1 \in R$ and

$$\lambda_1 < re\gamma_1 \qquad (6)$$

a single periodic orbit exists in an arbitrary neighbourhood of for small μ either positive or negative. This orbit is stable if $k_u = 1$.

The side of $\mu = 0$ for which the orbit exists is easy to determine by considering particular cases of (6) in more detail. The time spent in a small neighbourhood of 0 is large compared to the time spent during the global reinjection, so the period T of the periodic orbit tends to infinity as μ tends to zero from either above or below like

$$T = -(const.\ln|\mu|)/\lambda_1 + O(1). \qquad (7)$$

Thus for cases described by (6), maybe after time reversal, there is no chaotic behaviour in a neighbourhood of Γ, and the bifurcation is similar to that described in Fig. 1.

Symbol sequences provide a useful way of describing complicated dynamics. Let $\underline{s} = (s_i), i \in Z, s_i \in N$ be an infinite sequence of symbols. The shift, σ, is a map from the space of symbol sequences to itself defined by $\sigma\underline{s} = \underline{s}'$, where $s'_i = s_{i+1}$. Many complicated dynamical systems can be understood by showing that they are equivalent to the action of the shift on a space of sequences; thus Smale's horseshoe is equivalent to the shift on the space of sequences with two symbols, $s_i = 0$ or 1. This means that any sequence of 0s and 1s is equivalent to an orbit of some point under Smale's horseshoe and so there are infinitely many periodic orbits (corresponding to all periodic sequences of 0s and 1s) and uncountably many aperiodic orbits. Shil'nikov [26] uses this idea to characterise the behaviour of trajectories in the chaotic case.

Theorem 2 [26,29]

Suppose that the conditions of Theorem 1 hold with $n \geq 3$ with

$$re\lambda_1 > -re\gamma_1 \qquad (8)$$

and $im\gamma_1 \neq 0$. Then for $\mu = 0$ and in an arbitrary neighbourhood of Γ, there are trajectories are in correspondence with a subshift on arbitrarily many symbols $\underline{s} = (s_i), i \in Z, s_i \in N$, with $s_{i+1} \geq \alpha s_i, \alpha = -re\gamma_1/re\lambda_1$.

Thus we have an infinite number of periodic orbits, an uncountable number of aperiodic orbits and a countable number of horseshoes in the dynamics near Γ for $\mu = 0$. As noted earlier, all these complicated orbits are saddles, and the ensemble of invariant trajectories is not attracting. To understand some of the attracting behaviour observed in examples we need to consider the bifurcations in a typical one parameter family of systems containing a homoclinic system as described in Theorem 2, to see how the complicated trajectories are created. Here we shall consider the case $\lambda_1 \in R$, [13,14.15,18], the other case can be treated similarly [8].

For simplicity, consider flows in R^3 with linearisation near the origin

$$x' = -\rho x + \omega y$$
$$y' = -\omega x - \rho y$$
$$z' = \lambda z \qquad (9)$$

so (8) implies that $\lambda > \rho > 0$. Using standard techniques (linearising near 0 and approximating the global reinjection along Γ by an affine map) we obtain the model return map

$$x' = r(\mu) + axz^\alpha \cos(\xi \ln z + \Phi_1)$$
$$z' = \mu + bxz^\alpha \cos(\xi \ln z + \Phi_2) \qquad (10)$$

where $\xi = -\omega/\lambda, \alpha = -\rho/\lambda < 1$, for $z > 0$ on part of the plane $y = 0$ near the origin (see [14,15,18] for more precise statements). The point $(x,y,z) = (r(\mu),0,\mu)$ is the first intersection of the unstable manifold of 0 with the return plane $y = 0$. Thus when $\mu = 0$, this point lies on the local stable manifold of 0, $z = 0$, and a homoclinic orbit Γ exists.

Simple periodic orbits, which pass once through a neighbourhood of on each period, correspond to fixed points of (10). These have z-coordinates given (approximately) by

$$z = \mu + br(\mu)z^\alpha \cos(\xi \ln z + \Phi_2). \qquad (11)$$

The period T of these orbits is approximately

$$T \sim -(\ln z)/\lambda \qquad (12)$$

(obtained by considering the amount of time spent in a neighbourhood of the origin). Setting $z = \exp(-\lambda T)$ in (11) we find, ignoring higher order terms, using $\alpha < 1$

$$\mu = -br(\mu)e^{-\rho T}\cos(\omega T + \Phi_2). \qquad (13)$$

Solutions are sketched in Fig. 2. For $\mu = 0$, there is an infinite sequence of orbits with period (T_n) with $T_{n+1} = T_n + \pi/\omega$. Turning points of the right hand side of (13) correspond to saddle node bifurcations. These turning points are at $\mu_n \approx (-1)^n br(\mu_n)\exp(-\rho T_n)$ and hence

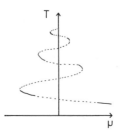

Fig. 2.   Locus of periodic orbits which pass once through a tubular neighbourhood of Γ on each period. The period T is plotted against the parameter μ, with solid lines denoting a stable orbit if $1/2 < \alpha < 1$ and unstable orbits if $\alpha < 1/2$, and dashed lines denoting nonstable periodic orbits.

$$\lim_{n\to\infty}(\mu_{n+1}-\mu_n)/(\mu_n-\mu_{n-1}) = \exp(-\pi\rho/\omega). \tag{14}$$

Figure 2 also gives the stability of these periodic orbits. Note that if $1/2<\alpha<1$ there are stable periodic orbits arbitrarily close to $\mu = 0$, and that these orbits lose stability via period-doubling bifurcations.

In the case $\alpha<1$ considered here there are also infinite sequences of more complicated homoclinic orbits in $\mu>0$. In particular, there is a sequence of homoclinic orbits which pass twice through a neighbourhood of Γ before tending to 0. These are called double-pulse homoclinic orbits [6,18,19], and occur when $'(r(\mu),0,\mu)$ is mapped back to $z = 0$ by the return map (10). This gives

$$-\mu = br(\mu)\mu^{\alpha}\cos(\xi 1n\mu+\Phi_2) \tag{15}$$

(c.f. (13)), and so there is a sequence of such homoclinic orbits which accumulate on $\mu = 0$ at the rate

$$\lim_{n\to\infty}(\mu_{n+1}-\mu_n)/(\mu_n-\mu_{n-1}) = \exp(-\pi\lambda/\omega). \tag{16}$$

Further details and examples can be found in [15,18].

III COMPLICATIONS

We shall consider two complications that can arise in examples: the presence of symmetry in the system (where the existence of one homoclinic orbit implies the existence of another which is the image of the former under the symmetry) and some codimension two situations, where two parameters are needed to describe perturbations from homoclinic systems with two homoclinic orbits.

Many equations encountered in physics and biology are invariant under some symmetry, which reflects a natural symmetry of the underlying problem. We shall only discuss simple symmetries of flows in $\mathbb{R}^3$, where the linear flow near the stationary point can be written either as (9) or

$$
\begin{aligned}
x' &= \lambda_1 x \\
y' &= -\lambda_2 y \\
z' &= -\lambda_3 z
\end{aligned} \tag{17}
$$

with $\lambda_i>0, i = 1,2,3$ and $\lambda_2>\lambda_3$. There are two symmetries which are encountered in many examples

$$(x,y,z)->(-x,-y,-z) \tag{18}$$

and

$$(x,y,z)->(-x,-y,z). \tag{19}$$

Note that homoclinic orbits to stationary points of linear type (9) cannot arise in systems with symmetry (19). The symmetry (18) is discussed in [16,20,28]. Typically, the homoclinic bifurcation at $\mu = 0$ creates two periodic orbits, which are the image of each other under the symmetry, on one side of $\mu = 0$ and a single symmetric orbit on the other side (modulo the results described in the previous section). However, symmetry (19) can lead to behaviour very different to that suggested by Theorem 1.

In these symmetric one parameter families the linear type of the stationary point must be described by (17), possibly after time reversal, and the bifurcations depend on the ratio $\alpha = \lambda_3/\lambda_1$. If $\alpha>1$ there is a cascade of more and more complicated homoclinic orbits on one side of the homoclinic bifurcation analagous to the standard period-doubling cascade in one-hump maps, leading to chaotic behaviour outside a neighbourhood of $\mu = 0$ [2]. If $\alpha < 1$, as in the Lorenz equations [21,27], there are no periodic orbits in a neighbourhood of the two homoclinic orbits on one side of the bifurcation, whilst on the other side the orbit structure is equivalent to the shift on two symbols (c.f. the remarks on Smale's horseshoe in section II).

Now consider systems with two homoclinic orbits which are not the images of each other under some symmetry. In this case we have the following theorem, analogous to Theorem 1.

Theorem 3 [10]

Suppose the conditions of Theorem 1 hold with two homoclinic orbits, Γ_0 and Γ_1, bi-asymptotic to 0 as $t \to \pm\infty$ and $k_u = 1$. Then if $\lambda_1 < \mathrm{re}\gamma_1$ perturbations of the homoclinic systems have nought, one or two periodic orbits in a tubular neighbourhood of $\Gamma_0 \cup \Gamma_1$, and these periodic orbits are stable.

These periodic orbits have a very simple symbolic representation [10], and given more information about the stationary point and homoclinic configuration this theorem can be refined to give interesting two parameter unfoldings [9,10,11,12]. It is more difficult to give general results about the remaining codimension two cases, and we refer the reader to [1,3,17] for more detail of these bifurcations.

IV THE FITZHUGH-NAGUMO EQUATIONS REVISITED

Many examples of ordinary differential equations with homoclinic orbits can be found in the literature [4,5,11,15,18], so we shall concentrate on applications to travelling wave solutions in reaction-diffusion equations such as the FitzHugh-Nagumo equations, (1). Hastings [19] shows that there are values of the parameters a, γ and σ at which the ordinary differential equation (4) has a homoclinic orbit which satisfies the conditions of Theorem 2, with linearisation of the form (9). Our aim here is to reinterpret the results of section II for these equations. Thus we suppose that for some given a, γ and σ there exists $c^*(a,\gamma,\sigma)$ such that (4) has a homoclinic orbit bi-asymptotic to the origin, where the linear flow is given by (9) with $\rho = \rho^*(a,\gamma,\sigma)$, $\lambda = \lambda^*(a,\gamma,\sigma)$ and $\rho^* < \lambda^*$. The parameter c, which represents the velocity of travelling solutions, is a natural parameter to use, and we assume (without justification) that for c sufficiently close to c^* we can set $\mu \approx d(c-c^*)$ in (10), for some $d \in R$. (Note that if $\mu \approx d(c-c^*)^k$ a similar procedure can be followed leading to the obvious modications of the results given below).

We begin by considering the double-pulse homoclinic orbits described by (16). These correspond to solitary waves of the FitzHugh-Nagumo equations with two peaks, as shown in Fig. 3. Assuming that the parameter dependence is as described above, we find that for fixed a, γ

Fig. 3. Sketch of a double-pulse solitary wave.

and σ which satisfy Hastings' conditions there is an infinite sequence
of velocities (c_n), n N for which there is a double-pulse solitary wave,
and this sequence accumulates on $c^*(a,\gamma,\sigma)$ at the rate

$$\lim_{n->\infty}(c_{n+1}-c_n)/(c_n-c_{n-1}) = \exp(-\pi\rho^*/\omega^*). \qquad (20)$$

Furthermore, by Theorem 2, there are infinitely many periodic travelling
waves and uncountably many aperiodic travelling waves with velocity c^*
and also with velocity c_n, for n sufficiently large. In particular,
there is an infinite number of periodic travelling waves with only one
large peak in each period with velocity c^* and wavelengths (λ_n) where
$\lambda_{n+1} = \lambda_n+\pi/\omega^*$ for sufficiently large n. Here we have used the fact that
the wavelength of these periodic travelling waves is given by the period
of the associated periodic orbits of (4).

Finally we use (13) to remark that one-peak periodic travelling
waves of this type also exist for values of c in a neighbourhood of c^*:
for all λ sufficiently large there is a velocity c near c^* for which
there is a one-peak travelling wave of wavelength λ.

I hope that some of these remarks are new, and leave it to the
experts to decide whether they are interesting in the context of the
FitzHugh-Nagumo equations. The same methods can be applied to similar
problems, such as fluid flow down an inclined plane [22]. Note, however,
that this analysis gives almost no information about the stability of
these solutions. This is totally different, and considerably more
difficult, problem.

V   A GAME

Consider a group of scientists sitting around a table (without loss
of generality we shall assume this to be in a bar). Each has a piece of
paper in front of him, and they begin to play the following game:

Step 1: each player writes down their favourite ordinary differential
equation, and some fact about it. The papers are then passed to the
left.
   e.g.   The Lorenz equations [21]
          $dx/dt = s(y-x), dy/dt = rx-y-xz, dz/dt = xy-bz.$
          Fact: there is very strong numerical evidence that for
          $b = 8/3$, $s = 10$ and $r = 13.926...$ there is a pair of
          homoclinic orbits, biasymptotic to the origin, which
          has real eigenvalues and satisfies the condition $\alpha<1$
          in the terminology of section III above.

Step 2: each player converts the o.d.e. in front of them into a partial
differential equation, bearing in mind the remarks of sections I and IV,
and passes their paper to the left.

   e.g.   Set $Y = s(y-x)$ in the Lorenz equations. After a little
          manipulation we consider the equation

          $\partial u/\partial t = \partial^2 u/\partial x^2 + au+buv, \quad \partial v/\partial t = -dv - u^2 + b^{-1}u.\partial u/\partial x$
          in $-\infty<x<\infty$, where a, b and d are constants.

Step 3: each player gives a description of a biological, physical or
chemical situation in which the pde might arise, and passes the paper to
the left again.

   e.g.   This is the difficult bit. Answers (postcards only) to me,
          please.

Step 4: now the results given at step 1 are reinterpreted for the system in step 2, and the papers passed to the left.

e.g. Let a = 129.26, b = 10 and d = 8/33. Then there is a pair of solitary waves with velocity c = 11, and for all c in some interval (11,c*) there exist an infinite number of travelling waves (equivalent to the full shift on two symbols for each c). For c greater than c* there is an infinite set of values of c for which pairs of multi-pulse solitary waves exist.

Step 5: (optional) the player reverses the order of steps 1 to 3 and chooses a relevant journal to submit the paper for publication.

This game is not totally without merit. Playing it a couple of times makes the various possible sequences of travelling and solitary waves which may exist in relatively simple pdes much clearer. Thus for another choice of initial ode (an example in [11]) we see that there are systems which have two different solitary waves at the same velocity, and these waves are not the images of each other under a symmetry.

REFERENCES

[1] V.S. Afraimovich and L.P. Shil'nikov, in "Nonlinear dynamics and Turbulence", G.I. Barenblatt, G. Iooss and D.D. Joseph, eds., Pitman (1983).
[2] A., Arneodo, P. Coullet and C. Tresser, Phys. Lett. 81A: 197-201 (1981).
[3] P. Collet, P. Coullet and C. Tresser, J. de Phys. Lettres 46:143-147 (1984).
[4] P. Coullet, J.M. Gambaudo and C. Tresser, C.R. Acad. Sci. (Paris) Serie 1, 299: 253-256.
[5] P. Coullet, C. Tresser and A. Arneodo, Phys. Lett. 72A:268-270 (1979)
[6] J. Evans, N. Fenichel and J.A. Feroe, SIAM J. Appl. Math. 42:219-234 (1982).
[7] J.A. Feroe, SIAM J. Appl. Math. 42:235-246 (1982).
[8] A.C. Fowler and C. Sparrow, Bifocal homoclinic orbits in four dimensions, preprint, University of Oxford (1984).
[9] J.M. Gambaudo, P. Glendinning and C. Tresser, C.R. Acad. Sci. (Paris) Serie 1, 299:711-714 (1984).
[10] J.M. Gambaudo, P. Glendinning and C. Tresser, The gluing bifurcation I: symbolic dynamics of closed curves, submitted to Comm. Math. Phys. (1985a).
[11] J.M. Gambaudo, P. Glendinning and C. Tresser, J. de Phys. Lettres 46:L653-L658 (1985b).
[12] J.M. Gambaudo, P. Glendinning, D.A. Rand and C. Tresser, The gluing bifurcation III: stable foliations and periodic orbits, preprint (1986).
[13] P. Gaspard, Phys. Lett. 97A:1-4 (1984a).
[14] P. Gaspard, Bull. Class. Sci. Acad. Roy. Belg., Serie 5, LXX:61-83 (1984b).
[15] P. Gaspard, R. Kapral and G. Nicolis, J. Stat. Phys. 35:697-727 (1984).
[16] P. Glendinning, Phys. Lett. 103A:163-166 (1984).
[17] P. Glendinning, Asymmetric perturbations of Lorenz-like equations, preprint, University of Warwick (1986).
[18] P. Glendinning and C. Sparrow, J. Stat. Phys. 35:645-696 (1984).
[19] S.P. Hastings, SIAM J. Appl. Math. 42:247-260. (19).
[20] P. Holmes, J. Diff. Equ. 37:382-403 (1980).
[21] E.N. Lorenz, J. Atmos. Sci. 20:130-141 (1963).
[22] A. Pumir, P. Manneville and Y. Pomeau, J. Fluid Mech. 135:27-50 (1983)

[23] L.P. Shil'nikov, <u>Sov. Math. Dokl.</u> 6:163–166 (1965).
[24] L.P. Shil'nikov, <u>Sov. Math. Dokl.</u> 8:54–58 (1967).
[25] L.P. Shil'nikov, <u>Math. USSR Sb.</u> 6:427–438 (1968).
[26] L.P. Shil'nikov, <u>Math. USSR Sb.</u> 10:91–102.
[27] C. Sparrow, The Lorenz equations: bifurcations, chaos and strange attractors, <u>Appl. Math. Sci.</u> 41, Springer (1982).
[28] C. Tresser, Homoclinic orbits for flows in R^3, preprint (1983b).
[29] C. Tresser, <u>Ann. Inst. H. Poincare</u> 40:441–461 (1984).

CHARACTERIZATION OF ORDER AND DISORDER IN SPATIAL PATTERNS

M. Markus, S.C. Müller, Th. Plesser and B. Hess

Max-Planck-Institut für Ernährungsphysiologie
Rheinlanddam 201
4600 Dortmund 1, FRG

ABSTRACT

Several algorithms are tested to characterize order and disorder in digitised images of spatial patterns. As source of images, the Belousov-Zhabotinskii reaction coupled to convective motion in a thin liquid layer is used.

I INTRODUCTION

Spatial self-organization is a ubiquitous feature in biological systems. A remarkable case is the occurrence of circular and spiral waves propagating in excitable media formed by cell aggregates and tissues. Examples are:

(a) Aggregation patterns of the social amoebae D. discoideum [4,8,28].
(b) Spreading depression around electrically stimulated areas, epileptic foci and lesions in the cortex of rats [15,27] and rabbits [24].
(c) Patterns during arrhythmias in porcine and canine hearts [14] as well as rabbit hearts [1].
(d) Spreading depression in isolated chicken retina [12,18].
(e) Zonation patterns in cultures of yeast and other fungi [5,16].

In some instances, biological waves may become disordered in the sense of being aperiodic, that is of having no recognizable period or wavelength. This is the case, for example, for the mutant Fr17 of D. discoideum [6,7], or for fungal culture zonation in the presence of chemical agents [16].

Aperiodicities may result from a sum of periodic waves having incommensurate frequencies (quasi-periodicity) or from deterministic chaos. In the latter case, there exists the problem that a theoretical description cannot be fitted to experimentally obtained patterns (e.g. by minimization of the sum of squares) even if the underlying mechanisms are comparable. This is due to the very nature of deterministic chaos, being unpredictable for given points in space and time. However, one may compare some global parameter obtained from a digitized image with the corresponding parameter obtained from a theoretical data array. In the present work, we test different methods to characterize ordered and disordered images by such a global parameter.

II SOURCE OF PATTERNS

As a model excitable system for pattern formation, we use the Belousov-Zhabotinskii reaction in a thin liquid layer in a petri dish. This system shares fundamental dynamic properties with the biological systems described above (see [30]). In particular, it displays circular and spiral waves. The experimental conditions for the reaction system used here are specified in [22]. Aperiodicities are generated by coupling these waves with convective motion in the layer. This motion can be evoked by temperature gradients due to evaporation (Marangoni effect [25]) and can be prevented by covering the petri dish with a plate. Thus, we can observe periodic patterns (order) while the dish is covered and aperiodic patterns (disorder) after uncovering the dish. When the dish is covered after the aperiodicities have developed, the structure returns

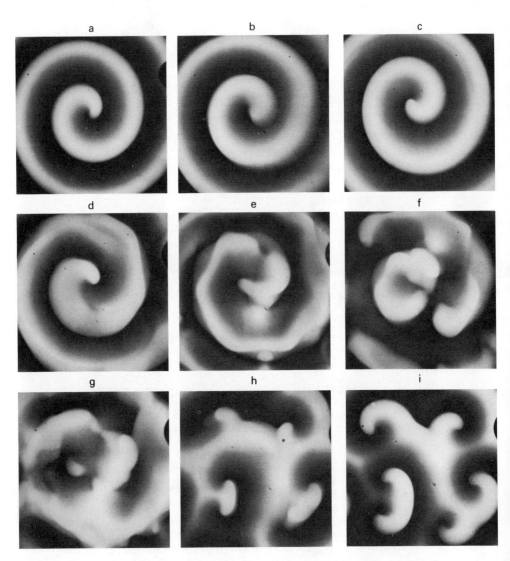

Fig. 1. Transition from order to disorder, and back to order for a spiral of the Belousov-Zhabotinskii reaction in a thin liquid layer.

again to a periodic one. In other words, this set-up allows switching from order to disorder, and vice versa.

Digitized images of transmitted light intensity are obtained by measurements using a two-dimensional spectrophotometer - a TV camera linked to a computer - with 512x512 picture elements (pixels) spatial resolution and 256 grey levels intensity resolution (for details, see [22]).

Figure 1 consists of a selection from one sequence of analyzed pictures (9 out of 12). Figure 1a shows a spiral of the Belousov-Zhabotinskii reaction without convective motion (covered dish). After uncovering the dish, the spiral becomes more and more distorted by convective motion (Figs. 1b to 1g). After covering the dish again, reorganization into periodic structures takes place (Figs. 1h to 1i). For a better visualization of the grey level distribution, we show in Fig. 2 three-dimensional representations of four images of Fig. 1 (Figs. 1a, e, g, i). The grey level is shown along a third coordinate. The graphics algorithm leading to these representations is given in [17].

The sequence of events is similar for the spiral pair shown in Fig. 3. In this figure, 12 out of 20 analyzed pictures are displayed. The dish is covered for Fig. 3a (ordered structure), uncovered for Figs. 3b to 3h (development of disorder), and covered again for Fig. 3i (reorganization).

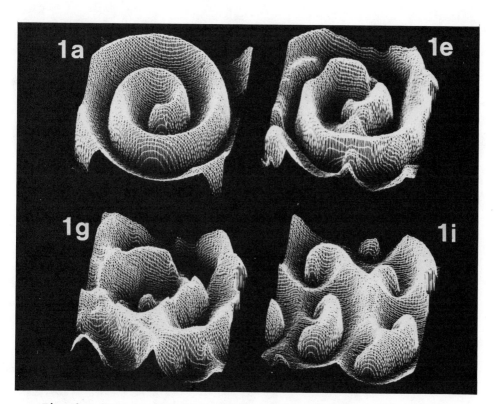

Fig. 2. 3D representation of the grey level distribution of four pictures of Fig. 1 (Fig. 1a, 1e, 1g, and 1i).

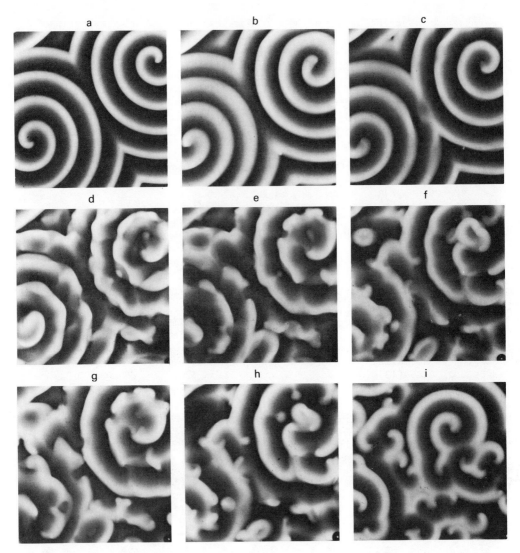

Fig. 3. Order and disorder transitions for a spiral pair, analogous to those in Fig. 1.

III METHODS OF ANALYSIS

The following methods were tested to characterize order and disorder in the digitized images:

1.   Determination of the so-called 'fractal signatures' [23]. These are fractal dimensions determined for just one degree of resolution. Pictures are evaluated pairwise by subtracting their fractal signatures and integrating over all resolutions.

2.   Determination of the grey level standard deviation, which has been used as a simple heuristic approach for describing glycolytic spatial patterns [21]. Also higher moments (skewness and kurtosis) of the grey level distribution were tested.

3. Methods based on comparing the grey levels of adjacent pixels. One method of this kind consists in the determination of the Sobolev norm [29] (integral containing the spatial derivatives of the grey levels).

4. Determination of the entropy of the Fourier spectra [26].

5. Characterization of curves (e.g. isointensity lines) using the method proposed by Mendès France [20]. This method is based on counting the intersections of the curves with a fixed set of straight lines. Disordered curves have more intersections than ordered ones.

6. Maximization of the autocorrelation function (ACF). The ACF for a pixel distance d is defined as

$$C(d) = \frac{1}{R(d)} \sum_{\substack{i=1 \\ |\vec{x}_i - \vec{x}_j| = d}}^{N} \sum_{j=1}^{N} \hat{g}(\vec{x}_i)\hat{g}(\vec{x}_j), \tag{1}$$

where $\hat{g}(\vec{x}) = (g(\vec{x}) - \bar{g})/\sigma$ (2)

$g(\vec{x})$ is the grey level matrix $(1 < g < 256)$, where the $\vec{x}$ are vectors of dimension 2 corresponding to the pixel locations. N is the number of pixels. $\bar{g}$ and σ are the mean and the standard deviation of the grey levels over the picture. In order to get the mean of the grey level product, we normalize by dividing through the total number of pixel pairs $R(d)$ (see [13]). Appreciable computation time can be saved by calculating $C(d)$ with the help of the fast Fourier transform.

For the characterization of pictures, the maximum C_{max} of $C(d)$ over all distances d is determined (discarding the trivial maximum at $d=0$). Order, resp. disorder, should be indicated by high, resp. low values of C_{max}.

7. Minimization of the 'autodifference function' (ADF) defined by

$$D(d) = \frac{1}{R(d)} \sum_{\substack{i=1 \\ |\vec{x}_i - \vec{x}_j| = d}}^{N} \sum_{j=1}^{N} |g(\vec{x}_i) - g(\vec{x}_j)| \tag{3}$$

We do not consider here the normalized grey levels $\hat{g}$, as in the ACF. In contrast to the product in equation (1), the best match between pixels is indicated by a zero grey level difference, independently of any normalization. The ADF has the disadvantage that no computational shortcut is known, such as the fast Fourier transform. In order to avoid excessively long computing times, we determined $D(d)$ after shrinking the pictures by transferring the average of a 5x5 pixel array into one pixel. For the characterization of pictures, the minimum D_{min} of $D(d)$ over all distances d is determined (discarding the trivial minimum $D=0$ at $d=0$). Order, resp. disorder, should be indicated by low, resp. high values of D_{min}.

8. The 'Birkhoff-Bense-Gunzenhäuser' method. This method is based on the introduction of the concept of order in theories of aesthetic quantification. As an aesthetic measure Birkhoff [3] proposed the quotient

$$M = O/\tilde{C} \tag{4}$$

where O is some measure of order and $\tilde{C}$ is some measure of complexity. Bense and Gunzenhäuser (see [2,19]) chose to set $\tilde{C}$ equal to the entropy

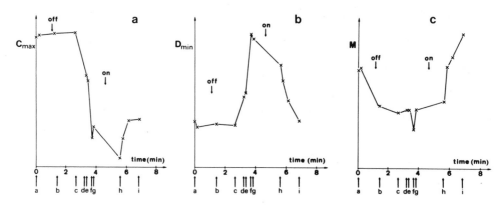

Fig. 4. Time dependence of three parameters for the identification
of order and disorder in the pictures of Fig. 1: (a) the
maximum C_{max} of the autocorrelation function, (b) the minimum
D_{min} of the 'autodifference function', and (c) the 'Birkhoff-
Bense-Gunzenhäuser' measure M. The arrows pointing to the
abscissae correspond to Figs. 1a to 1i. The arrows in the
upper parts of the graphs indicate uncovering ('off') and
covering ('on') of the dish. The ordinate units are arbitrary.

(S) and O equal to the redundancy (R), as given by

$$S = - \sum_{i=1}^{n} p_i \log_2 p_i \qquad (5)$$

and $R = (S_{max} - S)/S_{max}$ \qquad (6)

The p_i are the fractions of pixels having a quality i. S_{max} is equal to
$\log_2 n$. Setting O=R and C=S, equations (4), (5), and (6) lead to

$$M = \frac{1}{S} - \frac{1}{S_{max}} \qquad (7)$$

We call M the 'Birkhoff-Bense-Gunzenhäuser' measure. For the determination of M we consider the distribution of grey levels, setting p_i equal
to the fraction of pixels having grey level i (n=256). Order, resp.
disorder, should be indicated by high, resp. low values of M.

IV RESULTS

     The best correlation between the parametric description and the
impression of order and disorder elicited by mere visual inspection of
the pictures was obtained for methods 6, 7 and 8 described in section
III. The results with these three methods are shown in Fig. 4 for the
picture sequence corresponding to Fig. 1, and in Fig. 5 for the picture
sequence corresponding to Fig. 3. In Figs. 4 and 5, the times at which
the pictures on Figs. 1 and 3 were taken are indicated by arrows pointing
upwards. The arrows labelled 'on' and 'off' indicate the times when the
cover was put on and taken off the dish.

     From Figs. 4 and 5 we can infer that D_{min} is the most powerful
indicator, since it describes most clearly the transitions from spatial

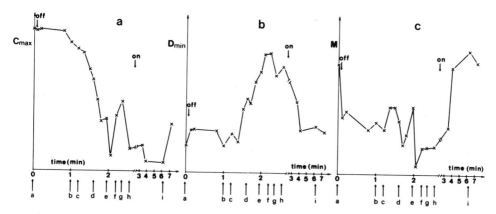

Fig. 5. Same parameters as functions of time as in Fig. 4, but for the pictures in Fig. 3. The arrows pointing to the abscissae correspond to Figs. 3a to 3i. The arrows in the upper parts of the graphs indicate uncovering ('off') and covering ('on') of the dish. The ordinate units are arbitrary.

order to disorder and vice versa, as perceived by the observer in Figs. 1 and 3. Although C_{max} describes well the transition from order to disorder in both picture series, it does not describe satisfactorily the reverse transition. In fact, one finds in Figs. 4a and 5a values of C_{max} which contradict the characterization of order and disorder by mere visual inspection. For example, according to Fig. 4a, Fig. 1h should be more disorganized than Fig. 1g, and according to Fig. 5a, Fig. 3i should be more disorganized than Figs. 3d to 3h. On the other hand, M describes well the transitions from disorder to order, but it fails in describing the transitions from order to disorder. In fact, a significant decrease of M is only obtained after the cover is taken off, even when the spirals are not yet visibly deformed. The change in M after the appearance of spatial disorganization is of the order of the scattering of M from one picture to the next.

V DISCUSSION

Methods 1 to 5 (see section III) were found to be unsuitable for the type of pictures treated here. In particular, methods based on the comparison of the grey levels of adjacent pixels failed because of the large noise at nearest neighbour distances, and because structures have longer characteristic lengths. Fourier analysis failed because of image boundary effects and the low number of spatial oscillations per picture. Characterization of curves extracted from the pictures by counting the intersections with a fixed set of lines failed because large portions of the curves drift out of the picture area as aperiodicities develop. Thus, the number of intersections is determined by this drift, and not by the degree of disorder.

The present work shows that the measure M should be discarded when spatial organization is considered, as it clearly 'overlooks' the state in which the spirals retain their shape after the dish is uncovered. This is because M reacts towards a decrease in grey level contrast which takes place while spatial order is maintained. Furthermore, the values of M scatter more than those of D_{min} during the disordered state.

The insensitivity of C_{max}, as compared to D_{min}, to some of the observed transitions may be explained as follows. Organization is a quality affecting all grey levels. However, the products in equation (1) give only a contribution to $C(d)$ as long as the g values are sufficiently far away from $\bar{g}$. Organization at $g \approx \bar{g}$ leads to contributions close to zero, instead of contributing to a higher C. Quite differently, organization at all grey levels renders a contribution to D_{min}. In particular, perfect matching in equation (3) leads to zero differences, irrespective of the grey level. Therefore, D_{min} is a more powerful indicator.

The diagnosis via D_{min} can be discussed in connection with the three-dimensional representations shown in Fig. 2. In fact, the minimization of $D(d)$ is related to the following procedure: (a) Draw all possible horizontal (bridging or tunnelling) linkages of equal length d between surface points corresponding to pixel locations, and (b) take the length d for which the number of linkages (N_L) is maximum. Pictures having large N_L for some d yield a large number of zeros in equation (3) and thus a low D. This is the case for Fig. 1a. In contrast, N_L is low for any d in Fig. 1g, and thus D_{min} is large. A quantitative comparison of the procedure of maximizing N_L and that of minimizing D is currently under investigation.

As compared to the ACF, the ADF is not only the algorithm with the optimal discrimination properties, but is also a more plausible biological mechanism. The biological advantage of the ADF over the ACF is its easy 'neural implementation', requiring only elementary properties of neurons: excitation, inhibition and summation. Furthermore, less effort in neural processing is required to maintain a low signal to noise ratio for differences as compared to multiplications [11]. Neural circuits of a difference comparator for spatially sampled patterns have been proposed [9,10].

ACKNOWLEDGEMENTS

We thank Mr. Jürgen Marsch for valuable computing assistance, Mr. Uwe Heidecke for fruitful assistance in the laboratory, Mrs. Gesine Schulte for enthusiastic photographic assistance, and Mrs. Angelika Rohde for efficient typing of the manuscript.

REFERENCES

[1] M.A. Allessie, I.M. Bonke and J.G. Schopman, Circ. Res. 33:54-62 (1973).

[2] M. Bense, Einführung in die informationstheoretische Ästhetik, Rowohlt, Reinbek b. Hamburg (1969).

[3] G.D. Birkhoff, Scientia 50:133-146 (1931).

[4] J.T. Bonner "The Cellular Slime Molds", Princeton University Press (1967).

[5] J.A. Bourret, R.G. Lincoln and B.H. Carpenter, Science 166:763-764 (1969).

[6] M.B. Coukell and F.K. Chan, FEBS Lett. 110:39-42 (1980).

[7] A.J. Durston, Develop. Biol. 38:308-319 (1974).

[8] G. Gerisch, Naturwissenschaften 9:430-438 (1971).

[9] H. Glünder, A. Gerhard, H. Platzer and J. Hofer-Alfeis in Proc. of 7th Int. Conf. on "Pattern Recognition", M. Wein, ed., IEEE Comp. Soc. Press, Silver Spring, MD, pp 1376-1379 (1984).

[10] H. Glünder, Biol. Cybernetics 55:239-251 (1986).

[11] H. Glünder, Human Neurobiol. 5:145-155 (1986).

[12] N.A. Gorelova and J. Bures, J. Neurobiol. 14:353-363 (1983).

[13] R.M. Haralick, K. Shanmugan and I. Dinstein, IEEE Trans. on Sys., Man and Cybernetics SMC-3, 610-621 (1973).

[14] M.J. Janse, F.J.L. van Capelle, H. Morsink, A.G. Kléber, F. Wilms-Schopman, R. Cardinal, C. Naumann d'Alnoncourt and D. Durrer, Circ. Res. 47:151-165 (1980).

[15] V.I. Koroleva and J. Bures, Brain Res. 173:209-215 (1979).

[16] G. Kraepelin and G. Frank, Int. J. Chronobiol. 1:163-172 (1973).

[17] W. Kramarczyk in: "Theoretical Foundations of Computer Graphics and and CAD", Springer Verlag, Berlin, Heidelberg, New York, Tokyo (1987 in press).

[18] H. Martins-Ferreira, G. de Oliveira-Castro, C.J. Struchiner and M.S. Rodrigues, J. Neurophysiol. 37:773-784 (1974).

[19] S. Maser, Grundlagenstudien aus Kybernetik und Geisteswissenschaft 6:101-113 (1967).

[20] M. Mendès France in: "Rhythms in biology and other fields of application", M. Cosnard, J. Demongeot and A. Le Breton, ed., Springer Verlag, Berlin, Heidelberg, New York, Tokyo, pp. 354-367 (1981).

[21] S.C. Müller, Th.Plesser and B. Hess in: "Temporal Order", L. Rensing and N.I. Jaeger, eds., Springer Verlag, Berlin, Heidelberg, New York, Tokyo, pp. 194-196 (1985).

[22] S.C. Müller, Th.Plesser and B. Hess, Naturwissenschaften 73:165-179 (1986).

[23] S. Peleg, J. Naor, R. Hartley and D. Avnir, IEEE Trans. on Pattern analysis and machine intelligence PAMI 6:518-523 (1984).

[24] H. Petsche, P. Prokasha, P. Rappelsberger, R. Vollmer and A. Kaiser, Epilepsia 15:439-463 (1974).

[25] J.K. Platten and J.C. Legros, "Convection in Liquids", Springer Verlag, Berlin, Heidelberg, New York, Tokyo (1984).

[26] G.E. Powell and I.C. Percival, J. Phys. A: Math Gen. 12:2053-2071 (1979).

[27] M. Shibata and J. Bures, J. Neurophysiol. 35:381-388 (1972).

[28] K.J. Tomchik and P.N. Devreotes, Science 212:443-446 (1981).

[29] A. Voigt and J. Wlocka "Hilberträume und elliptische Differential-operatoren", B.-I. Wissenschaftsverlag (1974).

[30] A.T. Winfree and S.H. Strogatz, Nature 311:611-614 (1984).

FRACTALS, INTERMITTENCY AND MORPHOGENESIS

Bruce J. West*

La Jolla Institute
Division of Applied
Nonlinear Problems
10280 North Torrey Pines Rd
Suite 260
La Jolla, CA 92037

*Permanent address:
Institute for Nonlinear
Science, R-002
University of California
San Diego
La Jolla, CA 92093

ABSTRACT

Until recently, there were no satisfactory models to account for complex physiological structures or processes that do not have characteristic scales of length and/or time. The concept of fractal offers new insights into multiple scaled structures such as the bronchial and coronary tree, His-Purkinje system and chordae tendineae as well as into the broadband, inverse power-law spectra associated with normal electrophysiological dynamics. In a broader biological context the notion of a fractal distribution may have implications regarding error-tolerance and evolution. These ideas are discussed and some supporting mathematical analysis and data are presented.

I INTRODUCTION

Nature is filled with geometrical shapes that arise from the imposition of some constraint on the evolution of a living system. For example, there are a large number of leafy plants that in order to maximise their exposure to sunlight arrange their leaves in a certain way. The pattern is that of a spiral in which the sequential leaves along the stem of the plant are rotated by a fixed amount so that no two leaves exactly coincide in the vertical direction, presumably the ideal condition. It is found that this fixed angle of rotation approaches the limit $2/(\sqrt{5}-1)$ which is $137°\ 30'28''$ [1]. This number is the limiting value for a Fibonacci series and is related to the fact that if the sum of any two successive values in a geometric progression equal the next term, then the values of the ratio are $(1\pm 5)/2$. The Fibonacci series describes the spiral growth pattern of the plant. The phyllo-taxis of other plants as well, such as cones, the trunks of date palm trees and flowers including the head of the giant sunflower all manifest spiral patterns [1,2].

Of course everyone is familiar with the swirling spiral of the conch shell and of other marine life. The logarithmic spiral shape is developed by the creature by adding material (usually carbonate of lime) to its shell. D'Arcy Thompson (1917) writes [3]

"In the growth of a shell we can conceive no simpler law than this, namely, that it shall widen and lengthen in the same unvarying pro-portion: and this simplest of laws is that which nature tends to follow. The shell, like the creature within it, grows in size *but does not change its shape*; and the existence of this constant relativity of growth, or constant similarity of form, is of the essence, and may be made the basis of a definition, of the equi-angular spiral."

D'Arcy Thompson was the first modern scientist to underscore the import-ance of scaling in biological structure and function. The concept of *similitude* emerges as the central theme of his studies.

In recent years the principle of similitude has been superceded by the concept of *self-similarity*. The principle of similitude was based on the idea that the underlying process is continuous and smoothly fills an interval. Such assumptions are not necessarily accurate [4]. The concept of self-similarity, however, does not require homogeneity on scales smaller than some characteristic size as does similitude. As an example consider the small scale architecture of the lung, far from being homogeneous, it is richly structured. At the same time, there is clearly a similarity between the bronchial branchings on the smaller levels and its overall tree-like appearance. Clearly any theory of self-similar scaling which is based on a multiplicity of scales would be an attractive candidate to test physiological structures and processes which are characterized by such variability and order [5].

Perhaps the most compelling feature of all physiological systems is their complexity. On a structural (*static*) level, the bronchial system of the lung serves as a useful paradigm for such anatomic complexity. In addition to static structure, we also seek to understand certain features of *dynamical* complexity so that the real time functioning and evolution of physiological systems can be explained. We have developed quantitative models that suggest a mechanism for the 'organized vari-ability' inherent in physiological structure and function. The essential concept underlying this kind of constrained randomness is that of *scaling*. But not the primitive scaling of D'Arcy Thompson, but one derived from the fast growing branch of the basic sciences called *non-linear dynamics* [6]. In particular, much of what we will have to say relates to the study of a ubiquitous class of irregular structures called *fractals*. In the decade since its introduction, the fractal con-cept has already had enormous impact on the physical sciences, with applications in the study of turbulence, meteorology, magnetization, polymer chemistry and so on. The first papers on the application of the fractal concept in a biomedical context have only recently been publish-ed [7-10]. Remarkably, a concept that promises to become an organizing principle in the physiological texts of the next decade remains unknown to the vast majority of medical practitioners and researchers [5]. This presentation is intended to indicate how one may apply nonlinear analysis and fractal constructs to the understanding of biological mechanisms.

Mandelbrot's concept of a fractal [11] extends our usual ideas of geometry beyond those of point, line, circle and so on into the realm of the irregular, disjoint and singular. Classical geometry deals with regular forms having integer dimensions. A line has dimension one, an area dimension two and a volume dimension three. However, structures in nature are perversely non-Euclidean, as anyone who has tried to trace the outline of a tree or watched the breaking of a wave can readily attest. Compared with smooth Euclidean forms, a fractal curve appears corrugated. On closer examination (magnification) each fold is seen to be composed of smaller folds, and these in turn of even smaller folds. At each successively smaller scale more and more levels of irregular structure are revealed. Clearly there can be no characteristic scale of length for such an irregular object: the smaller the ruler used to measure it, the longer the crumpled line appears to be. If ε is the unit of measure of a fractal curve of fractal dimension D then $N(\varepsilon) = 1/\varepsilon^D$ is the number of units required to measure the curve. The length of the curve is then $L(\varepsilon) = \varepsilon^{1-D}$ and since $D>1$ the length diverges as $\varepsilon \to 0$. In a volume of Euclidean dimension E the volume occupied by an object of fractal dimension D is given by $V(\varepsilon) \propto \varepsilon^{E-D}$.

The above example illustrates three related properties of fractal forms: *heterogeneity, self-similarity*, and the *absence of a characteristic scale of length*. The geometric features of heterogeneity, multiple scales and self-similarity are characteristic of a variety of seemingly unrelated biological forms: the tracheo-bronchial tree, the His-Purkinje system, the chordae tendineae, the biliary network, the vascular tree, and the urinary collecting system to name a few. The fractal dimension D has proven to be a useful way to characterize the geometric structure of a number of these biological systems [4,5].

The fractal concept is also useful for characterizing certain aspects of physiological dynamics. Consider a complex process that cannot be expressed in terms of a single characteristic rate, but instead is regulated over many order of magnitude in time (e.g. days to milliseconds) by a self-similar mechanism. The multiplicity in time scales will be reflected in a power spectrum with a broad profile of responses. The fractal (self-similar) scaling between variations on different time scales will lead to a frequency spectrum having an inverse power-law distribution [8,12].

We have mentioned geometric fractals and temporal fractals, now we turn to the third category of interest, namely statistical fractals. Such objects were investigated by P. Lévy in the 1920s and 1930s. He was interested in when the sum of a large number of identically distributed random variables have the same distribution as each of the contributors to the sum, i.e. when the statistical properties of the whole is the same as that of its parts. The class of distributions having this property are called the Lévy distributions. An important property of the Lévy distribution is that certain central moments diverge, e.g. the second moment is infinite [6].

If $X(t)$ is the dynamic variable of the interest and $P(x,t)$ is the probability density that it lies in the interval $(x,x+dx)$ at time t, then in terms of the characteristic function $\phi(k,t)$

$$P(x,t) \equiv \int_{-\infty}^{\infty} \exp(ikx) \, \phi(k,t) dk \qquad (1)$$

and for a Lévy process

$$\phi(k,t) = \exp(-\gamma |k|^{\alpha} t); \quad 0 < \alpha \leq 2. \qquad (2)$$

When $\alpha = 2$, $P(x,t)$ becomes the Gauss distribution and x^2 is finite. For general $0<\alpha<2$, $\langle x^p \rangle < \infty$ if $p<\alpha$ and $\langle x^p \rangle = \infty$ if $p>\alpha$. One consequence of this property is that the random variable $X(t)$ is intermittent, i.e. the time trace of $X(t)$ manifests bursting behaviour rather than a regular random fluctuation. The Lévy index α can be related to the fractal dimension of the underlying process, e.g., $D = 2-1/\alpha$ for the time trace of such a process.

III MAXIMUM ENTROPY AND FLUCTUATION-TOLERANCE

The preceding comments suggest that a biological structure may have the ability to maintain its integrity while allowing for a broad spectrum of variations both in space and time. This *fluctuation-tolerance* can be described by means of the maximum entropy formalism developed by Shannon [13]. This formalism is a technique for determining the least biased probability density consistent with a set of experimental constraints.

If $p(\alpha)d\alpha$ is the probability that the parameter α has a value in the interval $(\alpha,\alpha+d\alpha)$ then one can define a continuous state entropy (information) function H

$$H = -\int p(\alpha)\ln p(\alpha)d\alpha \tag{3}$$

with

$$\int p(\alpha)d\alpha = 1. \tag{4}$$

The probability density function $p(\alpha)$ that maximizes the entropy (3) is subject to a set of m auxiliary conditions

$$\int F_j(\alpha)p(\alpha)d\alpha = C_j , j = 1,2,..,m. \tag{5}$$

The functions $F_j(\alpha)$ and constants C_j are known for each of the m relations denoted by (5). The information measure introduced by Shannon gives a unique, unambiguous measure of the amount of uncertainty connected with the *unknown* distribution $p(\alpha)$. The distribution is determined by using the Lagrange multipliers $\lambda_j, j = 1,2,..,m$, to be chosen such that the variation of the functional $I(p)$

$$I(p) = H - \sum_{j=0}^{m} \lambda_j C_j \tag{6}$$

vanishes

$$\delta I(p) = -\int [\ln p(\alpha) + \lambda_0 + \lambda_1 F_1 + ... + \lambda_m F_m] \delta p(\alpha)d\alpha = 0. \tag{7}$$

Since the variation $\delta p(\alpha)$ is arbitrary the expression in brackets in (7) vanishes and we obtain the solution

$$p(\alpha) = N\exp \{-\sum_{j=1}^{m} \lambda_j F_j(\alpha)\} \tag{8}$$

where $N = \exp(-\lambda_0)$ is the normalization constant set by (4). Let us consider the constraint $F(\alpha) = \ln\alpha$ so that (8) becomes

$$p(\alpha) = \frac{N}{\alpha^{\lambda_1}} ; \alpha>0. \tag{9}$$

Thus the power-law distribution has the maximum diversity subject to the constraint that the average logarithm of the variate is known and fixed [14].

We note here that the Lévy distribution mentioned earlier has an inverse-power series representation, which asymptotically reduces to [15]

$$p(\alpha) = \int_{\infty}^{\infty} \exp(ik\alpha) \, \exp(-\gamma|k|^\mu) \, dk \sim \frac{N}{|\alpha|^{\mu+1}} \tag{10}$$

Thus identifying λ_1 with $\mu+1$ we can relate the asymptotic statistics of a Lévy process with those of the inverse power-law.

The efficacy of the power-law distribution is provided by the pulmonary tree where detailed measurements of the bronchial dimensions have already been made [16,17]. Traditionally, bronchial scaling from one level of branching to the next have been fit to a simple exponential curve: $d(z,\alpha) = d_0 \exp(-\alpha z)$ where $d(z,\alpha)$ is the average diameter of tubes in the zth generation, d_0 is the tracheal diameter and α is the scale factor between successive generations. However, this representation only satisfactorily accounts for data from the first ten generations. The data points from higher generations systematically deviate from the simple exponential regression (Figure 1). Weibel and Gomez [16] attributed this discrepancy to a change in the diffusive gas transport in the smaller tubes of the higher generations. However, if the pulmonary tree is a fractal structure, then there should be no single characteristic scale factor such as α.

For a fractal tree, a multiplicity of scale will contribute, each with a different weighting or probability of occurrence [7]. The power-law distribution of scales(9) is a candidate distribution to describe the variability in scales present in each generation. We denote the average tube diameter as

$$\overline{d(z)} = \int_0^{\infty} d(z,\alpha) p(\alpha) d\alpha \tag{11}$$

and note that the power-law distribution satisfies the scaling relation $p(\alpha/\beta) = \beta^{\lambda_1} p(\alpha)$. Thus if $d(z,\alpha) = d(\alpha z)$ and we denote $p = \beta^{1-\lambda_1}$ then the average diameter (11) satisfies the scaling relation

$$\overline{d(z)} = p\beta \, \overline{d(\beta z)} \tag{12}$$

independent of the form of $d(\alpha z)$. The solution to the functional equation (12) can be written

$$\overline{d(z)} = \frac{A(z)}{z^\alpha} \tag{13}$$

where by direct substitution we find

$$\alpha = \frac{\ln(\beta p)}{\ln \beta} = 1 - \ln(1/p)/\ln \beta \tag{14}$$

and

$$A(z) = A(\beta z) = \sum_{n=-\infty}^{\infty} A_n \exp\{i 2\pi n \frac{\ln z}{\ln \beta}\} \, . \tag{15}$$

Here p is interpreted as the probability that a scale of size β is present in the zth generation of the lung and $A(z)$ is a harmonic function in $\ln z$ with period $\ln \beta$ [5-7]. Therefore the multiple scales in the fractal model leads to the prediction (13) that the average bronchial diameter should decrease, with generation number, not as exponential, but as a type of modulated inverse power law. Data from human and animal lung casts show a good fit to this modulated power law for the twenty plus generations of branchings that were measured. A typical example of this fit is shown in Figure 2.

Now consider the action of a cardiac depolarization pulse. Each heartbeat is initiated in the sinus node of the right atrium and the activation wave spreads through the atria to the AV junction. Following activation of the AV junction, the cardiac pulse spreads to the ventric-

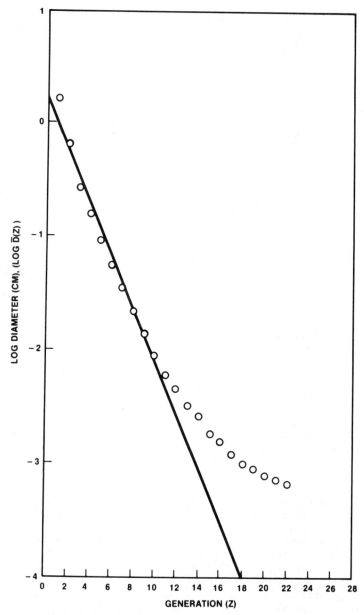

Fig. 1. A log-linear plot of Weibel's complete date [16,17] on the
average diameter of a bronchial tube as a function of
generation number is depicted. The data comprises 23
bronchial generations and shows a marked deviation from the
straight line (solid) predicted by an exponential function.

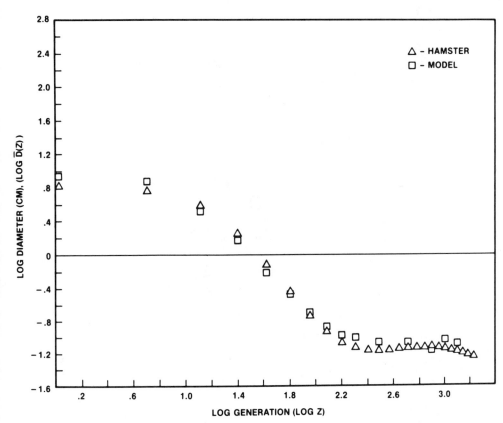

Fig. 2. Patterns of harmonically modulated power-law scaling were
obtained for average diameter measurements from Raabe et al.
[18] for dog, hamster and rat [see West et al. 7]. The data
for the hamster are fit with the theoretical points gener-
ated by $d(z) = [A_0 + A_1 \cos(2\pi \ln z / \ln \beta)]/z^{\alpha}$ with the parameter
values $A_0 = 0.32cm$, $A_1 = 0.11cm$, $\alpha = 0.86$, $\beta = 9$.

ular myocardium through the ramifying network, the His-Purkinje system
(see Figure 3). This branching structure of the His-Purkinje conduction
system bears a remarkable resemblance to the self-similar bronchial tree.
Elsewhere [8] we have conjectured that the repetitive branching of the
His-Purkinje system represents a fractal set in which each generation of
self-similar segmenting imposes a greater detail on to the process. At
each fork in this network, the cardiac impulse activates a new pulse
along each conduction branch, thus yielding two pulses for one. In this
manner, a single pulse entering the proximal point of the His-Purkinje
network with N distal branches, will generate N pulses at the interface
of the conduction network and the myocardium. These N pulses superpose
to yield the QRS waveform. In a fractal network, the arrival times of
these pulses at the myocardium will be staggered, i.e. the pulse generation process during the cascade along with the unequal distances
travelled by each of the pulses acts to decorrelate them. Now because
the conduction system is fractal, the process cannot be characterized by
a single decorrelation time (or rate).

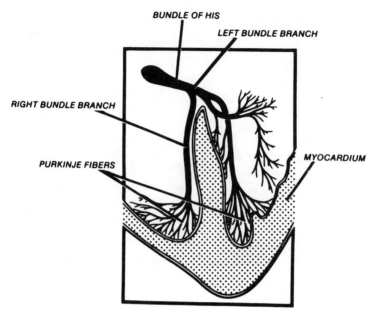

Fig. 3.   The irregular, but self-similar patterns of branchings seen
in the His-Purkinje conduction network of the heart depicted
here is characteristic of fractal structures.

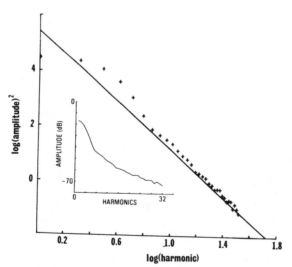

Fig. 4.   The power spectrum (inset) of normal ventricular depolarization (QRS) waveform (mean data of 21 healthy men) shows a
broadband distribution with a long, low amplitude high
frequency tail.   An inverse power-law distribution (solid
curve) is demonstrated by replotting the same data in log-
log format (crosses).   Fundamental frequency = 7.81 Hz
(adopted from reference [8]).

We can use the same argument we just applied to the average bronchial diameter to the correlation of pulses reaching the myocardium. We use a distribution of decorrelation rates of impulses by a direct correspondence to the distribution of a branch lengths in the conduction network. In this way, we obtain the relation for the correlation function $C(t)$ [5,6].

$$C(t) = aC(bt) \tag{16}$$

which if $a>1$ and $b>1$ then

$$C(t) = A(t)t^{\alpha} \tag{17}$$

with $\alpha = \ln a / \ln b$, $A(t)$ given by (15) and a is here not a probability. If $A(t)$ is a constant, or at least a slowly varying function of time, then the power spectrum for the QRS complex given by the Fourier transform of (17) is

$$S(\omega) \sim \frac{1}{\omega^{\alpha+1}} . \tag{18}$$

Spectral analysis of the QRS waveform (time trace) reveals a broad band frequency spectrum with a long tail corresponding to an inverse power law in frequency much as (18) [see Figure 4]. The actual data fits this model quite well [8].

IV MORPHOGENESIS

From a developmental perspective, the fluctuation-tolerant fractals provide a powerful new morphogenetic principle underlying the construction of many apparently unrelated highly complex structures by means of an appealingly simple algorithm. The basis of this algorithm is self-similarity, so a complex structure like the bronchial tree or His-Purkinje system can be generated by adding the same structure on progressively smaller scales. The multiple of scales at any one generation induces the self-similarity between generations. This fractal algorithm tends to suppress the effect of any error arising during morphogenesis since it relies on a self-similar, iterative mechanism. Cohn [18] in the early fifties, introduced the notion of an 'equivalent bifurcation system'. The equivalent bifurcation systems were examined to determine the set of rules under which an idealized bifurcating system would most completely fill space. The analogy was based on the assumption that the branchings of the arterial system should be guided by some general morphogenetic laws enabling blood to be supplied to the various parts of the body in some optimally efficient manner. The branching rule in the mathematical system is then to be interpreted in the physiological application of the self-similarity idea. Such physiologic structures are only static in that they are the 'fossil remnant' of a morphogenetic process. It would seem reasonable therefore to suspect that morphogenesis itself could also be described as a fractal process, but one which is time dependent [7].

In the pulmonary tree, the decrease in mean bronchial diameters between generations follows a type of inverse power law. From a functional view point, the fractal geometry of the lung appears to provide an optimal solution to the problem of maximizing surface area for diffusing O_2 and CO_2. Gas exchange in the lungs over this broad surface area is effected by the interleaving of three fractal networks: pulmonary arterial, pulmonary venous and bronchial alveolar. Similarly the findings are consistent with our conjectured link between the fractal structure of the His-Purkinje branchings and a fractal time process which leads to the inverse power-law spectrum of the QRS.

313

It is interesting to note that the general principles of minimum entropy production[20], minimum use of mass and minimum energy expenditure [21] all lead to the exponential decrease of the average measure with generation number. The modulated inverse power law, on the other hand, is derived here from a maximum entropy argument and the assumption of no characteristic scale. The question that arises is: 'What is the mechanism of such inverse power-law spectra and averages?'. Traditionally in physiology the term mechanism applies to the linear interaction of two or more (linear or non-linear) elements which *causes* something to happen. Receptor-ligand binding, enzyme substrate interactions, and reflex-arcs are all examples of traditional physiological mechanisms. The 'mechanisms' responsible for power-law behaviour in physiological systems, however, is probably not a result of a linear interactive cause-effect chain, but more likely relates to the kinds of complex scaling interactions we have been discussing.

ACKNOWLEDGEMENT

Financial support was provided by La Jolla Institute Independent Research and Development funds.

REFERENCES

[1]   T.A. Cook, "The Curves of Life", Dover, N.Y. (1979);   first publ. (1914).

[2]   H.E. Huntley, "The Divine Proportion", Dover, N.Y. (1970).

[3]   D'Arcy Thompson, "On Growth and Form", Cambridge University Press, (1961);   first publ. (1917).

[4]   B.B. Mandelbrot, "The Fractal Geometry of Nature", W.H. Freeman and Co., New York (1982).

[5]   B.J. West and A.L. Goldberger, Am. Sci. (in press).

[6]   B.J. West,"An Essay of the importance of being nonlinear",Lecture Notes in Biomathematics 62, Springer-Verlag, Berlin (1985).

[7]   B.J. West, V. Bhargava and A.L. Goldberger, J. Appl. Phys. 60:1089 (1986).

[8]   A.L. Goldberger, V. Bhargava, B.J. West and A.J. Mandell, Biophys.J. 48:525 (1985).

[9]   M. Sernetz, B. Gelléri and J. Hofmann, J. Theor. Biol. 117:209 (1985).

[10] A. Babloyantz and A. Destexhe, Proc. Natl. Acad. Sci. USA 83:3513 (1986).

[11] B.B. Mandelbrot, "Fractals, Form, Chance and Dimension", W.H. Freeman, San Francisco (1977).

[12] M.F. Shlesinger, Ann. N.Y. Acad. Sci. (in press).

[13] C.E. Shannon, Bell System Tech. J. 27:379 (1948);   ibid. 623 (1948)

[14] B.J. West and J. Salk, Eur. J. Oper. Res. (in press).

[15] E.W. Montroll and B.J. West, in "Fluctuation Phenomena", eds. E.W. Montroll and J. Lebowitz, North-Holland, Amsterdam (1979);   2nd ed. (1987).

[16] E.R. Weibel and D.M. Gomez, Science 137:577 (1962).

[17] E.R. Weibel, "Morphology of the Human Lung", Academic, New York (1963).

[18] O.G. Raabe, H.C. Yeh, G.M. Schum and R.F. Phalen, "Tracheobronchial Geometry: Human, Dog, Rat, Hamster", Lovelace Found., Albuquerque, New Mexico (1976).

[19] D.L. Cohn, Bull. Math. Biophysics 16:59 (1954);   ibid 17:219 (1955).

[20] P. Glansdorff and I. Prigogine, "Thermodynamic Theory of Structure, Stability and Fluctuation", Wiley, New York (1971).

[21] N. Rashevsky, "Mathematical Biophysics Physico-Mathematical Foundations of Biology", Vol. 2, 3rd rev. ed., Dover, New York (1960).

TEMPERATURE STABILITY OF DAVYDOV SOLITONS

L. Cruzeiro, P.L. Christiansen, J. Halding, O. Skovgaard
and *A.C. Scott

Laboratory of Applied
Mathematical Physics
The Technical University
of Denmark
DK-2800 Lyngby, Denmark

*Department of Electrical and
Computer Engineering
The University of Arizona
Tucson, Arizona 85721

ABSTRACT

The stability of Davydov solitons is investigated numerically for different values of the nonlinearity parameter, the temperature and the initial conditions.

I INTRODUCTION

Energy transport is a central phenomenon in biological processes. A possible mechanism for energy transport in proteins has been suggested by Davydov [1,2,3] consisting of soliton formation and propagation in the protein α-helices.

Here we follow Davydov's approach and study the evolution of solitons in α-helices at different temperatures. The dependence of the stability of the soliton on the initial conditions, nonlinearity parameter and temperature is investigated. A window in the nonlinearity parameter value for soliton propagation is found which decreases with temperature and is also critically dependent on the initial conditions.

II THE DAVYDOV MODEL

Let us consider a one dimensional molecular lattice with N sites, located a distance a, apart. The Hamiltonian of the system, averaged over the phonon states, is

$$H_T = \sum_n ((\varepsilon|\phi_n|^2 - J(\phi_n^* \phi_{n-1} e^{W_{nn-1}} + \phi_n^* \phi_{n+1} e^{W_{nn+1}}) -$$

$$-|\phi_n|^2 (\frac{1}{N})^{\frac{1}{2}} \sum_q (F(q) e^{iqna} (\beta_{qn} + \beta_{-qn}^*)) +$$

$$+|\phi_n|^2 \sum_q (\hbar\Omega_q (\bar{\nu}_q + |\beta_{qn}|^2)) \tag{1}$$

where ε is the intramolecular excitation energy, ϕ_n is the probability amplitude for an excitation in site n, $-J$ is the dipole-dipole interaction energy, and

$$W_{nn\pm1} = \sum_q ((\bar{\upsilon}_q+1)\beta^*_{qn}\beta_{qn\pm1}+\bar{\upsilon}_q\beta_{qn}\beta^*_{qn\pm1}-$$
$$-(\bar{\upsilon}_q+\tfrac{1}{2})(|\beta_{qn}|^2+|\beta_{qn\pm1}|^2)) \tag{2}$$

where the summation is over all phonon wave numbers q and $\bar{\upsilon}_q$ is the average number of phonons with wave number q, given by $\bar{\upsilon}_q = 1/[\exp(\hbar\Omega_q/kT)-1]$, and β_{qn} characterizes the displacement of site n from its equilibrium position due to a phonon with wave number q; finally,

$$F(q) = (\frac{\hbar}{2M\Omega_q})^{\tfrac{1}{2}}[\chi(\cos qa-1)+i\chi\sin qa] \tag{3}$$

where $\hbar$ is the normalized Planck's constant, M is the mass of each site, Ω_q, is the frequency of a phonon with wave number q and the <u>nonlinearity parameter</u> χ is the change in ε per unit extension of the <u>following</u> hydrogen bond.

From the above Hamiltonian (1), we derive the following dynamical equations

$$i\hbar\frac{\partial\phi_n}{\partial t} = \frac{\partial H_T}{\partial\phi^*_n} = \varepsilon\phi_n-J(\phi_{n-1}e^{W_{nn-1}}+\phi_{n+1}e^{W_{nn+1}})- \tag{4}$$
$$-\phi_n(\tfrac{1}{N})^{\tfrac{1}{2}}\sum_q(F(q)e^{iqna}(\beta_{qn}+\beta^*_{-qn}))+\phi_n\sum_q(\hbar\Omega_q(\bar{\upsilon}_q+|\beta_{qn}|^2))$$

and

$$i\hbar\frac{\partial\beta_{qn}}{t} = \frac{\partial H_T}{\partial\beta^*_{qn}} = -J[\phi^*_n\phi_{n-1}e^{W_{nn-1}}((\bar{\upsilon}_q+1)\beta_{qn-1}-(\bar{\upsilon}_q+\tfrac{1}{2})\beta_{qn})+$$

$$+\phi^*_{n+1}\phi_n e^{W_{n+1n}}(\bar{\upsilon}_q\beta_{qn+1}-(\bar{\upsilon}_q+\tfrac{1}{2})\beta_{qn})+$$

$$+\phi^*_n\phi_{n+1}e^{W_{nn+1}}((\bar{\upsilon}_q+1)\beta_{qn+1}-(\bar{\upsilon}_q+\tfrac{1}{2})\beta_{qn})+$$

$$+\phi^*_{n-1}\phi_n e^{W_{n-1n}}(\bar{\upsilon}_q\beta_{qn-1}-(\bar{\upsilon}_q+\tfrac{1}{2})\beta_{qn})]+$$

$$+|\phi_n|^2(\tfrac{1}{N})^{\tfrac{1}{2}}F(q)e^{-iqna}+|\phi_n|^2\hbar\Omega_q\beta_{qn}. \tag{5}$$

In the next section, we present some results of the numerical integration of these equations.

III   RESULTS

The integration of equations (4) and (5) can lead to different dynamical behaviour, according to the conditions specified. In figure 1, five characteristic situations are shown: figure 1a shows a situation in which the initial excitation is completely DISPERSED after 10 picoseconds; figure 1b shows another situation in which part of the initial excitation is dispersed, while the rest is built into a 1 MOVING SOLITON; figure 1c is similar to 1b, but instead of 1 soliton it shows 2 solitons (1 MOVING AND 1 PINNED); figure 1d shows yet another situation in which the non-dispersed part of the initial excitation is HOPPING between the bonds around the initially excited ones; and, finally, figure 1e, shows a situation in which almost all of the initial excitation is PINNED.

316

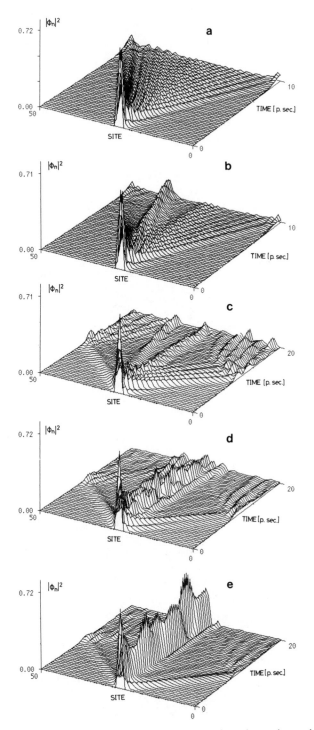

Fig. 1. Variation of the probability of excitation with site and time
(picoseconds). The following values were used: number of
sites = 51; mass of each site = 114. a.m.u.; $J = 7.8$ cm^{-1},
which are the alpha-helix parameters[2]. (a) $T = 10°K$, $X = .13$
10^{-10}N; (b) $T = 10°K$, $X = .23$ 10^{-10}N; (c) $T = 310°K$,
$X = .20$ 10^{-10}N; (d) $T = 310°K$, $X = .29$ 10^{-10}N;
(e) $T = 310°K$, $X = .41$ 10^{-10}N.

317

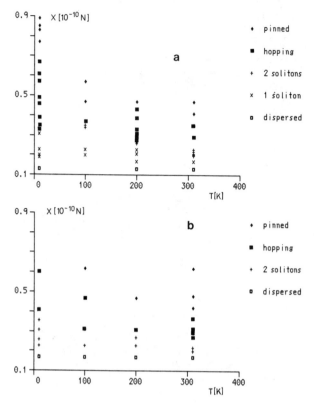

Fig. 2.   Dependence of dynamics on nonlinearity parameter χ and
temperature for M = 114 amu and J = 7.8 cm^{-1}. Initial con-
ditions (a) $|\phi_{24}(0)|^2 = |\phi_{25}(0)|^2 = 0.5$, (b) $|\phi_{25}(0)|^2 = 1$.

Figure 2 shows how the dynamical situations presented in figure 1
depend on the value of the nonlinearity parameter χ, on the temperature
T and on the initial conditions in the case of two bonds excited
(figure 2a) and one bond excited (figure 2b). We see that, at each
temperature, for values of χ below a critical value, there is total dispersion of the initial excitation (figure 1a), while above that critical
value there is a range in which there is soliton propagation (figure 1b
and 1c). Increasing χ further leads to hopping (figure 1d) and finally
pinning (figure 1e). Figure 2 shows that the window of χ in which there
is propagation decreases as temperature increases, and that propagation
of 1 soliton is gradually replaced by propagation of 2 solitons. Also,
figure 2b shows that the more spread out the initial excitation is, the
easier it is to have 1 soliton propagation and that the dynamics of a
1 initially excited bond at T = 10°K resembles that of 2 initially
excited sites at 310°K.

IV DISCUSSION

Questions have been raised concerning the stability of the Davydov
solitons at biological temperatures [4,5]. Figure 2 shows that the
critical value of χ, below which there is dispersion, is roughly 0.17
10^{-10} N in the case of one bond excited initially and 0.13 10^{-10} N in the
two bond case. Since theoretically determined values of χ are 0.265

10^{-10}N-0.364 10^{-10}N [6] the stability of the Davydov solitons, even at biological temperatures, is assured. Figure 2 also shows that several cases are possible, from storage of energy, equivalent to pinning and hopping, to energy propagation, in the 1 soliton or 2 soliton modes. The specific behaviour is dependent on the initial conditions, i.e., on how the energy is delivered to the system and on the respective value of χ. Therefore, the evolution of the initial excitation is regulated by the biochemistry of the system, which is related to the proteins' structure and dynamics. In a general way, storage of energy requires strong values of χ and/or excitations in more restricted regions, while energy propagation requires smaller values of χ and/or spatially broader excitations.

ACKNOWLEDGEMENT

The financial support of the Danish Council for Scientific and Industrial Research is acknowledged.

REFERENCES

[1] A.S. Davydov and N.I. Kisluskha, Phys. Status Solid (b), 59:465 (1973).
[2] A.S. Davydov, J. Theor. Biol. 38:559 (1973).
[3] A.S. Davydov, "Biology and Quantum Mechanics," Pergamon, Oxford (1982).
[4] P.S. Lomdahl and W.C. Kerr, Phys. Rev. Lett. 55:1235 (1985).
[5] A.F. Lawrence, J.C. McDaniel, D.B. Chang, B.M. Pierce and R.R. Birge, Phys. Rev. A 33:1180 (1986).
[6] V.A. Kuprievich and Z.G. Kudritskaya (unpublished).

INDEX

DATE DUE

JUL 27 1990			
NO 25 92			
MR 8 '95			
FE 14 96			

DEMCO 38-297